CANDID SCIENCE

Conversations with Famous Chemists

CANDID SCIENCE

Conversations with Famous Chemists

István Hargittai

Edited by Magdolna Hargittai

ICP

Imperial College Press

Published by

Imperial College Press
57 Shelton Street
Covent Garden
London WC2H 9HE

Distributed by

World Scientific Publishing Co. Pte. Ltd.
5 Toh Tuck Link, Singapore 596224
USA office: Suite 202, 1060 Main Street, River Edge, NJ 07661
UK office: 57 Shelton Street, Covent Garden, London WC2H 9HE

István Hargittai
Budapest University of Technology and Economics
Eötvös University and Hungarian Academy of Sciences
H-1521 Budapest, Hungary

Magdolna Hargittai
Eötvös University and Hungarian Academy of Sciences
H-1518 Budapest, Pf. 32, Hungary

British Library Cataloguing-in-Publication Data
A catalogue record for this book is available from the British Library.

First published 2000
Reprinted 2002

CANDID SCIENCE
Conversations with Famous Chemists

ISBN 1-86094-151-6
ISBN 1-86094-228-8 (pbk)

Printed by Fulsland Offset Printing (S) Pte Ltd, Singapore

Foreword

This collection of interviews by Professor Hargittai is a timely celebration of the remarkable century of chemistry, which has just passed. Those interviewed are some forty chemists, who were fortunate enough to live during the momentous developments in chemistry, which have occurred during the last decades of the millennium.

These historic events will be retold for centuries to come, often with more understanding and appreciation than we can have today. But, in one respect, what is written here will never be repeated because it is the personal experiences of the chemists who took part, each of them a unique individual.

Scientists are sometimes seen as automata without human feeling or sentiment (unlike artists for example!). The scientists themselves, and those who live with scientists, know how false this view is. But, unfortunately, most people know very little of science or scientists personally and often see them as living in a different world from "ordinary" people.

I hope that the interviews recorded here will do something to dispel this increasing divide in our understanding. They tell how chemists of today have lived and worked, why they chose chemistry as a career and how they were led to make the discoveries for which they are best known. The informality of the interviews gives the impression that one is sharing in a conversation rather than listening to an autobiographical lecture.

Children are born scientists but, if they are to retain their love of science, they need role models to follow and encouragement from those who have been successful. This book will be enjoyed by all who have some interest in science and it will be of special value to the young people whom it may encourage to follow those, whose stories are told here.

London George Porter

Preface

At the dawn of the twenty-first century and the new millennium, it is fascinating to review the achievements of chemistry during the past decades of the twentieth century, so extraordinary in scientific progress. It is also fascinating to get acquainted with the stories behind these achievements directly from some of their most important participants. The subject matter in this collection of interviews covers a broad range of chemistry, with emphasis on the following loosely identified areas: structural chemistry, medicinal chemistry, natural products chemistry, stereochemistry, theoretical and computational chemistry, inorganic chemistry, physical organic chemistry, NMR spectroscopy, fullerenes, kinetics and reaction mechanisms, atmospheric chemistry. There is, of course, considerable overlap between these areas, and even more so in the research activities of most of the scientists interviewed. There are also important missing fields. Future collections, already under preparation, will considerably expand the scope of this volume.

This volume contains the first selection of my conversations with famous scientists. In addition to the interviews, we present here a few auxiliary entries. They include a brief chapter on Odd Hassel to whom I never posed questions in the way I did to the others, but my interactions with him made me think about him as one of my interviewees. I prepared two auxiliary entries of quotations by Erwin Chargaff and John Cornforth to augment their interviews. I did this because I wanted to share the intellectual pleasures I have experienced from their writings. Another entry augments the brief interview with Linus Pauling. It is about the controversy of the resonance theory in the Soviet Union at the beginning of the 1950s. I attended the Pauling memorial session of the ACS meeting in 1994 in Anaheim and was surprised by the scarce knowledge of this story among our colleagues who have been interested in Linus Pauling's life and works. Yet another entry is on the beginnings of the multiple metal-metal bond studies. The relevant systems were produced and examined by scientists in Moscow, although the multiple bonds in them were not recognized until Albert Cotton's studies. I compiled a brief segment on Buckminster Fuller because of his conspicuous, if indirect, role in the fullerene story. This was greatly helped by a conversation I had recorded with Ed Applewhite, Fuller's assistant in creating his opus magnum, *Synergetics*. Finally, there is a brief entry on Paul de Kruif because his book, *Microbe Hunters*, was at least as important as the chemistry set in turning interested children's attention to chemistry for the generations that are so prominently represented among my interviewees.

Although my first conversation with a famous chemist, Nikolai Semenov, was recorded back in 1965, launching the quarterly magazine, *The Chemical Intelligencer* by Springer-Verlag, New York in 1995 gave a great incentive for these conversations. I am grateful to Springer-Verlag for their gracious permission to reproduce published material here from *The Chemical Intelligencer*. Some interviews are reproduced in full, others with some changes, saving some space or extending the description of the circumstances of the interview. Quite a few of the interviews are presented for the first time in this volume. The relevant bibliographical data are given in each respective entry.

The general procedure of preparing an interview is the following. I contact the interviewee, we set up a date for our meeting, and have a fairly loose conversation recorded on audiotape. I then prepare and slightly edit the transcripts and send them to the interviewee for correction, deletion, and addition. This procedure is repeated until the interviewee feels absolutely comfortable with the material. Before publication, there is usually some additional light copyediting. The conversations are illustrated with snapshots made during the conversation and by photos provided by the interviewee. In a few cases, the interview was by correspondence.

In all the conversations I am trying to bring out some important aspects of chemistry as well as to learn about the life and thoughts of the interviewee. Since I am a fellow scientist rather than an investigative reporter, I never try to go after a particular problem with which the interviewee seems uncomfortable. From the start of our conversation I am asking my interviewee to ignore any of my questions that he or she does not care to discuss.

Thus the reader may at places feel that I should have pressed for more answers or more details and would never learn about my unanswered questions: I may and then, I may not have asked that particular question. My approach has limitations in this respect. On the other hand, I see benefit in my approach in that the interviewee may open up more to a friendly colleague than to an investigative reporter. In any case, the purpose of the interviews was to learn about chemistry and bring great scientists in human proximity. I have myself learned a great deal from these conversations, in terms of both chemistry and human behavior and personal philosophy. I am grateful to all my interviewees for their time, their contribution, and for their interest in and support for the project.

This book was edited by Magdolna Hargittai, herself a chemistry Ph.D. and Research Professor, who is my partner in my ventures in chemistry and all other aspects of my life, and who often knows my emphases and preferences before I could even articulate them.

For my doctoral work I did not have a supervisor, not even in a formal way. I never liked authority. Yet I always felt blessed when I had a good teacher and somehow I always had one at the right moment, although not necessarily in physical proximity. Having good teachers is the greatest thing that can happen in one's career. A good teacher need not be appointed to be your teacher; a good teacher is a good example, and that may come in a great variety of ways. I feel that my interviews have given me another experience of benefiting from good teaching, and my greatest hope is that they will be received as such by my readers, too. It is in this sense that I would like to dedicate this volume to the coming generations of students, for whom much of the material presented here will be science history.

Budapest, Fall of 1998 — István Hargittai

Contents

Linus Pauling (photo courtesy of the late Linus Pauling).

1

Linus Pauling

The life and work of Linus Pauling (1901–1994) have been documented in great detail. His fame and broad interests are expressed by his being unique in having received two unshared Nobel Prizes. He received one in chemistry in 1954 "for his research into the nature of the chemical bond and its application to the elucidation of the structure of complex molecules." He then received the Nobel Peace Prize for 1962, by the decision of the Norwegian Parliament in 1963.

From time to time I had correspondence with Linus Pauling on a variety of issues, including symmetry and quasicrystals. I met him in person in the early 1980s at the University of Oslo. He came for a brief visit at the invitation of the rector of the University who was also my host and long-time friend, Otto Bastiansen. Pauling gave a lecture to a packed auditorium on structural chemistry, and in particular he discussed at length the hybridization of transition metals, a topic he had published on extensively. During the lecture he was deriving complicated expressions without as much as a scrap of paper. He was marching back and forth in front of a very long blackboard, and he covered it with his formulas. It was impressive and I felt sorry that the blackboard was to be erased after the lecture. Pauling kept his enthusiastic Norwegian audience in awe and he was visibly enjoying it. Only gradually did it dawn on me that the sophisticated derivations were superfluous to the understanding of the subject matter, that he could have just as well explained what he wanted to teach us with words. In any case, this was a rare exhausting lecture that I did not wish to end after one hour. During the luncheon following the talk, Pauling stayed fresher and more alert than any of us, he discussed disarmament and teaching, made publicity for his *General Chemistry* co-authored with his son Peter, and asked us about our research, too.

As I was preparing to launch *The Chemical Intelligencer*, I had some correspondence about it with Linus Pauling, in the fall of 1993. He was very enthusiastic about a magazine devoted to the culture of chemistry. This would have sufficed for support, but he added that his other works prevented him from writing a contribution for the magazine. This gave me the idea of posing some questions to him that would not need too much of an effort to answer. I suspected that he was rather ill by then.

Given that his views had been expressed copiously about a wide range of issues, I decided to focus on four questions that I personally found most interesting. They referred to the long-standing validity of his teachings about structural chemistry; the controversies his ideas on resonance and electronegativity had generated in the Soviet Union; the discoveries of quasicrystals and fullerenes; and the ever growing importance of computations in chemical research.

In addition to what he said, I am quoting two paragraphs from his paper that he referred to in his answer to my last question. I am also giving a brief account of the 1951 Moscow meeting on the resonance theory followed by a few comments in the next chapter. Regarding quasicrystals, I had some correspondence with Pauling at the time of his contributing an article to one of my edited symmetry books. The article had a very telling title, "Interpretation of So-called Icosahedral and Decagonal Quasicrystals of Alloys Showing Apparent Icosahedral Symmetry Elements as Twins of an 820-Atom Cubic Crystal".[1] I was curious whether he had changed his rather rigid opposition to the notion of quasicrystals and, as it turned out, he did not. Since then, *The Chemical Intelligencer* has carried a set of interviews with the major players of that discovery.[2] As for the fullerenes, there is a set of interviews in this volume with the major participants of their discovery, including those who had predicted the stability of C_{60}. Apparently, Pauling was not aware of such predictions.

Now the questions and answers:*

*This interview was originally published in *The Chemical Intelligencer* **1995**, *1*(1), 5

It has been estimated recently[3] *that at the time of the first edition of your* The Nature of the Chemical Bond, *only about 0.01% of today's structural information was available. Nevertheless, your observations and conclusions concerning molecular structure and bonding have withstood the test of time. Would you care to comment on this and, in particular, on the value of comparative methods in chemistry?*

I think that the structural information available in 1939, when my book *The Nature of the Chemical Bond* was first published, was enough to permit reliable general conclusions about chemical bonds to be made, and I am not surprised that the generalizations have withstood the test of time. I think that my own crystal-structure work and that of R. W. G. Wyckoff and Roscoe G. Dickinson involved a good selection of crystals to answer questions that at that time needed to be answered. Also, our electron diffraction work in the 1930s involved many simple molecules of such a sort as to permit further general conclusions to be drawn.

In the early fifties, at the time of your difficulties with the U.S. State Department, because of your allegedly leftist views, you were also sharply criticized in the Soviet Union because of the resonance theory, which was found to be ideologically hostile. How closely did you follow the debates in the Soviet Union? What was your reaction about a decade later to the nonacceptance of the electronegativity concept by official Soviet chemistry?

It took several years, from about 1949 to 1955, for the chemists in the Soviet Union to get a proper understanding of the resonance theory. I may say that I do not know about the nonacceptance of the electronegativity concept in official Soviet chemistry.

Recent discoveries such as the quasicrystals and the fullerenes seem to have caught the solid state and chemical communities by surprise. Were these greatly exceptional events or should we be getting prepared to seeing more of these kinds of unexpected findings in the future?

I am rather surprised that no one had predicted the stability of C_{60}. I might have done so, especially since I knew about the 60-atom structure with icosahedral symmetry, which occurs in intermetallic compounds. It seems to be difficult for people to formulate new ideas. An example is that from 1873 to 1914 nobody, knowing about the tetrahedral nature of the bonds of the carbon atom, predicted that diamond has the diamond structure.

As to the quasicrystals, you know that I contend that icosahedral quasicrystals are icosahedral twins of cubic crystals containing very large icosahedral complexes of atoms. It is not surprising that these crystals exist. The first one to be discovered was the MgZnAl compound reported by my associates and me in 1952. We did not observe quasicrystals of this compound, but they have been observed since then.

According to some, the ever improving computational techniques are overtaking the physical experimental techniques, at least in structural chemistry. What's your expectation of the near future in this respect?

I do not think that quantum mechanical calculations of molecular structure or crystal structure will ever make the sort of chemical arguments about structure presented in my book obsolete. The quantum mechanical calculations are made for one substance, and perhaps then for another somewhat similar substance. My arguments about the chemical structure are very general and pertain to all molecules and crystals.

Simple quantum mechanical calculations have great value, as pointed out in the early chapters of my book and in a paper that I published in *Foundations of Physics* a couple of years ago.

Poster on the wall of the Linus Pauling Institute of Science and Medicine in Palo Alto, California. I gave a seminar at the Institute on February 28, Linus Pauling's birthday, in 1996, not long before the Institute was dissolved. The symbolism of the poster refers to Pauling's two Nobel Prizes and displays conspicuously a pentagonal structure. Both the quasicrystal and fullerene discoveries were intimately related to pentagonal symmetry.

Two paragraphs are quoted here from Linus Pauling's paper that he referred to in his response to the last question. The paper was titled "The Value of Rough Quantum Mechanical Calculations."[4]

"In thinking about the history of science in the period around 60 years ago, I have come to the conclusion that much of the progress was the result of carrying out approximate quantum mechanical calculations. It is my impression that in recent years the effort has been made to carry out quantum mechanical calculations that are as quantitatively accurate as possible. Instead of making calculations of energy levels and other properties of a system with use of a simple approximate wave function corresponding to some simple model, the effort of many physicists is to formulate as complicated a wave function as can be handled by the computers. I remember reading a paper in which the author reported that his wave function contained a million terms. There is little doubt that such calculations will give results in agreement with experiment. With a complicated wave function, however, it is essentially impossible to formulate an interpretation in terms of a model of the system." (p. 830)

...

"Chemistry has probably been fortunate in that a great amount of empirical knowledge about the properties of chemical substances had been obtained before the development of quantum mechanics. Chemists strove to understand these properties, and as a result the classical structure theory of chemistry was developed. If the accumulation of large amounts of information about the properties of substances had not been gathered before quantum mechanics was formulated, it may well be that chemical structure theory would not have developed. In fact, at the present time, in 1991, little use is made by chemists of quantum mechanics, except to the extent that the principles of chemical bond formation that were formulated on the basis of quantum mechanical principles are extensively used. Some use is made by chemists of accurate quantum mechanical treatments of molecular structure, but much more use is made of the chemical structure-theory model." (p. 834)

References

1. Pauling, L. in Hargittai, I. Ed., *Symmetry 2: Unifying Human Understanding*, Pergamon Press, Oxford, 1989, pp. 337–339.
2. Hargittai, I. *The Chemical Intelligencer* **1997**, *3*(4), 25–49.
3. Murray-Rust, P. in Goodfellow, J. M.; Moss, D. S. Eds., *Computer Modelling of Biomolecular Processes*, Ellis Horwood, New York, 1992.
4. Pauling, L. *Foundations of Physics* **1992**, *22*, 829–838.

2

The Great Soviet Resonance Controversy

This is a brief review of the minutes of a meeting in Moscow, which took place on June 11–14, 1951. They were published in Russian, in a 440-page volume, entitled *Sostoyanie Teorii Khimicheskogo Stroeniya v Organicheskoi Khimii* (*The State of Affairs of the Theory of Chemical Structure in Organic Chemistry*), published by the Izdatel'stvo Akademii Nauk SSSR (Publishing House of the Soviet Academy of Sciences), Moscow, 1952.

The meeting was organized by the Chemistry Division of the Soviet Academy of Sciences. The subject of the meeting was the structure theories of organic chemistry. Four-hundred-and-fifty chemists, physicists, and philosophers attended, representing major centers of scientific research and higher education from all over the Soviet Union. There was a report on "The status of chemical structure theory in organic chemistry" compiled by a special commission of the Chemistry Division, followed by 43 oral contributions. An additional 12 contributions were submitted in writing. The conference adopted a resolution and sent a letter to I. V. Stalin.

The letter to Stalin expressed self-criticism for past deficiencies in appreciating the role of theory and theoretical generalizations in chemical research. This has resulted in the spreading of the foreign concept of "resonance" among some of the Soviet scientists. This concept is an attempt

Statue of A. M. Butlerov (1828–1886) in front of the Chemistry Department of Moscow State University (photo courtesy of Dr. A. A. Ivanov, Moscow). Butlerov was professor of chemistry at Kazan' University when he introduced the term "chemical structure" ("khimicheskoye stroyeniye").

to liquidate the materialistic foundations of structure theory. However, Soviet chemists have already started their struggle against the ideological concepts of bourgeois science. They have unmasked the falseness of the so-called "theory of resonance", and would cleanse Soviet chemical sciences of the remnants of this concept.

During the meeting, there has been repeated reference to Stalin's teachings on the importance of the struggle between differing opinions and of the freedom of criticism.

The Report of the Chemistry Division was submitted to the meeting by Academician A. N. Terenin on behalf of the special commission. The names of the members of this commission and those of many of the speakers in the subsequent discussions read like a *Who's Who* in Soviet chemistry. There were many academicians among them and also many future academicians.

The report consisted of the following chapters:

1. Butlerov's teachings and their role in the development of chemistry
2. The development of structure theory during the second half of the 19th century and the first half of the 20th century
3. Advances of Soviet organic chemistry
4. Quantum chemistry and structure theory
5. About the so-called "theory of resonance"
6. On the mistakes of some Soviet scientists
7. Current state of Butlerov-Markovnikov's teachings on the intramolecular interactions of atoms and on reactivity
8. Perspectives of further development of structure theory.

These are telling titles and I mention a few points from Chapter 6 only. Here we learn that G. V. Chelintsev has criticized actively the concept of "resonance" in the open press. It was owing to him, to a great extent, that the Soviet scientific society has turned to this question. The basis of his criticism was his own "new structure theory" which, however, completely contradicted the modern theory of chemical structure and was contrary to the experimental facts and theoretical foundations of quantum physics.

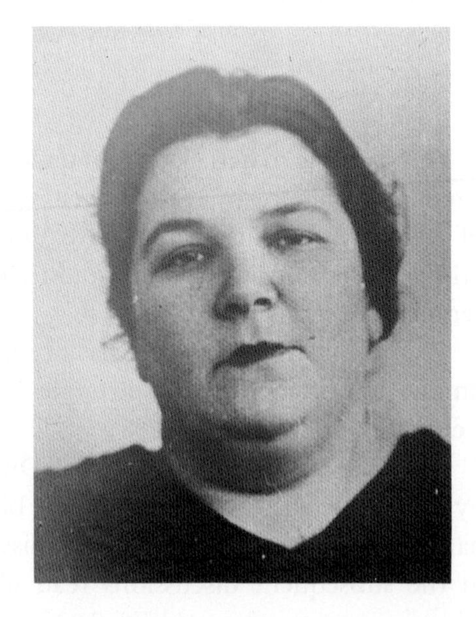

Professor Ya. K. Syrkin (1894–1974) and Professor M. E. Dyatkina (1914–1972). Photographs courtesy of Professor L. V. Vilkov, Moscow State University.

Ya. K. Syrkin and M. E. Dyatkina were named as the main culprits of disseminating the theory of resonance in the Soviet Union. They were accused of having even further developed the erroneous concepts of Pauling and Wheland, of ignoring the works of Soviet and Russian scientists, of idolizing foreign authorities, and of quoting even works of secondary importance by American and English authors.

Others were also mentioned, among them the organic chemist A. N. Nesmeyanov. He, along with R. H. Freidlina, interpreted the diverse reactivity of chlorovinyl compounds of mercury and other quasi-complex compounds by the "resonance" between their covalent and ionic structures. These lesser sinners, along with many others, however, have eventually repented and have themselves become critics of the application of "resonance".

The report, as the subsequent contributions, was followed by a questions-and-answers period. Perhaps the most important question was about the idealism of the concept of resonance. The answer to this question started with a quotation from V. I. Lenin, according to which philosophical idealism is a one-sided exaggeration of an insignificant feature of the cognition process. Such a feature was then detached from the matter and from nature, and made into something absolute. The answer then turned to the concept of resonance, where the insignificant features of the cognitive process were specified to be the individual components of the approximate computational techniques employed in the calculation of the molecular wave function. They are made into something of primary importance, as if objectively existing in the molecule, and as if determining *a priori* the molecular properties. In reality the resonance structures and their resonance are torn from the matter, and the theory of resonance becomes an absolute above the matter. Obviously, more was read into the resonance theory than what Linus Pauling may have ever wanted it to mean.

Only a small, though very vocal, group attacked blindly the theory of resonance and attacked especially viciously the alleged proponents of the theory. Quantum chemistry and all of the science of the West were also under attack. The return to historical Russian results was advocated and the attackers' own theories were proposed for general acceptance. These theories, however, had been shown to be worthless nonsense by many. However, all the participants painstakingly dissociated themselves from the theory of resonance. At times the self-criticism of some excellent scientists was humiliating in the extreme.

It was characteristic of the atmosphere of the meeting that a philosopher (B. M. Kedrov) declared Schrödinger to be a representative of modern "physical" idealism which made Schrödinger a relative of Pauling's. Furthermore, he stated that Dirac's superposition principle was as idealistic as Heisenberg's complementarity principle and even more idealistic than Pauling's theory of resonance.

A writer (V. E. L'vov) criticized the report for a serious political error, namely, that the protagonists of the theory of resonance were equated with the greatest Soviet scientists. These protagonists had been unmasked as spokesmen of the Anglo-American bourgeois pseudoscience by the Press and Soviet society. According to L'vov, the report was vague about the main thrust of the ideological struggle taking place in theoretical chemistry. He also quoted, as a positive example, the criticism of Mendel by T. D. Lysenko who proved that Mendel's work had nothing to do with the science of biology. Furthermore, L'vov attacked fiercely the theories of Heisenberg as well as those of Heitler and London. He protested the report's view of quantum mechanics as a development of Butlerov's teachings. The most important political task of Soviet chemistry, he declared, was the isolation and capitulation of the insignificant group of unrepenting proponents of the ideology of resonance.

The last entry in the minutes of the meeting is a dissenting opinion in the form of a short letter by E. A. Shilov, member of the Ukrainian Academy of Sciences. He is critical of the report and the resolution of the meeting for looking so much backward rather than forward. He suggested concentrating on new results and new teachings instead of conducting scholastic debates about questions such as "Where does resonance end?" and "How does the 'healthy' mesomerism of Soviet authors differ from Ingold's 'erroneous' mesomerism?" and "How can ideal structures be considered real at the same time?" The result of ending such debates would be, Shilov added, that the efforts and time of Soviet organic chemists could be devoted to valid and productive work. Professor Shilov's contribution was not delivered as an oral presentation during the meeting.

***

The significance of the Moscow meeting can be evaluated at a number of levels. At one level, the records speak for themselves and provide an excellent demonstration of orwellian double speak, but even George Orwell's *1984* pales by comparison. They also show the fear of the Soviet political system of everything coming from the West, even if it is only a chemical theory. The inferences of the theory of resonance appear exaggerated beyond any reasonable limit and the statements reflect the atmosphere of a staged trial rather than that of a scientific discussion. There are chemists who do not like the description of molecular structure by a series of resonance structures. What is frightening and mind boggling is that such a dislike was made into an official dogma with philosophical justification. The story of resonance, however, should not be viewed in isolation from the rest of Soviet life in the early fifties, the last years of Stalin's reign. To me the question that is most telling was that asked of M. E. Dyatkina: "How do you explain that you are so conspicuously familiar with the teachings of foreign scientists? May it be that you, along with Professor Syrkin, are intentionally bowing to foreign scientists?"

According to a different evaluation, the Moscow meeting had a distinctly positive significance. It was governed by a healthy mechanism of self-defense of the higher echelons of the Soviet chemical establishment. Rather than letting harsher outside interference crashing it, as it had happened to some other fields of science, this establishment organized itself a milder purge, and, by the same stroke, it ridiculed and marginalized the pseudoscientific extremists in their own ranks, whose most vocal representative was Professor Chelintsev. It is true that jobs were lost in chemistry, but not lives as in less fortunate areas, due to scientific controversy.

There may have been long-term negative consequences of the resonance controversy for Soviet chemistry. Many of the brightest young scientists in the Soviet Union may have stayed away from theoretical chemistry, if not from chemistry, for decades. It was just not the field to be associated with. Roald Hoffmann mentions this in the interview with him in this Volume.

Pauling's response to my question about resonance in Soviet chemistry makes the whole problem appear rather innocent. This may just be a reflection of his magnanimity after so many years. I would find it hard to imagine that he did not follow the developments in Moscow in the 1950s with somewhat more penetrating interest than what his response seems to reveal. On the other hand, he, indeed, may have known much less about the problems of the electronegativity concept in Soviet chemistry. They culminated in the mid-1960s when Soviet science was less rigidly molded by central directives. Besides, any criticism of Linus Pauling by then was muted by the appreciation of Soviet politics for his well publicized activities in the peace movement.

Erwin Chargaff, 1994 (photograph by I. Hargittai).

3

Erwin Chargaff

Erwin Chargaff (b. 1905, in Czernowitz, Austria-Hungary, now the Ukraine) is Professor Emeritus of Biochemistry, Columbia University. His foremost research contribution has been his seminal work for our understanding of the nucleic acids and their role in genetics. According to what is known today as Chargaff's rules, DNA has equal numbers of adenine and thymine residues and equal numbers of guanine and cytosine residues. Dr. Chargaff was educated in Austria and received his doctorate from the University of Vienna. He was Professor and at the later part of his career chairman of the Biochemistry Department of the College of Physicians and Surgeons of Columbia University, New York.

Dr. Chargaff's numerous honors and awards include the Pasteur Medal (Paris, 1949); Carl Neuberg Medal (1958); Gregor Mendel Medal (Halle, 1968); the Charles Leopold Mayer Prize (French Academy of Sciences, 1963); the H. P. Heineken Prize (Royal Netherlands Academy of Sciences, 1973); the National Medal of Science (1975); and the Johann-Heinrich-Merck Prize from the German Academy of Language and Literature (1984). He has been a member of the National Academy of Sciences, fellow of the American Academy of Arts and Sciences, the American Philosophical Society, the German Academy of Sciences, and many other learned societies. Our conversation took place in the Chargaffs' home, Central Park West, New York City, in November, 1994. It first appeared in *The Chemical Intelligencer*.*

---

*This interview was originally published in *The Chemical Intelligencer* **1998**, *4*(1), 4–9

Recently I found a set of birthday cards in a Manhattan store. There was a separate card for each day of the year, listing several names of important people who were born on that particular day of the year. Among those listed on the card of my birthday, I was familiar with one name only, Erwin Chargaff's. Appearing on such a card is a sure sign of popularity. And yet I know that your writings lately have increasingly appeared in German rather than in English because you found a diminishing audience in America for what you have to say.

The American Chemical Society also publishes a yearly calendar and whenever there is a birthday of what they consider a famous scientist, it is mentioned. So my name appears under August 11. I always got along very well with my chemistry colleagues, better than with my biology colleagues, because I always emphasized the importance of chemistry. This is especially so now when so many people, who had never had any training in chemistry, especially M.D.'s, are being let loose on what they call molecular biology and what I still call biochemistry. It is important to keep alive the idea of what chemistry really was, can be, and is. Linus Pauling is a shining example. In general, however, chemistry has had a lot of a bad press as they connected it with poison gases and with poisoning the soil and water. In this, however, most chemists are completely innocent; it was the chemical industry that has used or misused the findings of chemistry. Chemists have learned a little bit from nature and nature is, of course, the greatest chemist. In contrast to other scientists, chemists have not tried to improve nature. Chemistry is more a descriptive science than an exploratory and agressive science, and I prefer descriptive sciences.

The reason why I stopped publishing in English is very simple, because I couldn't find a publisher. The funny thing is that we live in the greatest democracy of them all, and there are no black-lists in this country but when you look more closely you see that certain people are not *persona grata*. I believe in my paranoid way that I belong to them. When I wrote *Heraclitean Fire*, I shopped around for a publisher, and some didn't even want to receive me although I was already a member of the National Academy and so on. Finally I had to go to a very small university press where I was treated very well. The Rockefeller University Press published the book in 1978. With *Serious Questions*, which is a book of essays, I also had trouble. It's not easy since I am old and impatient and lose my temper easily and get discouraged and become sleepless. All this I didn't use to be when I was younger. So finally I said, the hell with it; I write at least as well in

German, so let's try it. So far I have produced 12 books and the thirteenth is in press, in German. There I have an audience and lots of reader's mail, so I continue it. I have not given up publishing in English out of malice. I love the English language; it was my third language. I started with German, then came French and only when I was 17 or 18, did I start English in Vienna. I admire English, it's a great language. It has some drawbacks; it is an ugly-sounding language with a crazy pronunciation and crazy spelling. It is also a very comfortable language and it is easier to write and speak bad English than other languages and this is an advantage too.

When did you come first to the United States?

I first came to the United States in 1928 when I was 23 years old after receiving my Ph.D. in chemistry in Vienna. I got a fellowship in Yale University, and worked with Rudolph Anderson. I stayed for two years. I worked on the chemistry of tubercle bacilli on which Anderson was a great entrepreneur at that time. We worked with a lot of cultures, tens of liters at a time.

Nobody got infected?

Not that I know of, and it was not handled in a microbiological way; it was handled by chemists at that time.

Erwin Chargaff in 1931 (courtesy of Erwin Chargaff).

What happened when your two years expired?

I got an offer to be assistant professor from Duke University in North Carolina. But the condition was that I would have to work on tobacco because the support came from the tobacco industry. Though I smoked a lot I still didn't want to devote my entire life to tobacco and had second thoughts.

You still smoke?

Oh, yes. I've always smoked a lot. I also thought that if I'd run out of money, I could hire myself out for advertisement to the tobacco industry. I have smoked cigars, cigarettes, and pipe. I'm not afraid and don't even believe quite in it. There are probably some dangers in tobacco but it came about the time when the American tobacco industry started treating the cigarettes with various ingredients. I would like to know whether there are some statistics from the last century when people mostly smoked Egyptian and Turkish tobacco. America is a crusading country, and it makes crusades in favor of everything. When we get wild, we get awfully wild. Prohibition was an example. Of course, alcohol is dangerous if you overdo it. I first came to the United States at the hight of the prohibition. You could almost see how crimes originated due to ethical pressures to drink some innocent cocktails. You couldn't help meeting bootleggers and maffias on the streets of New York, and it was easier than ever to buy alcohol.

Did you then return to Vienna?

Yes, I declined the assistant professorship at Duke, and returned to Vienna. At that time I was already married to my wife. She had also studied chemistry, started two years after me but she didn't finish her studies. We met at the University in Vienna and got married in New York in 1929. Thank God, we are still together. So, following my return to Vienna, I went to Berlin and got a job in the chemical laboratory of the Institute of Hygiene and Bacteriology of the University. I met there Julius Hirsch, a microbiologist. He knew our work on tubercle bacilli and this was a marvelous introduction. Dr. Hirsch arranged my job with the chief of the Department Martin Hahn, a coauthor of the famous Buchner and Hahn work on yeast from the 1890's. He appointed me with money from outside by the Deutsch-Österreichische Forschungsgemeinschaft and some extra money from the University. I stayed until Hitler came to power. After that, within two months I was in Paris.

Were you Jewish?

Yes, but not awfully, but for Hitler we were anyway. We had an Austrian passport and that was marvelous. Mussolini and Hitler were not friends at that time yet, and Mussolini kept his hands over Austria. I remember April 1, 1933, the famous boycott day, when I had to go to the state library in order to look up something. I was stopped at the entrance by two husky students and they asked me, "Sind Sie Jude, Herr Kollege?" I said I was even better, I was Austrian.

You did not consider to return to Vienna?

No, no, no, no one ever could get a job in Vienna on his merits.

Didn't you think about the job situation before you chose your profession?

On the contrary, I gave it careful consideration. I was interested in languages, also in history, cultural history, philosophy, all kinds of things. I also thought that I ought to be able to help my parents, and certainly to help myself. It was clear to me that for a physicist it was difficult to get a job and for mathematics I was not bright enough. Chemistry, I said, was the only occupation that I knew of where at the age of 23 I could already support myself. For possibilities, of course, I considered not only Austria but Germany as well.

The first year, I went to both the University and the Technical University. I started out with chemistry at the Technical University and at the same time linguistics at the University. After one year, I recognized that it was too much for me to be in two places and I moved to Chemistry at the University. Eventually my thesis was on organic silver complexes. My sponsor was Fritz Feigl. Biochemistry barely existed at that early time. It was more medical chemistry, such as diagnostic tests and urine analysis. Although I had gone into chemistry to support myself, I grew to really like it during my studies and work. At home I could find chemistry only in the dictionary. My father worked in a bank as deputy director.

Places like Vienna and Budapest have ejected many people who then became great scientists and who might not have become great had they stayed.

True, but at the time I was too young to think about greatness; I was only thinking of where my sausage would come from the next day. When

I left Berlin in April 1933, I still had my assistantship. I lost it the next summer when they started throwing out the state employees. I was then dismissed, four-five months after I'd left.

That I could get out of Berlin so promptly, two months after Hitler's taking over, with the assurance of a job in Paris, was due to more than a lucky coincidence. When I came as a chemist to the Bacteriology Institute of Berlin University I started work on the chemistry of diphtheria bacteria, the results of which were published later in *Z. physiol. Chem.* after I had left Germany. I had not planned to work again on tubercle bacilli, but I was pushed back into the field by the circumstances.

A. L. C. Calmette (1863–1933), French bacteriologist on Soviet stamp.

At that time the medical community in Germany was very much excited by the oncoming Lübeck Trial. In the Children's Hospital of that northern city a very large number of children had died after receiving treatment with the Bacillus Calmette-Guérin (BCG). That was a strain of tubercle bacilli originally prepared at the Institut Pasteur by these two French bacteriologists. One distinguishes three principal strains of tubercle bacilli according to the host for which they are virulent: human, bovine, and avian. BCG had been derived originally from *bovine* tubercle bacilli by subjecting cultures to an elaborate weakening process until virulence was said to be lost. It was at that time a much used vaccine. Considering the poisonous atmosphere prevailing in Germany in the early thirties it was to be expected that the Lübeck catastrophe would be the cause of much anti-French and Nazi propaganda, the more so, since some of the physicians involved in the trial were Jewish.

My chief, Martin Hahn, was to function as one of the experts at the coming trial. He came to me and said "Dr. Chargaff, could you think of something to help me to make a decent scientific statement?" I proposed to make a thorough comparison between the strains isolated from the several autopsies of the Lübeck victims and authentic virulent strains as well as BCG. The results, published later in *Biochem. Z.* **1932**, *255*, 319, showed without little doubt that the strains isolated from the pathological materials resembled closely the human strain and not the BCG vaccines. As it turned out, one of the assistants at the hospital who did research on various tubercle bacilli had bottles of various strains on one shelf of the incubator, on another shelf of which a technician had kept some of the BCG used in the treatment of patients. Somehow the bottles must have been mixed up, and so many children had to die.

In any event, Calmette, who was Deputy-Director of the Pasteur Institute, was highly gratified by my results, and when the Nazi pest arrived, in 1933, he wrote me a letter, inviting me to the Pasteur Institute in Paris. He must have guessed without knowing anything about me, how happy I should be to get out of Germany.

Thus on April 15, 1933, we were on the train to Paris. I was rather unhappy though in the laboratory; I was used to much better chemicals and the *Institut Pasteur* was not to my taste as a chemist at that time. They have considerably improved since. Microscopy and bacteriology were marvelous but the chemical work was different. I wanted to work on polysaccharides of bacteria and not lipids. Nothing was known about the role of sugars at that time. Rather soon, however, I determined that I had to leave Paris. It was at the time of the beginning of fascist disorders. So I came eventually in the Fall of '34 to New York. Initially I went to Mount Sinai Hospital because they had created a fund for taking care of refugee scientists. They offered me a temporary position and for months I could go around looking for a job. By then I had more papers because I was working on my habilitation in Berlin. It happened so that in the Summer of 1933 I was being promoted to Privatdozent and dismissed at the same time. By then I was already in Paris anyway. My habilitation work was presented in Berlin just one week after Hitler came to power. Many things kept moving though for a while because even the nazis needed some incubation time.

Have you ever been back to Berlin?

I've been back twice. First in '73 and visited both West Berlin and East Berlin. I went because I had to go to Halle to receive the Mendel Medal from the Leopoldina which is the old Science Academy there.

Erwin Chargaff in the late 1940s (courtesy of Erwin Chargaff).

When you left for New York in '34, did you leave family behind in Vienna?

My father had died before the nazis came to Vienna. My mother was taken by the Germans to Poland and killed. I had tried desperately to get her out of Austria and bring her to America. It is well known that the American consul in Vienna was a nazi. My mother was 65 and had some infection of the eye, and the consul sent her to a nazi doctor who claimed that she had trachoma, a very dangerous eye disease and the consul immediately refused to admit her. I have fought and fought and went to the greatest ophthalmologist in New York and they had the information and sent back their expert opinion that the disease was not infectious but all this did not influence the American consul. Then she was deported. All my books remained in our apartment and have never reappeared. I also have a sister whom I could get out in '38. I have heard since that the actions of the American consul in Vienna at that time showed a characteristic pattern and what happened to my mother had been repeated with many other people.

Have you been back to Vienna?

Yes, many times. The first time it was in 1958 because I was to attend a biochemistry meeting, and I was asked to give the opening lecture. I like Vienna, but with reservations. All I can say is that without the Germans the Austrians would have never become virulent. They have always been anti-Semitic but they would have never created Theresienstadt, not to speak of Auschwitz. They are pleasant, unorganized, but they don't fabricate murders on an industrial scale.

Let's now have a very different question. For many young scientists when they enter Academia nowadays, they start as an assistant professor, and it is very difficult to get funding. How do you feel about this problem?

The story goes back to Humboldt when he founded Berlin University stating that one can only teach by doing research. The Forschungsuniversität (research university) was founded at the time of Napoleon. The first such research university was in Berlin, or maybe Giessen predated it somewhat, where Liebig was the main driving force. The Ph.D. dissertation itself was created as a consequence of that everyone, including the students were supposed to do research. The Americans have taken over this model in a certain mindless way and believe that the only way to grow up in a university is by doing a certain amount of teaching with simultaneously doing research. It is difficult for a young man who does not have collaborators, doctor students. Once you are a professor you can teach because other people do the experiments for you. You have 6 to 8 Ph.D. students and 2 or 3 postdocs. The assistant professor's situation is very different and he really has to tear himself apart. I don't know whether there is a remedy but I have the feeling that what we really need now is the opposite, a diminution in the production of researchers. There should be a healthy proportion maintained. When I was in Germany in the early '30s, there was a good balance, and most of the new doctors went into the chemical industry and only a small fraction stayed on at the universities. Except for the political situation, anti-Semitism, inflation, and bad economy, the balance was pretty good.

How did they decide who should get support?

It was very different. They didn't believe in peer review, and neither do I, for that matter. I can tell you about my case. I was told by Martin Hahn that if I wanted to supplement what the Department paid me, I should make an application to the Deutsche Notgemeinschaft for a grant. A certain amount would then be branched off to supplement my salary, and I could buy chemicals, etc. I made an application and got an invitation in writing to see, and this was incredible, the chief of the Notgemeinschaft who wasn't even a scientist. His name was Schmidt-Ott and he was an orientalist. I was let into the office of His Excellency, and he made me sit down and we started talking about what books I was reading at that time. Then he asked me about my plans and I told him that I had written a proposal about my work on polysaccharides of tubercle bacilli and so on. He asked me what was a polysaccharide. I explained and then he said that I would soon hear from him. Three days later I had the grant. This is not as stupid as it may seem. If you get the right people they don't have to spend too much time on it. I find it silly when they call this proposal valuable and the other proposal not valuable. Most proposals are half so and half so because one doesn't know yet, and if it is good it may still not work. So the peer reviews are completely useless except that they grow into old boys networks. I think the most important is to get the general behavior, the general way of thinking of the person rather than to decide that this is a marvelous problem. They are not marvelous except in lucky hands, in very gifted hands. The hands you can't look at in a proposal.

Could we look back a bit to your research. Was base-pairing your most important contribution?

Yes, it was.

Did you feel it at the time?

Yes, I did. I've often been asked about the double helix and this is related to it. In a way the double helix is a gimmick because there is no evidence that it occurs in the body. It occurred in X-ray diagrams. What was important in Crick and Watson's work was that they proposed something that was *double*. The shape of helix is not so much of consequence. The base-pairing seems to be a novel law of nature that had not been dreamed of because the amino acids and the fatty acids, none of them show this gregariousness, this steric fit. Therefore I always say that of the three Chargaff rules, as they are called, the most important is the one stating that six amino equals six oxo. That is true not only for intact DNA but also for each of its separate strands, whereas the other rules are not true, and it is true also for the complete RNA in the body. We published this but it has not been remarked by anybody, nor acted upon it. My last few papers could have been drowned before publication.

Why?

I don't know why, perhaps because I am an old man. The other thing is that Crick and Watson made this terrific noise; this was an advertising company. I really think that the base-pairing which we discovered goes even beyond the structure of DNA. For instance, we tried to show that base-pairing was even important in completely different things. ATP has certain effects in activating enzymes transferring phosphate groups. They are very nice systems. We showed that when we added polythymidine to

Erwin Chargaff in 1980 (courtesy of Erwin Chargaff).

such a system, ATP was suppressed. In other words the thymine captured the A of the ATP system. We published these observations in the *Proceedings of the National Academy* and there was no echo whatever. We did also good work on blood coagulation; we also worked on hydroxyaminoacids and others. I have never sat on one thing only. These were mostly works with doctoral students at Columbia, and they could each work on what they liked most. I have worked in most fields.

Who supported your research?

I got money from the NSF, NIH, sometimes from Merck, from Hoffman-LaRoche.

You are also a literary man. Were you part of the literary society here in New York?

No.

Do you think that scientists and writers can get together easily?

No, I don't think so, but this is probably quite characteristic for America. I am just reading Kafka's letters to his fiancé who never became his wife. He was not very much in favor of intellectual activities for he was a peculiar man. But you should see how much goes on and whom he sees and so on in Prague of his time. Maybe this is no longer so there either but it was so in the precommunist time.

Cannot it then be that you have a nostalgia for something that no longer exists?

Oh, sure, and I know Europe very well. We used to spend our summers for 30 years in Switzerland, not only sitting in the mountains but we used to go to the cities as well, and to read the Swiss and German papers there. No, no, you are right. I call this a progressive Americanization of Europe. But the deterioration is home-made, of course. If you have 12 years of the nazis and then communism, this is all more than enough. There is then this new ideology recently all over the places. I call it consumentia. The man is consumer. Whether it is a work of art or a work of science, everything has to be consumable, has to be digestible somehow. For this reason Rimbaud or Kafka would have to commit suicide.

Don't you think that the large number of Europeans who came to the United States, like yourself, could have had some Europeanization effect too?

No, quite the contrary, the Europeans have been changed, and I am one of the few who still speaks his mother tongue. I still speak better German than English. I've met lots of students, of course, but usually the Americans have a one-track mind. They can be very good but it's in one thing. But speaking about America, I have already stated this repeatedly, especially when I am saying some very unfriendly things about this country, that there is no other country where I'd prefer to live.

Do you have children?

One son. He is a police detective and lives in Los Angeles. He got a master's degree in administration and then went into the police.

You told me that you were packing up your books.

It's in the office. I used to have a real big one at Columbia. Then after retirement I had to move to a smaller office so I picked out the best of my library, but it's still about 4,000 volumes. Now they are throwing me out because I am too old. I received a letter that I had to go in two months.

Erwin Chargaff with Magdolna Hargittai in 1998 (photograph by I. Hargittai).

What are you going to do now with those 4,000 books?

I am shipping them to Vienna, to the National Library. All these books that you see in my home will go to Vienna too after my death. I don't want books to be junked. Here they would go into the incinerator. Don't misunderstand my gesture. It's not my love for Vienna, it's my love for books. They need be preserved because even if an idiot reads good books maybe he'll be less of an idiot. I have a good collection of books in all the languages which I read in which is German, English, French, Italian, and Russian. My wife and I learned Russian too and read all the classics in original. We started it around the 1950's. And we still read.

How did you find time for all this?

I found time for everything. Also, it was never difficult to drag me out of the laboratory. Science was a hobby, a wonderful hobby but a hobby which was not enough to fill my life.

Quotable Chargaff

I compiled this sampler from three of Erwin Chargaff's books with Professor Chargaff's gracious approval (August 1998) of the selection. [Page numbers are given in brackets.]

Erwin Chargaff, *Voices in the Labyrinth: Nature, Man and Science*
The Seabury Press, New York, 1977

The Ph.D. is essentially a license to start unlearning. [2]

Chemists do not have to bother about the sociology of molecules. [5]

The cell is more than a chemical slum. [5]

Science is wonderfully equipped to answer the question "How?" But it gets terribly confused when you ask it the question "Why?" [8]

With very few exceptions, it is not the men that make science; it is science that makes the men. [12]

It is true of every scientific discovery that the road means more than the goal. [17]

Each pioneer is *eo ipso* an outsider. [18]

Unappreciated geniuses in the natural sciences remain unrecognized; for them there is no posterity. [19]

We can explain it all, but understand only very little. [21]

What counts in science is to be not so much the first as the last. [24]

To the scientist nature is like a mirror that breaks every thirty years; and who cares about the broken glass of past times? [24]

The natural sciences are furiously writing on second volumes of which there exist neither the first nor the last. [26]

In the end, we know nearly everything about nearly nothing. [26]

Humanity has an enormous capacity to disregard the incomprehensible. [31]

We humanize things, but we reify man. [31]

Through each of the great scientific-technological exploits, the points of contact between humanity and reality are diminished irreversibly. [33]

What members of an academy of science have in common is a certain form of semi-parasitic living. [41]

The kind of questions we ask is conditioned by the kind of answers we expect. [43]

In hell, everybody works for the devil! [51]

Science is a way to investigate, not to define, reality. [52]

The sciences, like other professions, cannot endure if their practitioners are unable to know more than an ever smaller portion of what they must know in order to function properly. [53]

The extreme dislike, and therefore ignorance, of chemistry I have often encountered among molecular biologists are puzzling. [57]

In a rotten society even the saints will carry a slight aroma of rottenness. [58]

The road counts more than the destination. [60]

Our period, just because it is so intellectually weak, is given to extraordinarily strong assertions. [61]

There is nothing in the world that does not begin to appear ridiculous if you look at it for too long a time, especially, if you know nothing about it. [65]

Only those who have no use for their knowledge can acquire it. [66/67]

The DNA may well predetermine the shape and composition of the "bio-piano," but not the music that is being played on it. [72]

There are few things in the world before which the biochemist feels as uncomfortable as when he has to deal with life itself. [73]

The Hegel of biology has not yet arisen. [73]

No scientist's lifework is really continued when he is dead. [74]

Never say no to an experiment. [81]

Editors are even more afraid of polemics than of originality. [82]

Scientific fashions last longer than women's fashions but not as long as men's. [85]

The Oscars of science do not set trends, but follow them. [86]

A classic in science is a man who no longer has to be quoted. For the pickpocket, the man with the widest pockets is a classic. [99]

The practitioner of a science must know much more than he can know. Ignorance is no handicap in the arts — Renoir did not have to have seen all the nudes that were painted before him. [149]

Life and all its functions have become a spectator sport. [151]

Science has become an eye without a head, a desperate attempt to fill holes with gaps. [151]

Erwin Chargaff, *Heraclitean Fire: Sketches from a Life before Nature*
The Rockefeller University Press, New York 1978

A good teacher can only have dissident pupils. [7]

Language separates man from man. It is the most faithful mirror of growth and decline. [19]

A teacher is one who can show you the way to yourself. [45]

It is hoped that our road will lead to understanding; mostly it leads only to explanations. [56]

Often the regularities of the "laws of nature" are only the reflection of the regularity of the method employed in their formulation. [57]

Their [successful scientists'] books mostly are accounts of a career, not a life. [62]

The great predicament of the scientist: that what he leaves behind is his experiment, not his experience. [79]

Karl Kraus (1874–1936) on Austrian stamp. Kraus was a satirical and polemical writer to whom Chargaff referred to in the following way: "He was the deepest influence on my formative years; his ethical teachings and his view of mankind, of language, of poetry, have never left my heart. He made me resentful of platitudes, he taught me to take care of words as if they were little children, to weigh the consequences of what I said as if I were testifying under oath. For my growing years he became a sort of portable Last Judgment. This apocalyptic writer ... was truly my only teacher." [Erwin Chargaff, *Heraclitean Fire*, p. 14.]

What counts in science is to be not so much the first as the last. [83]

Great scientists are particularly worth listening to when they speak about something of which they know little; in their own specialty they are usually great and dull. [85]

Most students no longer study nature; they test models. [106]

Knowledge and wisdom are far from being communicating vessels, and the level of one has no bearing on that of the other. [110]

More people have gained wisdom from unknowledge, which is not the same as ignorance, than from knowledge. [111]

We take from others only what we already have in ourselves. [111]

Like all good things in life, we seem to have noticed the environment only when it began to deteriorate. [112]

In science there is always one more Gordian knot than there are Alexanders. [116]

Most sciences are predictive and most of their results are predictable. … for me the real interest begins when these attributes no longer apply. [116]

If instead of the IQ, the intelligence quotient, somebody succeeded in working out the HQ, the humanity quotient, I believe the latter would yield surprising test results. [120]

The vested interest in a scientific subject compresses as it intensifies; it restricts as it deepens. [137]

As happens always in science, it is the facts that pull the thoughts to the bottom of the sea of oblivion. [139]

The definition of molecular biology is the practice of biochemistry without a license. [140]

Man is only strong when he is conscious of his own weakness. [155]

Science is the attempt to learn the truth about those parts of nature that are explorable. [156]

Never before has science become so alienated from the common man, and he, in turn, so suspicious of science. [158]

[On biology:] In no other science is the span so wide between what it ought to understand and what it can understand. [163]

We require a minimum of ever-ready information without which fruitful analogies and even completely original ideas are impossible. [164]

Great scientific concepts often have an entirely noninductive, dream-like quality. [164]

We always cram the newest into our students: lost souls teaching the young how to lose theirs. [164]

Unpredictable associations and the free play of imagination are no less important in science, that is, in real science, than they are in writing. [166]

The arts create their own truth, whereas the sciences are said to reveal the truth that is hidden in nature. [166]

The sign over the door of the laboratory reads: "There is no hurry; there never is any hurry." [167]

The manner in which questions are asked, i.e., experiments designed, is either completely random or conditioned by our ideas of a preestablished harmony, a harmony that we seldom recognize as a contract with God that He has never signed. [169]

I am not sure whether one can say that man — just as he possesses an almost instinctive sense of symmetry — is governed by an equally elemental desire for simplicity. [169]

The longing for simplification has been one of the intellectual driving forces during the growth of modern science. [169]

The attempt to find symmetry and simplicity in the living fabric of the world has, however, often given rise to false conclusions or to anthropomorphic short-cuts. [169/70]

The world is built in many ways: simple for the simple-hearted, deep for the profound. [170]

The ideal state which we approach asymptotically is to know everything about nothing. [170]

The edifice of the animated world rests on two pillars: one is the unity of nature, the other is its diversity. [170]

The availability of a large number of established methods serves in modern science often as a surrogate of thought. [170]

The hope is that all this shattered universe of knowledge will eventually coalesce into a total vision; but it has never done so, nor is it likely to happen in the future, for the more we divide, the less we can integrate. [170]

Every few years the techniques change; and then everybody will use the new techniques and confirm a new set of facts. [171]

Models — in contrast to those who sat for Renoir — improve with age. [171]

Our inability to comprehend life in its reality is due to the fact that we are alive. [172]

Only a pessimist can make a good prophet. [175]

Even the shortest text of any value is untranslatable. [176]

A balance that does not tremble cannot weigh. [179]

A man who does not tremble cannot live. [179]

The caricatures of the past become the portraits of the present. [179]

Most people are wise and applaud the inevitable; but I, inexplicably love to be on the losing side. [181]

Winning and losing are not the same as good and bad. [181]

I have been searching for the third face of the coin. [181]

Nothing that has been done or thought is ever lost. [181]

The principal one [of my struggles] was my quixotic attempt to maintain science with a human face. [182]

My life has been marked by two immense and fateful scientific discoveries: the splitting of the atom, the recognition of the chemistry of heredity and its subsequent manipulation. It is the mistreatment of nucleus that, in both instances, lies at the basis: the nucleus of the atom, the nucleus of the cell. In both instances do I have the feeling that science has transgressed a barrier that should have remained inviolate. As happens often in science, the first discoveries were made by thoroughly admirable men, but the crowd that came right after had a more mephitic smell. [183]

A life is the heaviest investment a man can make. [183]

Erwin Chargaff, ***Serious Questions: An ABC of Skeptical Reflections***
Birkhäuser, Basel, 1986

There are people who occasionally engage in thinking when they have nothing better to do. [vii]

Everybody who has lived long enough knows that at the end the quiet word prevails, the word that promises less than it gives. [viii]

If the world can still be saved, it will be saved by the amateurs. [1]

Death seems to be the only human function that has not been overtaken by progress. [3]

A professional association of amateurs is unthinkable. [5]

The revolt against expertism has long been brewing, with more and more people beginning to realize how much worse than ignorance it is to know the wrong thing at the wrong time. [6]

Meanness is contagious, not nobility. [17]

Books last longer than empires. [21]

Illuminated darkness is not light. [31]

Fervent belief may be stronger than truth, but it is not identical with it. [31]

Those who can speak loudly of their beliefs or nonbeliefs are mostly phonies. [32]

Language is impotent: when it cannot name, we cannot think. [34]

America is in danger of becoming a totalitarian democracy. [51]

Most of what can be done ought not to be done. [79]

I prefer the search for the truth to its possession. [111]

The newest is a deadlier enemy of the new than of the sturdy old. [113]

How many books could there be, if every writer had to make an enforceable vow of originality? [117]

Rome did not have to scream that it had to be Number One: it was. [118]

The peculiar thing about human life is that, while there may be peaks that cannot be surmounted, one can always fall deeper. [136]

As concerns the creations of the human mind, the prevailing taste of a period is not only a tyrant and a slave driver, it is also an abortionist. [146]

Language is the greatest gift conferred on humanity. [163]

If you suspect yourself of otherness, join the Elks. [165]

Reading is no longer state-of-the-art data retrieval. [178]

What scientific research has become in our time is an attempt to change the destiny of man, to undo what millions of years have achieved, to improve on creation. [186]

All languages are as clear as the minds that use them. [191]

We have learned not to be afraid of what experts fear, but to be frightened by what they do not. [223]

Ours is the first generation that has dared to pawn the future. [224]

Frank H. Westheimer (by Koby-Antupit Studio, courtesy of Frank Westheimer).

4

Frank H. Westheimer

Frank H. Westheimer (b. 1912, in Baltimore) is Professor of Chemistry Emeritus, Department of Chemistry, Harvard University, Cambridge, Massachusetts. He received his A.B. from Dartmouth College in 1932 and his M.A. and Ph.D. from Harvard University in 1933 and 1935, respectively. He did postdoctoral work at Columbia University in 1935–36. He joined the University of Chicago in 1936 as Instructor of Chemistry and rose to full professor in 1948. During World War II he worked in the Explosives Research Laboratory of the National Defence Research Committee in 1944–45. He has been at Harvard since 1953 and served as Chairman of the Chemistry Department in 1959–62. In 1964–66 he was Chairman of the Committee of the U.S. National Academy of Sciences that surveyed chemistry and compiled the famous Westheimer Report.

Dr. Westheimer's numerous distinctions include the James Flack Norris Award in physical organic chemistry (1970); the Willard Gibbs Medal (1970); the Theodore W. Richards Medal (1976); the Award in the Chemical Sciences of the National Academy of Sciences (1980); the Rosenstiel Award (1981); the Nichols Medal (1982); the Robert A. Welch Award (1982); the Cope Award (1982); the Ingold Medal (1983); the National Medal of Science (1986); and the Priestley Medal of the American Chemical Society (1988). He has been member of the National Academy of Sciences, the American Academy of Arts and Science, the American Philosophical Society, and Foreign Member of the Royal Society (London).

My wife and I spent a lovely day, July 23, 1995, in the Westheimers' summer house at Owl's Head on Squam Lake, Center Harbor, New Hampshire, and what follows is taken from the conversation we recorded on that occasion. It has also appeared in *The Chemical Intelligencer*.*

I often see references to a chapter of yours on the calculation of steric effects in the 1956 book, edited by M.S. Newman, Steric Effects in Organic Chemistry.[1] *Is that your most important contribution to chemistry?*

It may well be. That was the beginning of what has become known as Molecular Mechanics. But I am also proud of several other projects, such as the work that Harvey Fisher, Birgit Vennesland, and I did on the direct and stereospecific transfer of hydrogen in enzymatic oxidation-reduction reactions. That work began and stimulated many studies that depend on distinguishing between enantiotopic hydrogen atoms, and initiated the use of deuterium as a stereochemical, rather than simply a chemical, tracer, especially in biochemistry. Much later, Ed Thornton, Ajaib Singh, and I invented photoaffinity labeling, which has taken off in a big way. I am very happy, too, with my work with Ed Dennis that showed the impact of Berry pseudorotation on phosphate ester chemistry. But I plead guilty to scattering my efforts among too many projects and failing to develop any of these fields adequately.

How did you become a chemist?

I went to Dartmouth College with the expectation of going to business school and then going into my father's business. He was a stockbroker. I graduated in 1932, at the bottom of the depression — a great time to be a stock broker. But I really enjoyed science and wrote my parents in my junior year that I would like to become a chemist. They took the next train north from Baltimore, where they lived, and came to Hanover. But they didn't come to dissuade me; they came to ask my professors whether this was realistic or just the dream of a nineteen-year-old boy. My professors told my parents that I was actually talented in chemistry and that if I worked hard and was reasonably lucky, they saw no reason why, some day, I might not earn as much as five thousand dollars a year

---

*This interview was originally published in *The Chemical Intelligencer* **1996**, *2*(2), 4–11

as a chemist. Well, corrected for inflation and taxes, I never did; perhaps I didn't work hard enough. This was 1931, five thousand dollars was fairly good money then, and my compulsory retirement from Harvard came a year or so before the large increases in academic salaries. I was an only child, and my father was terribly disappointed that I didn't continue his business, but he never pressured me to do it. His only stipulation was that I should work hard.

How did your attention turn to steric effects?

This was just after World War II. Before the war, I had been fortunate enough to get a junior position on the staff of the University of Chicago. During the war, I worked at the Explosives Research Lab of the National Defence Research Committee (NDRC) in Bruceton, Pennsylvania. After the war, on returning to Chicago, I had no ongoing research, no graduate students, and had to start over again. What could I do, just by myself? I sat down and tried to think, which is a difficult and painful process. I thought through the idea of calculating the energy of steric effects from first principles and classical physics, relying on known values of force constants for bond stretching and bending, and known values of van der Waals constants for interatomic repulsion. I applied this idea to the calculation of the energy of activation for the racemization of optically active biphenyls. Minimizing the energy of a model for the transition state leads to a set of n equations in n unknowns, one for each stretch or bend of a bond in the molecule. It seemed to me that, to solve these equations, one needed to solve a huge $n \times n$ determinant.

Fortunately for me, Joe Mayer came to the University of Chicago at the end of World War II. Joe was an outstanding physical chemist; he and his wife Maria (who won a Nobel prize in physics in 1963) wrote *the* outstanding text in statistical mechanics. During the war, he had been working at Aberdeen, Maryland, using the world's first digital computer to calculate artillery trajectories. Perhaps Joe could have access to that computer and could show me how to solve my determinant on it. So I went to him and asked him to help me. He didn't know about optically active biphenyls, so I made some molecular models and explained the stereochemistry to him, and showed him my mathematical development, up to the determinant. Then, in something like half an hour, he found a mathematical trick that we used to solve my equations without needing the determinant. That's how the solution of real problems in molecular mechanics got started. It has become big business since. Furthermore, it turns out that my instinct for computerizing was correct, since that is the way in which the field has since been developed.

The history of molecular mechanics must include — in fact perhaps begins with — a publication by Terrell Hill that presented the same general method I had invented of expressing the energy of molecules in terms of bond stretching, bond bending, and van der Waals interactions, and then minimizing that energy. Hill published the method,[2] but with no application, no "reduction to practice". I hadn't known that we had a competitor, or that one could publish a bare research idea. After Hill had published, I immediately wrote up the work that Mayer and I had already done, theory *and* successful application to determining the activation energy for the racemization of an optically active biphenyl, and submitted it for publication.[3]

Really, the work on direct and stereospecific transfer of hydrogen is also related to steric effects, but of a very different sort. Fisher, Vennesland, and I showed that, in the enzymatic oxidation of ethanol, alcohol dehydrogenase could distinguish between the two hydrogen atoms on the methylene group. Monodeuteroethanol, CH_3–CHD–OH, consists of two enantiomers, and the two hydrogen atoms of ethanol are enantiotopic. The preference that the enzyme shows for removing one or the other of the isotopic atoms of monodeuteroethanol is not an isotope effect; it is not a question of hydrogen versus deuterium, but of position in space. This is obvious today, but was novel in 1951, when we began this work.

Being an organic chemist, you also served at one time on the Editorial Board of the Journal of Chemical Physics. *This is an interesting combination.*

I drifted from being half a physical chemist through being a physical-organic chemist to becoming a biochemist. When I was in my first academic job at the University of Chicago, I realized that my training in physics was inadequate, so in 1936/7 I took the course on electricity and magnetism with the physics majors. I really took it; I did all the problem sets and took all the hour exams along with the undergraduates. That course sensitized me to the importance of electrostatic effects in organic chemistry. I found Bjerrum's paper in which he calculated the ratio of the first to the second ionization constants of dibasic acids; the difference arises mostly from the electrostatic effect of the negative charge of the monoanion that results from the first ionization. Then there was Eucken's paper in which he made a similar electrostatic calculation for the effect of dipoles such as the carbon-chlorine bond on the ionization of substituted acids. A major difficulty with those papers arose, however, because Bjerrum used 80, the dielectric constant of water, in his calculations, whereas Eucken used 1, the dielectric constant of empty space. Neither dielectric constant worked very well. One can make a logical argument for either, but one simply cannot have both. I was greatly puzzled by this, and could not resolve the problem.

However, in 1937, J. G. Kirkwood came to the University of Chicago. He had been an Assistant Professor at Cornell, came to Chicago as an Associate Professor, and returned to Cornell after only a year as a full Professor. But the one year he spent at Chicago provided me with a wonderful opportunity. I had seen some of Kirkwood's papers on electrostatic effects, but couldn't understand them. When he arrived in Chicago in 1937, I took him my problem, which was just down his alley. He gave me Byerly's book on Fourier's series and spherical harmonics and told me to master it and come back. That wouldn't wash today, when young Instructors and Assistant Professors are under such pressure to produce immediately. I took perhaps a month to learn the mathematics (which, shamefully, I have largely forgotten) and returned to Kirkwood. Then we adapted the equations he had previously published — the ones I hadn't understood — to my problem of electrostatic effects in organic chemistry.

What we did was make a model where we placed charges or dipoles in a sphere or prolate ellipsoid of dielectric constant 2 that was imbedded in a solvent of dielectric constant 80. Then — using Kirkwood's equations — we worked out the classical electrostatics for the interactions of the charges. This is, of course, a terribly crude model, but compared to Bjerrum's model, where the charges were placed directly in water, or compared to Eucken's model, where there was no solvent, it was sophisticated. It accounted for acid strengths of dibasic and dipole substituted acids remarkably well, and it used only a single model to account for the effects of both charges and dipoles.[4] This was real physical chemistry, as was the work on steric effects, and accounts for my serving on the Editorial Board of *J. Chem. Phys.* But I also carried out studies on the mechanisms of reactions in organic chemistry and enzymology, and so worked my way into biochemistry.

I read that you were active during World War II.

I had to do something for the war effort; it was a moral imperative. This was the time of the Metallurgical Project at the University of Chicago. No one was supposed to know what it was about, but somehow many of us at Chicago had a general idea. Someone — I've forgotten who —

asked me if I'd like to join. I said absolutely not, not on any moral principle, but because I thought it was wild, totally unrealistic. I needed to do something to help in *this* war. Instead, I went to the Explosives Research Laboratory in Bruceton and worked on rocket propellants. I made some contributions to the war effort, but regrettably they were minimal.

Did many chemists contribute to the war effort?

Yes, indeed. It was mostly a physicist's war, and the contributions of physicists, such as radar and the atomic bomb, are well known, but many chemists did participate. The most important — and vital — contribution of chemistry was in the synthetic rubber program. The United States was cut off completely from the natural rubber of the East Indies, and would have been helpless without synthetic rubber; the program involved a lot of chemists and a lot of chemical engineers.

There were other contributions of chemistry. The implosion device used for igniting the Nagasaki bomb involved lenses of chemical explosives. New antimalarials were modestly important. Penicillin was developed at that time.

Let me mention a few names. E. Bright Wilson became Director of the Woods Hole Underwater Explosives Laboratory for developing weapons for antisubmarine work. He was really a theoretical physical chemist, later known for microwave spectroscopy, but he managed a very practical laboratory. George Kistiakowsky, also a Harvard Professor, was Chairman of Division 8, the explosives division of the NDRC; the laboratory at Bruceton where I worked was one of those under his supervision. The Director of that laboratory was Louis Hammett. In the Bruceton lab, John Kincaid invented an important rocket propellant.

What was the Westheimer Committee?

In the 1960s, the National Academy of Sciences set up a series of committees to survey the state of the various sciences in the U.S., with the object of presenting sort of lawyers' briefs — honest ones — to the federal government in the hope of encouraging the financing of pure research. The first of these was in astronomy; and then physics and chemistry were more or less simultaneous, followed by biochemistry. A group, which included me, was invited to Washington to discuss writing a chemistry report. I was unable to attend, and probably that is how I got to be selected as chairman, although of course I should prefer to believe that my research in several different fields of chemistry had something to do with it. The report is sometimes nicknamed for me, but is properly titled *Chemistry: Opportunities and Needs.*

It was a splendid committee, with excellent physical and organic chemists, although we didn't have enough industrial representation, or any at all from the government laboratories. Our report had several features that distinguished it from the astronomy and physics reports, and it served as a model for the subsequent biochemistry report. First, our report was written from the point of view of chemistry in the service of the nation. The nation should support chemistry because chemistry benefited the nation; it should support chemistry with the purpose of improving national health, industry, agriculture, and defence and should support it to the extent that it accomplished these objectives. The report was also distinguished for being inclusive. A good many chemists were exclusionists; for example, they didn't want to include any biochemistry. Biochemists worked with impure compounds; it was dirty work and not proper chemistry. As you know, I was doing perfectly clean stereochemistry with coenzymes and enzymes, and my view was a very different one. Our report was inclusive, and properly predicted the future direction of our science.

Writing a report about physical science isn't itself physical science; it is social science. But we tried to set a real scientific standard for our efforts. We tried to back up our statements with an analysis of research papers and of citations from those papers. For example, we wanted to show that a fabulous increase in instrumentation had occurred and was occurring in chemical research, a change that was making chemistry a highly instrumented science. When I started my graduate work, the main difference between Harvard's lab and Emil Fischer's laboratory in Germany at the end of the nineteenth century was that we had Pyrex glass and he didn't. In 1935 we didn't have any of today's important tools: no IR, no electronic UV, of course no NMR, no paper chromatography or column chromatography, no VPC, no mass spectrometry. Essentially none of the instruments we use today existed. We wanted to show the increase in the use of instruments in chemistry, so what we did was to go through the leading chemical journals from Germany, Britain, Russia, and the United States, more or less at random, and note the numbers of papers, and the fraction of papers that presented research that used each of these various instruments. We did this for several different years, so as to show the growth of the use of instrumentation over time, as well as the international nature of their use.

In another study, we analyzed important chemical inventions to show their roots in pure research. We selected about 20 of the most important industrial and about 20 of the most important pharmaceutical inventions of the previous decade and found the first papers published in the scientific literature where the inventors described their work. Those papers were footnoted; so we could see what the inventors themselves had considered the important underpinnings of their own inventions. The vast majority of the references came from fundamental, academic research.

It sounds like an enormous project.

We had very few paid staff. Most of the work was done and most of the areas of chemistry were written up by members of the Committee. We were not paid; members of the National Academy may only be reimbursed for out-of-pocket expenses, although some students who could read foreign languages were paid to sample the foreign journals, and the National Academy contributed some help, and especially contributed the services of Martin Paul. He joined me in condensing and rewriting what the various members of the Committee contributed, and fortunately he and I had similar prose styles, so the final report seems all of a piece. I spent most of a sabbatical on the report.

How did you choose your methodology?

We just tried to do our work as best we could. I was absolutely convinced that it was in the nation's interest to increase its investment in chemistry. So I went to a prominent and wise social scientist and asked him for the best, most honest criteria by which to judge our case. I was confident that they would show that chemistry was grossly underfunded. He threw back his head and laughed at me. There aren't any accepted criteria. There isn't any absolute way to show that the nation would do well by putting more, or less, money into chemistry. So that wasn't much help; we were on our own.

What was the basic message of the report?

That chemistry is vital to the health, prosperity, agriculture, and defense of the United States and that basic research is essential for practical results. That chemistry was (and is) growing in a spectacular fashion, so that the United States would be well advised to continue and to increase the financing of basic research in chemistry. In particular, chemistry was becoming a highly instrumented science, so that the nation should support an increase in funds for instrumentation. That was the message, and I believe it worked.

This was 30 years ago. How did it work?

The first and honest part of the answer is that there is no way of knowing for sure. There is no way of telling how much influence the report had. The federal financing of chemistry increased dramatically, but so did the federal financing of physics and biochemistry. The report did at least one thing: the NSF set up a line item in its budget for chemical instrumentation. I also believe that the report had some influence in inclining the NIH toward more support of chemical research on drugs.

Has there been any similar report since?

The biochemistry report, issued in 1970, was modeled on ours. And about ten years ago the NAS supported a second report on chemistry, entitled *Opportunities in Chemistry*, by a committee chaired by George Pimentel. Pimentel actually asked me to serve on his committee. That was a high compliment, but of course I refused; can you imagine anything worse for him than having someone looking over his shoulder and saying, in effect, "Well, of course you *could* do that, but *we* did it this way"? I found the report disappointing, not because Pimentel didn't do a good job, but because chemistry didn't seem to me to have advanced as much as I had hoped and expected it to.

Can we determine the proper amount of support for science?

No, not in any precise, scientific way. What scientists have to do is to show that the American people is getting its money's worth from its support of science. That can be mainly in intellectual terms as in astronomy, where the public seems quite content with spending on projects with no financial return, such as putting a man on the moon, or it can be in practical terms, but there must be a return in one way or another. Naturally, we chemists think the return from chemistry has been wonderful, but it's a bit difficult to look at this objectively if you're a chemist. Even if you could look at it completely objectively, how do you tell?

Well, you could tell in World War II. The return was so great as to leave little doubt. The investment in radar was extremely successful. The investment in atomic weaponry was extremely successful. The investment in synthetic rubber was extremely successful. Over the years, investment in antibiotics has been extremely successful.

Ask yourself the question as to what scientific advances have benefited the nation, intellectually or practically, in the last 20 years. That's a hard question. Forty years ago, Watson and Crick announced the structure of nucleic acids, and that was a fantastic intellectual breakthrough; but it has taken almost until now before it has had a practical effect; now a huge biotechnological industry has arisen that is based on that discovery and the multiple discoveries it has spawned. About the same time — early 1950s — we had the great advance made by Fred Sanger, who showed how to determine the sequence of amino acids in proteins, and demonstrated that proteins have structures in the same sense that other organic compounds do. That was an extremely important intellectual advance. The biochemical and molecular biological industries that finally arose from that discovery are reasonably important now, and will almost certainly become even more important as we get new pharmaceuticals from the drug companies. We have the additional impact of advances in synthetic organic chemistry on the drug and plastic industries. Where does chemistry pay back to the American people the investment that the American people makes in chemistry? Certainly, in part it is in these industries.

Frank Westheimer in the lab in the pre-instrument days, early 1950s (photograph by MINOT, courtesy of Frank Westheimer).

I think we've been fairly financed, but then you see all the good projects that are submitted to the NSF and NIH and rejected, and you wonder. One of the least well-defined words in the English language is "should". What "should" the American people do? I don't know. What are the criteria? I think we (the NAS Committee) did the right thing thirty years ago in defining these criteria in terms of aid to agriculture, medicine, industry, and the national defense. But we have to add the true intellectual understanding of the world. In 1965, it was reasonable to say that we didn't know how much more financing was justified, but more. In 1995, I'm not sure that we can say, "more". Maybe that's because I'm old and crotchety.

What kind of research do you find most exciting today?

Much of what I find exciting is in the interface between chemistry and biology. On the intellectual front, I am wildly enthusiastic about finding out more about RNA catalysis. There are good arguments that the first living organisms on Earth used RNA and not DNA, and we know why evolution would progress from RNA to DNA.

Life consists of the interplay between proteins and nucleic acids. DNA is the blueprint for making RNA, and RNA is the blueprint for making proteins, and in particular making enzymes, and the enzymes in turn catalyze the formation of DNA. This is the circle of life: DNA makes RNA and RNA makes proteins and proteins make DNA. But it seems very unlikely that life could have originated that way. It's just too complicated. Manfred Eigen, one of the great chemists of all time, had a scheme like that, but it was so complicated that it couldn't have been right. Thomas Cech and Sidney Altman discovered independently that RNA can serve as a catalyst as well as a blueprint. So you didn't need two (or three) starting materials; life can start just with RNA, where you have a blueprint and a catalyst in the same molecule.

A lot of work in intermediary metabolism has shown that there is no direct pathway for the biosynthesis of DNA; DNA is always made by the biochemical reduction of RNA. This is another reason to believe that the origins of the living world must have been in RNA.

Incidentally, it was consideration of the chemistry of RNA that led my coworkers and me to discover the pseudorotation that accompanies the hydrolysis of some cyclic phosphate esters. This is related to an understanding of the reason why evolution led from RNA to DNA: RNA undergoes fairly rapid degradation, and so is not really suitable as a genetic material for long-lived organisms, whereas DNA is much more stable. It doesn't have the chemical mechanism for decomposition that RNA has.

The chemistry of RNA provides fascinating intellectual advances. We can predict that they will lead, in the future, to practical advance. At present, the intellectual advances of the past provide the basis for rational chemotherapy, and that seems to me another of the most exciting developments in chemistry. Most chemotherapeutic agents come as modifications of naturally occurring drugs, and from the serendipitous discovery of their effects. However, more recently the drug industry has come up with at least a few drugs based on rational chemotherapy. One of the earliest of these is the discovery of allopurinol for gout, a discovery for which I am personally grateful.

Gout is a disease where the calcium salt of uric acid crystallizes in the muscles. The pain comes from scraping nerves over the sharp edges of the crystals. The objective of a cure for gout is to get rid of the uric acid. Uric acid arises from the oxidation of a purine that is, in turn, a degradation product of nucleic acids. The enzyme that catalyses the oxidation is called xanthine oxidase. If you could inhibit this enzyme, you would shut off the production of uric acid, stop the production of calcium ureate in the muscles, and cure gout. George Hitchings and Gertrude Elion, who later shared a Nobel prize for other work, looked for inhibitors for xanthine oxidase and found allopurinol, which is structurally very similar to xanthine but is not oxidized so readily and which inhibits the enzyme. There are a few other examples of successful products of rational chemotherapy and lots of research today where one tries to develop new drugs by inhibiting enzymes that are involved in the disease process in one way or another.

Please tell us about your family.

My mother's family came to the U.S. in the 1820s, and settled in Baltimore. My paternal grandfather emigrated from Germany in 1845 or thereabouts as an adolescent, with $20 sewn in his jacket pocket, no relatives to greet him in America, and little if any English. He settled in St. Joseph, Missouri, where he raised eight sons, all of whom, at least for a while, went into the whiskey business with him. My father, who was one of those eight sons, moved to Baltimore with one of his brothers to establish a whiskey business there. They were of course deprived of that business by the Volstead Act of 1917 but later became established in the brokerage business.

I married Jeanne Friedmann, of Philadelphia, and we have now celebrated our 58th wedding anniversary. Our elder daughter is a practicing psychiatrist

Jeanne and Frank Westheimer at Owl's Head on Squam Lake in 1995 (photograph by I. Hargittai).

in Worcester, Massachusetts; our younger daughter was, for many years, a computer programmer, but has another job now.

Can you single out any one person who had the strongest influence on your career?

Aside from my father, that would probably be James Bryant Conant. I went to graduate school at Harvard in order to work with him. But Conant became President of Harvard in 1933, when I had been with him for less than a year, so I transferred to Elmer Peter Kohler, who assigned me a research problem that turned out quite differently than he had expected. Kohler came by once a week, and asked me, regularly, "What did you do last week? What are these set-ups on your lab bench? What are you planning to do next week?", and I'd tell him. Then he would literally snort, turn, and walk out. I never really knew what that was all about. But whatever it was, he didn't instruct me what to do next; he let me work out my own salvation. I owe my independence largely to Kohler because he didn't press me to do what he thought would be best.

Near the end of my Ph.D. research, I applied for a National Research Council Postdoctoral Fellowship. There weren't many postdoctoral fellowships in those days, no government financing of research, and not much industrial support either. But the National Research Council offered about a dozen postgraduate fellowships in chemistry, and I won one of them to work with Louis Hammett, one of my heroes in physical organic chemistry, in his laboratory at Columbia.

About that time, Conant called me into his office. He said that he knew I was getting my doctorate and was interested in my career. What was I going to do? I told him I'd won this fellowship and explained with great pride the problem I'd submitted and was going to work on. Conant had the habit of putting the tips of his fingers together and rocking back and forth while he thought, and he put the tips of his fingers together and rocked back and forth, and then he said, "Well, if you are successful with that project, it will be a footnote to a footnote in the history of chemistry". As I walked out of his office, I realized what he had told me.

Professor James B. Conant (from the Harvard University Archives, courtesy of Frank Westheimer).

Really, it was two things. One was, of course, that my project wasn't very important. The other was — and it may have been pretty stupid that I had never thought this until that moment — that I was supposed to do important things. Chemistry was a lot of fun; it was great entertainment, and I was going to be paid — or at least so I hoped — for entertaining myself with it. Yet Conant had essentially told me that I was expected to do things that were scientifically important. The interview with Conant provided a vital kick in the pants for me. It changed the way I thought about my future.

At Columbia, I did the project that I had proposed, and it worked out beautifully. But it was exactly what Conant had said it was, a footnote to a footnote in the history of chemistry.

Then I set my sights higher — much too high, as a matter of fact. As a physical-organic chemist I was concerned with general acid-general base catalysis and had decided that enzyme catalysis was probably caused by simultaneous general acid-general base catalysis. I was going to demonstrate this in my next piece of research. Amino acids, with their combination of acid and base in the same molecule, should prove especially active catalysts themselves. So I tried their catalysis of the mutarotation of glucose, but it turned out that there was nothing special about them. The project was obviously enormously ambitious, and although I was fundamentally correct about enzymes, demonstrating it was much too big a project for me at the time. That attempt came to nothing, but at least Conant wouldn't have been able to object that the attempt was directed at a footnote to a footnote.

Eventually I settled down to things that were more important than footnotes to footnotes, but not as grandiose as the youthful project that I just mentioned. I never discussed my research with Conant again, but I did restrict myself to things that he might have approved of.

Many years later, after I was a Professor at Harvard, and after Conant had retired from his many careers, I was working in my office one Saturday when someone knocked on my door. I opened it, and there he stood. He looked at me and said, "Do you remember me?" Needless to say, I did.

References

1. Westheimer, F. H. in Newman, M. S., Ed., *Steric Effects in Organic Chemistry*, Wiley, New York, 1956, p. 523.
2. Hill, T. *J. Chem. Phys.* **1946**, *14*, 465.
3. Westheimer, F. H.; Mayer, J. E. *J. Chem. Phys.* **1946**, *14*, 733.
4. Westheimer, F. H.; Kirkwood, J. G. *J. Chem. Phys.* **1938**, *6*, 513.

Gertrude B. Elion, 1996 (photograph by I. Hargittai).

5

Gertrude B. Elion

Gertrude B. Elion (1918–1999) was Scientist Emeritus and Consultant of the Glaxo Wellcome Inc. in Research Triangle Park, North Carolina at the time of the interview. She retired as Head of the Department of Experimental Therapy of the Company (then Burroughs Wellcome) in 1983. In 1988, Gertrude B. Elion shared the Nobel Prize in Physiology or Medicine with Sir James W. Black and George H. Hitchings "for their discoveries of important principles for drug treatment."

She earned a Bachelor's degree in chemistry from Hunter College and a Master's degree in chemistry from New York University. Her research resulted in the dicovery of important drugs, known to the general public as Purinethol®, Thioguanine®, Zyloprim®, Imuran®, and Zovirax®. Although she had to abandon her doctoral studies at the Brooklyn Polytechnic Institute for financial reasons, she was awarded a host of honorary doctorates, the first from George Washington University in 1969. She was a member of the National Academy of Sciences and many other learned societies.

Our conversation was recorded on November 9, 1996, in Dr. Elion's office in what was recently named the "Elion-Hitchings" building at Glaxo Wellcome in Research Triangle Park, North Carolina. The interview appeared in *The Chemical Intelligencer.**

---

*This interview was originally published in *The Chemical Intelligencer* **1998**, *4*(2), 4–11

What was the most important influence that led to your becoming a chemist?

Early on I read the book the *Microbe Hunters* by Paul de Kruif. It had a tremendous influence on me. I'm sure even today's children would be influenced by it. It should be required reading. My main motivation, however, was the death of my grandfather from cancer when I was 15 and about to enter college. I decided then that I wanted to become a chemist so that I might be able to find a cure for cancer.

Where did you go to College?

I went to Hunter College and my brother went to the City College of New York. They were free of charge and we could not have gone to college otherwise. Hunter was an excellent school at that time and it is still a very good school today. In our time there was not open enrollment. You had to maintain an average of 85% in high school and you had to live in New York City. Hunter was the women's college and City College the men's college. Since then they have become coeducational and also have open enrollment. That is to give minorities and other underprivileged an opportunity for higher education.

Gertrude Elion graduating from high school (courtesy of Gertrude Elion).

Little Trudy Elion with her mother (courtesy of Gertrude Elion).

How much did your parents witness your success?

Well, they witnessed my coming to work at Burroughs Wellcome. My mother died in 1956, so during my first 12 years she knew that I was doing well. My father died two years later, in 1958. My father came to the United States as a small boy from Lithuania and graduated from New York University School of Dentistry in 1914. My mother came from Russia in 1911 at the age of 14. They were doing well until they lost everything in the stock market at the very beginning of the Great Depression. Yet they would never have thought of me or my brother not getting a College education. As immigrant Jews they knew education was the key to success, and they wanted their children to be educated.

My first outside recognition was my medal from the American Chemical Society, in 1968. My first honorary degree was in 1969, so 10 years after

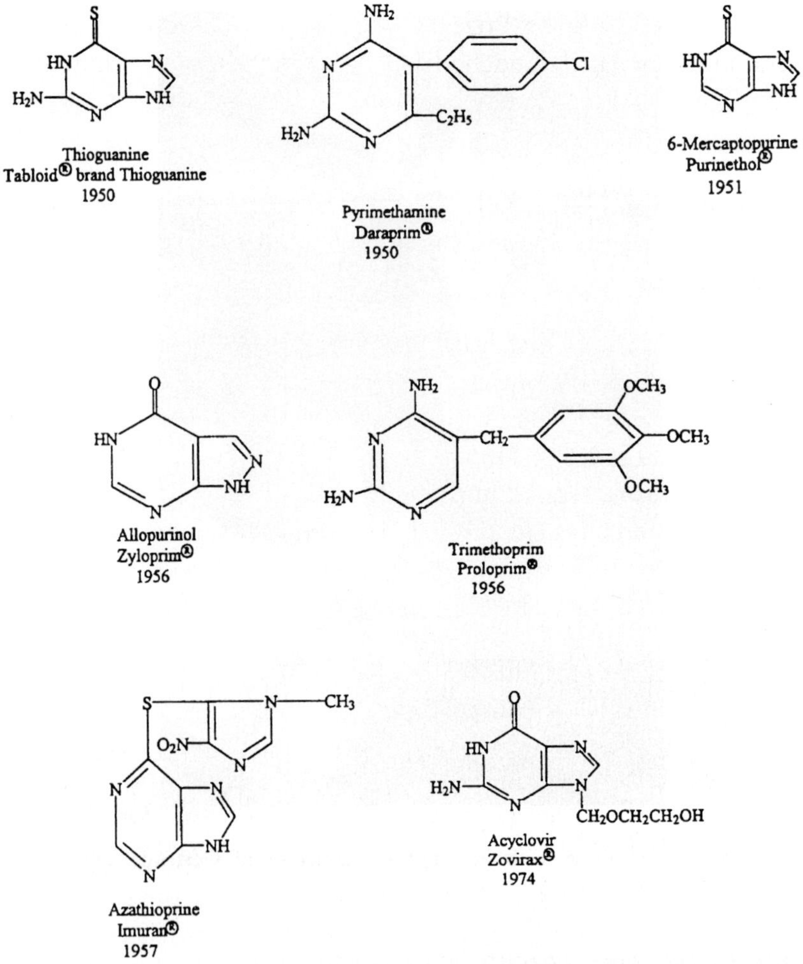

Thioguanine
Tabloid® brand Thioguanine
1950

Pyrimethamine
Daraprim®
1950

6-Mercaptopurine
Purinethol®
1951

Allopurinol
Zyloprim®
1956

Trimethoprim
Proloprim®
1956

Azathioprine
Imuran®
1957

Acyclovir
Zovirax®
1974

Structural formulae of Gertrude Elion's drugs.

my parents died. But they knew that I loved my work and they did know that I had had some successes with 6-mercaptopurine, for example, in the early fifties. But they did not know about the rest.

I have read that you took grandnephews and grandnieces to the Nobel ceremonies in Stockholm.

Oh, yes, and they behaved very well. I insisted that the children were going to be at the dinner. The whole family came to Stockholm, including five children. Unfortunately, my brother was ill and could not come. Generally by the time people get the Prize their children are already grown up and they seldom bring the grandchildren. I did not have children myself so I brought my brother's family, his four children and their spouses, and his five grandchildren.

I would like to ask you about George Hitchings. Obviously there must have been a complex relationship between the two of you, going through different stages.

Absolutely. When I was first interviewed by him in 1944 and he told me that he was working on antimetabolites, nucleic acid derivatives, I really did not know what a purine or pyrimidine was. People did not know much about DNA in those days. However, I was very intrigued by what he was telling me. He was interested in the biochemistry of nucleic acids. Hitchings graduated from the University of Washington and got his Ph.D. from Harvard University. He joined the company when he was 37, in 1942.

Gertrude Elion and George Hitchings (1906–1998) in their laboratory; Elvira Falco is seen at the end of the lab (courtesy of Gertrude Elion).

It was during the war that women were suddenly able to get a job in chemistry, which they had not been able to do before. I started as Hitchings' assistant. He had another assistant, also a young woman, Elvira Falco. She's been retired now for a long time. She left us during the fifties, got married and had a child. Later she worked as a chemist at the Sloan-Kettering Institute in New York. She now lives in Maine.

Hitchings wanted certain compounds to be made. Since I had my Master's degree in chemistry and had all good grades, he figured I could probably do it. I could read German and I could follow the German literature. A lot of it was Emil Fischer's work. I had learned Yiddish at home and then I took German at college. Chemists had to know German. You could not read *Beilstein* if you did not know German.

Our department consisted of the three of us at the beginning. We were focused on making antimetabolites of nucleic acid bases. Although Avery, MacLeod and McCarty had just discovered that DNA is the carrier of genetic information, we really did not know the structure of DNA at that time. The double helix was described only in 1953.

What was your first research?

Hitchings was very intrigued by the effects of sulfonamides on bacteria. By the time we started our work together it was understood that the reason that sulfonamides were so effective against bacteria was that they were antagonists of the metabolite paraaminobenzoic acid. They were very close in structure except that the carboxylic acid was replaced by a sulfonamide group. We started out with the idea that if we made things that were very similar to the naturally occurring purines and pyrimidines, they might actually antagonize the formation of DNA. The idea was to interfere with the division of things like bacteria, tumor cells and malaria parasites since anything that had to multiply in order to survive had to make DNA.

You wanted to interfere with the DNA of the bacteria. Was there any danger that you would interfere with the human DNA?

Of course, we had no way of knowing whether this would work or not. We did not even know which enzymes might convert these bases to nucleotides. The best way was to make the compounds and test them. We had antibacterial tests going. We had one bacterium, *Lactobacillus casei*, which was ideal because when we gave it a natural purine and thymine, it would grow and divide. It was a way of determining whether we were interfering with the utilization of the purine, or were we interfering with the utilization of the pyrimidine thymine. Were the compounds getting into the DNA and fouling it up so that it did not function? To help sort out these several possibilities, we had to have some other systems to look at. We made an agreement with the scientists at Sloan-Kettering Institute who had animal tumors growing in mice and asked them to test the compounds that we thought were promising. If they worked in *Lactobacillus casei* and inhibited growth they might inhibit the growth of tumors. At the same time, if you are testing a compound in a mouse, you'll see whether it is toxic to the mouse, and whether you can affect the tumor without affecting the mouse. A young biologist, Samuel Bieber, came to work in our lab and he worked on frog embryos. He would fertilize the frog eggs in a big pan and watch them divide. He could then put compounds into the water and see whether they prevented cell division and, if so, at what stage. That was another very useful test.

Since we had colleagues in England at the Wellcome Research Laboratories who were working on malaria, we sent them compounds to see if they would inhibit the growth of malaria parasites. What we were really doing was spreading the net widely to see in what kind of system these compounds would work. It was very interesting because soon we began to find that certain types of compounds worked on malaria and bacteria, and a different set of compounds worked on tumors. The purines separated themselves out as being effective on tumors and even on some viruses, whereas the pyrimidines were definitely antibacterial and antimalarial. The reasons for their being antibacterial and antimalarial we did not understand at first, but we soon found out that in order to be active they had to be diaminopyrimidines, and, then, they could have different side-chains. The diaminopyrimidines were folic acid antagonists. But the structure of folic acid was not known at that time. It was known that if you gave the *Lactobacillus casei* folic acid, you did not have to give it purines or pyrimidines. So we had in that one organism a way to find out whether the compound was a folic acid antagonist or purine antagonist or thymine antagonist. That's how we chose the compounds to go to Sloan-Kettering for testing in the tumors. It was really amazing that in the same year, in 1950–51, we found the first good antimalarial and also the antileukemic compounds, mercaptopurine and thioguanine. Elvira was making pyrimidines and I was making purines.

The choice of focusing on DNA proved to be very timely and very lucky.

Of course, it was because George had done his doctorate on DNA, on analytical work to find out how to measure the amounts of adenine, guanine, and so on. In those days there was no HPLC, there was not even good ultraviolet absorption spectrometry. So he isolated these compounds as picrates and did quantitative analysis. When we got our first Beckman DU Spectrophotometer in 1946, it was a great event.

As we began to have some successes, we got additional people and then, as we had further successes, we got more people. The department grew and I became head of a little section of the department. Eventually, I got a department of my own with about 20 people. When we moved here, to North Carolina, in 1970, my department grew to about 55 people. But it was not just chemistry, it was also biochemistry, immunology and pharmacology. When one compound began to show activity in gout, for example, I had to have pharmacologists doing metabolic studies. Eventually we also had a virology section. People used to tease me that I didn't have a department, I had a mini-institute because I had all the people in that department that I needed to bring a drug to development.

Did being in charge interfere with your research?

For about 15 years I did syntheses with my own hands, but I also helped to analyze the results from the *Lactobacillus casei* screen and analyze the results from Sloan-Kettering. I had to make decisions about what needed to be made. I was also interested in the mechanism of action, so I became involved in biochemistry. Then I began to work in pharmacology, investigating drug metabolism. At that time it was unusual to study the metabolism of drugs in people.

Do you think researchers today may have a similar latitude as you did in those times?

Probably not, and especially not in large companies which are usually divided up into departments. It may still be possible for small companies. However, we always had to do everything ourselves in those early days. Although there were some pharmacologists there, they were involved with antihistamines and muscle relaxants. I could not go to them and say, "Could you help me find out what happens to 6-mercaptopurine in the body?" So I decided to make radioactive 6-mercaptopurine, put it in the mouse, collect the urine, and find out what happened to the drug. I synthesized the metabolites, tried to find out how the compounds worked, and analyzed whether they were converted to nucleotides. It took me years to follow the metabolism in animals and then in man because the technology to separate and purify the metabolites was just not there.

Did the company support you all the time?

Oh, yes. This was because the company was very different. Burroughs Wellcome was owned by a Trust. Originally Silas Burroughs and Henry Wellcome were two American pharmacists who went to England to make their fortune. They remained in England and formed a company in 1880. Burroughs died fairly early and, because it was a partnership, everything was left to Wellcome. Wellcome became very interested in tropical medicine. He also became very wealthy. Since his wife had left him, he did not leave her anything in his will. He left everything to a Trust which was to have five trustees, including a scientist, a lawyer, and a businessman. They had to decide how to spend the money that the company made. There was nobody looking to make a personal profit but only for ways to give the money away in philanthropy. Thus, there was nobody looking over our shoulders.

The company, when I joined it, made most if its money from a drug for headaches. Then one of the chemists in England made an antihistamine. They made a lot of over-the-counter drugs, they made ointments, they isolated digoxin from digitalis leaf, they got into antibiotics, they were very heavily into vaccines. They were the suppliers of vaccines for the British Army during the war. They also made polio vaccine after the war. These were their main products; they did not have a lot of synthetic organic chemistry then.

Our compounds were medically important but they were not making a lot of money. You could not make a lot of money on 6-mercaptopurine; it was very useful for treating childhood leukemia but that was not a major product. It was not really till we got into antibacterials, for example, trimethoprim, which was one of the diaminopyrimidines, that the company began to see some profit for what we were doing. Then with allopurinol, which was a direct result of trying to prevent 6-mercaptopurine from being oxidized, gout became a major market.

Elvira Falco synthesized allopurinol. We tested it on *Lactobacillus casei* and as an antitumor agent but it was not active. So it was put aside. We also had a woman in our lab named Doris Lorz who worked on xanthine oxidase because it was one of the few enzymes which acted on purines and she could isolate it from milk. She tested many purines and pyrimidines as well as antimetabolites on that enzyme. So we knew allopurinol had activity as a xanthine oxidase inhibitor. We really did not proceed with that until the time came when we needed a xanthine oxidase inhibitor. We had at least a dozen compounds which inhibited xanthine oxidase.

We picked allopurinol to work on, and it turned out to be the right one. We could have picked several others and, in fact, after we found that allopurinol did work, we did look at the others. None of them was as good as allopurinol for various reasons. Some were not well absorbed, some were metabolized or excreted very rapidly. Allopurinol was unusual for two reasons. One, it was a very good inhibitor, and two, it was oxidized into a very good inhibitor. The second product, oxipurinol, was the xanthine analog and had a much longer half-life. Allopurinol itself had a half-life of only an hour and a half, whereas its product, oxipurinol, had a half-life of 18 to 24 hours. So we had a drug that could reach a steady state and really continue to inhibit xanthine oxidase.

How about gout itself, the cause is not yet eliminated?

What we have is only a treatment. We aren't able to cure people. They can go on for the rest of their lives taking allopurinol. It just controls the amount of uric acid formed so they do not have accumulation of uric acid in their joints or kidneys. In this sense it has solved the problem. However, about 3% of people get rashes from allopurinol and have to take something else.

Which is the single most important drug you discovered?

I knew you would ask me that. Everybody asks that. But I do not answer that question, I always say that I don't discriminate among my children. In a way I feel that these are my children. They came along at different times and each one was very exciting and wonderful and so was the next one and the next one. You do not forget about the oldest because a new one is born.

Which drug brought the most money to the company?

Acyclovir. It's an antiherpes drug. It was the last one and it was, and still is, a major drug for the company.

I understand the company is no longer a nonprofit organization.

No, it's not. What happened was the following: Some time ago, and this was after I retired, the Wellcome Trust who owned all of the stock of the company, decided that they were not making enough money to support biomedical research on a large scale. They got permission from the Charity's Commission in the U.K. to sell some of their stock. At first they said that the Trust should not lose control of the company, but could sell up to 49% of the stock. Later they were given permission to sell more, since the money would be given away for biomedical research. By that time the company was doing very well. About two years ago, Glaxo offered to buy the rest of the stock and to merge the two companies. The Wellcome Trust accepted the offer.

What happened to the name Burroughs?

It sort of disappeared with the merger. Actually in England it was the Wellcome Foundation which managed all of the Burroughs Wellcome companies and they were all over the world. After the merger the name was changed to Glaxo Wellcome.

There is a beautiful book about women scientists who had made to the top but this is still very rare. How do you view this?

I think women will catch up eventually. We were kept back until the Second World War. You couldn't get a job in a chemistry lab; they just didn't want you. Also, there were very few women Ph.D.s, very few women teaching in higher education. This is gradually changing. In my day, if I had wanted to go to medical school I couldn't have gotten in. Duke Medical School did not take women; very few medical schools in this country took women. Women had to go abroad if they wanted to go to medical school. So it is taking a while for us to catch up. There were also very few women in the National Academy of Sciences because there weren't enough who had made a real scientific contribution. But all this is changing. Now women are getting into the Academy. Medical schools now have 50% women. It is going to take a while though before a lot of women reach the top.

There are indeed few women at the very top in chemistry but those who make it there seem then to be in great demand, overburdened by memberships and assignments.

I think you are right. That question you had asked me before is important. Once you become a department head, can you still do real science? I think you can but only up to a point. You are no longer at the bench. If you really keep your hands in, you must do it in a way that the people working for you want you involved. They may want to come and talk to you about it, they may want you to come and ask what happened yesterday, but you have to sense the proper degree of such an involvement. I tried to do that, I really did. For me it was a family. I did want to know what was going on and in fact I became involved in trying to help them if they ran into a snag. I tried to develop a team effort so that people wanted to work together. I acted as a conductor of an orchestra. I wanted everyone to be a good musician but I wanted them to play together. I think I succeeded in that.

George Hitchings made a discovery by employing you. Did you employ anybody who became a great success?

Yes, though not in the same way. I think almost all the people I employed were very successful. They became senior scientists, section heads, department heads, and one of them became a research director. They were with the company for 25 to 30 years. Many took early retirement after the merger. The character of the place, however, has changed. It is big, it is subdivided in different ways than it used to be. The chemists, biologists and pharmacologists still work together but not in the same atmosphere. They are separated by buildings, they are separated by different departments, it is just different. I know they think I am very old fashioned, because I like people to work closely together when they are working on a project, to learn from one another, the chemist learning from the biologist and the other way around. That is much more difficult in a large company.

Does it make any difference for the research to be here in North Carolina and not in New York as far as manpower is concerned?

No. Because we have many good universities nearby and people are mobile now. If they want to come to work with you, they come across the country; they don't care where you are.

Did your life change much when you moved to North Carolina?

It made it more difficult to go to the opera. I am a great opera fan. I go up anyway but I don't go very often. The move came at the right time in my life. If I had been younger, it would have been much more difficult. New York has a lot to offer, but at the time I came down here, I was not going into New York that often anymore. I lived in the suburbs and I went in to the opera, but I did not use New York as much as I did when I was growing up. Then I really used New York, the museums, the theaters, the opera — everything. In fact, if anything, it improved my life when we moved here. I became more closely associated with the universities than I had been in the New York area, where there are a lot of companies and the universities do not need you. I received adjunct appointments at Duke and at the University of North Carolina.

Did the Nobel Prize change your life?

Oh, yes. I thought it would not, but exactly the thing you pointed out happened to me. I am so much in demand to be on committees, to be on scientific advisory boards, and to give lectures. Also, it came when I was already retired, so that it did not interfere with my research. It added one thing which I very much enjoy. I take on a third year medical student at Duke each year to train in research and that has been very satisfying. Up until now there have been 6 women and 6 men, just as their percentage is at Duke Medical School. They work in the lab at Duke with Dr. Henry Friedman who is in pediatric oncology. We deal mainly with chemotherapy of brain tumors. I act as a mentor, I go to Duke every couple of weeks, spend several hours, discuss the data, talk about where we go from here, what it means. I try to get them to think because most doctors are not really trained to think. They learn a lot of facts but they don't learn how to solve a problem. If nothing else, they know how to approach and to solve a problem by the end of the year they spend on research. They usually write two good papers at the end of the year. And then it's up to them. I am not necessarily trying to train researchers, I want to teach doctors how to solve problems. If they want to do clinical work that's fine, or if they want to do research that's fine too.

Of the 12 students I have had so far, the first two have gone into research in cancer centers; both were men. Most of the women have gone into practice, usually not private practice, but hospital work in universities. It is hard to know what will happen to all of them. After they have been with me they complete their fourth year of medical school, then three years of residency. They are just coming out now and beginning to make decisions. I saw one of them in Chicago last week, she is working in a hospital as a gynecologist. She wrote me a very nice note, saying: "I hope you are not disappointed that I am not in research." But I never have tried to influence them. They must do what is best for them.

You seem to have been concerned with not having earned your doctorate originally, a sad story. But was it so important?

It turned out not to be. In fact, in some ways it was a good thing because it made me more visible that I could do this without a doctorate.

However, for a woman, believe me, it would have been much easier with a Ph.D. For one thing, at the time I came to work at Burroughs Wellcome I was offered a job at another company. When I was interviewed there, they said: "Oh, you only have a Master's degree, therefore we should tell you now, this is as high as you can go." I thought it was a very strange thing to tell me. They had no idea how good I was or if I was any good; I did not ask them how far I could go, it was not even on my mind. That was not what I was working for. I wanted to do something important, I really wanted to cure cancer.

I have to say that George Hitchings said to me, when I had to make the decision: "You won't need a doctorate." I knew that as long as I stayed with him, he knew what I could do and that was fine. But if I had tried to go to somewhere else, I am not sure that would have been the case.

The first honorary doctorate, George Washington University, 1969, with George Hitchings (courtesy of Gertrude Elion).

George Hitchings is still alive; he is 91; he has had a couple of strokes, and he is in rather frail condition. I have not seen him for about six month but he lives nearby. I did go there for his 91st birthday but I found it difficult to communicate with him. He was a wonderful man. He gave me an opportunity to do as much as I could do, he never said: "It is not your business to do that." He never said: "You are not a pharmacologist, not a virologist, not an immunologist," so I became all of them.

This comprehensive approach to drug research must have been quite different from today's race for new drugs.

Today people give preference to high-throughput screens. They can test 30 000 compounds in a month. I never thought that way. When I found some activity, I was happy to pursue improving that activity by making progressive changes in the compounds. I think that must still go on but it is not as popular any more. It is popular to screen 30 000 and find some "hit" compounds and then pursue structure-activity relationships. What we did was, as George used to call it, a biochemical approach to chemotherapy. Try to find out what are the systems that you want to interfere with, what enzymes you want to inhibit, and then concentrate on that. Of course, we started out not knowing the enzymes but we often discovered the enzymes as a result of finding compounds that interfered with them. It was a circle. When we found an active compound, it taught us something about the biochemistry that we were able to utilize to make other compounds.

People now want to do it the other way. They want to know what the target is, and then make and screen thousands of compounds against that target. That may be one way to discover new drugs. Then you have a target that everyone already knows about. How about a new target that nobody knows about yet? You are not going to find it that way. So there has to be another way. You can identify a target by finding an active compound by screening, not knowing ahead of time how it works. Find out what it is doing and then you have a new target. We did not know the exact target when we began the antiviral work. We were screening compounds against the viruses, which were potential antimetabolites of nucleic acids. That was true with acyclovir which was made by another chemist, by Dr. Howard Schaeffer. He was studying something entirely different. He was not looking for antivirals, he was looking to see how much of the ribose moiety of adenosine he could eliminate and still have a compound that is acted on by the enzyme adenosine deaminase. He found out that he only needed a portion of the sugar, an acyclic side chain with a terminal hydroxyl group. This suggested that the other enzymes might be fooled in the same way. When we put this adenine derivative into the antiviral screen, it had activity against herpes viruses. Then began the effort to improve on it. When he put the acyclic side chain on a guanine instead of an adenine, this was acyclovir. What we did not know, but we found out, was that there was a very specific enzyme coded for by the virus, but not present in uninfected cells, that was required to activate the compound to be an antiviral. When we found the antiviral activity, the first thing we asked was: "Why is it an antiviral?" Then we found this activating enzyme, and found that acyclovir had to be converted to a triphosphate to inhibit the viral DNA polymerase.

I am trying to get people to realize that we do not know all the targets yet; there is still a lot to find. To solve a problem, you don't need to know everything to start with. You need to find out why things work. I like to think that all the people I trained developed the attitude of "WHY?" That was the big question one always had to ask; why does it work? If you are a doctor, how do you make a diagnosis if you don't understand what is happening? This is the most important thing a doctor can learn, how to diagnose. This is complex. People may have two or three different diseases at the same time.

I know that if you are a woman and you want to have both a family and a career, it is very difficult. I did not have that problem, I did not have a family although it is not the way I had planned it. My fiancé died in 1940 of a disease that could have been cured by penicillin that was discovered two years after he died. I think that encouraged me to stay in research. It was a feeling that there were still so many things to do, that research is worthwhile and that here you can really make a difference.

You can imagine how I feel when somebody comes to me after I have given a lecture somewhere and says: "I have a kidney transplant thanks to you and I have had it for 25 years." I get many "Thank you" letters and I keep every one of them. They represent the real reward, making sick people well.

Sample of quotations from letters from patients
(courtesy of Gertrude Elion)

D.S.: "I'm sure there are thousands of people saved or helped by your drugs, but I just wanted to tell you that my daughter, Tiffany is alive and well today because of you and your research."

N.K.C.: "I now know that herpes-induced encephalitis accounts for only 10% of all encephalitis cases but causes 50% of encephalitis death. I feel very strongly that your work (particularly the development of acyclovir) helped me beat these odds."

S.C.: "I am one of the grateful individuals who has directly benefited from your research that resulted in the discovery of Imuran. I received a kidney transplant from my brother almost 7 years ago and my quality of life is superb."

B.O'C.: "I received a cadaveric kidney transplant ... I have been on a regimen of Imuran and prednisone ... I owe you a very special thanks."

J.E.P.: "I have a little boy who was diagnosed some two years ago with ALL. Since that time, he regularly takes #6-Mercaptopurine. ... with inexpressible gratitude for having contributed to the savings of one human life so very dear to me, and so many human lives, that I write to you ..."

K. and J.B.: "In 1984, our then 5-year-old daughter was diagnosed with Acute Lymphocytic Leukemia and was put on a chemotherapy protocol including, among other drugs, 6-mercaptopurine. Our family went through many difficult times, both physically and emotionally. We are thrilled to tell you that our daughter has been in remission for over 5 years, with no relapses, and has been completely off medication for 2 years, 3 months, and 24 days! She will celebrate her 11th birthday in May and is a delight to her parents. When we see the adoration in the media of overpaid sports figures and entertainment figures with inflated egos, we can only think how much more you and your colleagues have contributed to society, with little or no recognition. You are truly a hero."

Carl Djerassi, 1996 (photograph by I. Hargittai).

6

CARL DJERASSI

Carl Djerassi (b. 1923 in Vienna, Austria) is Professor of Chemistry at Stanford University. He received his education at Tarkio College and Kenyon College (A.B., 1942) and completed it with his Ph.D. in chemistry from the University of Wisconsin in 1945. He had worked for Ciba Pharmaceutical Company in New Jersey and for Syntex, S. A., in Mexico City before his first academic position at Wayne State University in Detroit, Michigan, in 1952. Concurrently with his academic career, he was involved with chemical companies throughout most of his career. He has been at Stanford University since 1959. He has published over 1200 articles and seven books on the chemistry of natural products and the application of physical measurements for structure elucidation. In medicinal chemistry, he is best known for oral contraceptives.

He is a member of the National Academy of Sciences and the American Academy of Arts and Sciences, and has received the National Medal of Science (1973); the National Medal of Technology (1991); the first Wolf Prize in Chemistry (Israel); the Perkin Medal; the Priestley Medal of the American Chemical Society (1992); and many other honors and awards. Lately, he has also become a writer and has published novels, short stories, autobiographies, and essays. His web page address is www.djerassi.com.

The following conversation was recorded on February 26, 1996, in Professor Djerassi's San Francisco home. It first appeared in *The Chemical Intelligencer.** I am grateful to Dr. Djerassi for letting me use his poems to illustrate our conversation in this edition.

---

*This interview was originally published in *The Chemical Intelligencer* **1997**, *3*(1), 4–13

What do you consider to be your most important achievement in science?

Importance is a subjective term. You may mean something that is the most complicated, or something that impressed your colleagues the most, or something that had the most societal benefit.

From a societal point of view, the first synthesis of an oral contraceptive is without any doubt the most important. This was almost 45 years ago, October 15, 1951. It's a very specific date recorded in our lab notebook when we synthesized the steroid that was the active ingredient of oral contraceptives eventually taken by hundreds of millions of people. There are very few synthetic drugs that have been around for forty-some years. I don't really know if there has been any more important discovery from a societal standpoint during the postwar period. For example, genetic engineering may bring its most powerful societal benefits and impact in the next century.

If I'm talking about my achievements as a scientist, I consider myself more as a degradative chemist rather than a synthetic chemist, because what has always interested me was structure elucidation. That, in a way, is becoming somewhat anachronistic. Structure elucidation is nowadays just a pseudonym for applied spectroscopy. We can now determine the structure of most compounds either by a combination of mass spectrometry and NMR or X-ray crystallography. I'm talking about organic compounds, of course. You may also call it analytical chemistry. But historically, that was not the case in natural products chemistry which really created all the important advances in organic chemistry. Synthetic chemistry came only afterwards. Thus, even many of the subtle synthetic reactions were discovered in steroid chemistry. There was no other field of organic chemistry where the impact of physical methods was as noticeable as it was in natural products chemistry in general and steroids in particular.

First came ultraviolet spectroscopy, and for 10–15 years, it was the only spectroscopic method in the organic lab. Fieser and Woodward were the first who carried out real generalizations about ultraviolet absorption, relating specific structures to UV. The next was infrared, and most of the early applications of infrared spectroscopy started in the steroid field. We used infrared for identification. Before that it was mixed melting point. Yet another 10 years later came NMR, and a little bit later mass spectrometry. In between came the chiroptical methods and some others like ESR, but these were more specialized.

I, as a structure elucidation chemist, was emotionally receptive to new physical methods. Synthetic chemists are usually not so receptive. You almost had to drag them onto the use of physical methods. So I was different from most of my colleagues, in that I also became interested in the physical methods themselves. First it was optical rotatory dispersion and then mass spectrometry. I published hundreds of papers on them. The only reason I could be doing this so fast was that I was a steroid chemist first, and steroids were the absolutely ideal templates for all of these methods. In this context, this may have been my most important contribution to science at that time.

Having this arsenal of physical techniques, what kind of tool are we still missing, if any?

In organic chemistry, there may not be any, and this may not be a very clever answer. I don't think we'll have a paradigmatic shift or quantum jump in what we can do in experimental organic chemistry with new physical methods unless you categorize the methodologies of combinatorial synthesis as "physical methods." We really don't need these any more. We can basically elucidate the structure, easily and quickly and with small amounts, of any molecule, except for some bizarre complicated ones, with the applications of the existing methods, for example, by electrospray mass spectrometry, and by coupling two or several methods. Further, couple this with X-ray crystallography which can handle now giant molecules, and you may ask the question what type of molecule do you want to look at which we haven't already looked at? So I believe that the new methods will only be improvements and cannot have the enormous impact that the earlier ones had. That the intellectual challenge of structure elucidation is not around anymore may

István Hargittai and Carl Djerassi during the interview (photograph by M. Hargittai).

be good for science, but it is surely bad for the individual chemist. The natural products scientist is not a chemist anymore but an analytical chemist or a spectroscopist. It's not chemistry anymore; it's black boxes. Today we have specialists, and the giants of the past, like Sir Robert Robinson and R. B. Woodward, were generalists. That is what really made them so great.

Natural products chemistry seems especially important in less developed places.

That's absolutely true. I've always felt that natural products chemistry was one of the best forms of organic chemistry that people in the so-called third-world countries should do. It gives them an advantage which people in advanced countries don't necessarily have. It is not only the availability of plant products and sometimes even animal products, but the knowledge of their utilization. This is true for Chinese and Indian indigenous medicine or what had originally interested me, the Mexican and Latin American medicine. There is a great deal of knowledge out there and I worked a lot on this in Mexico and in Brazil. A lot of it was based on indigenous use of these materials. But even in this kind of work the new methodologies have changed not only structure elucidation but the methods of screening as well. Before, you studied a whole animal and were observing the behavior of a mouse, blood pressure, etc., whereas now we use receptors for testing biological activity and we need minute amounts of the material. Instead of chopping down entire trees, we merely need a leaf or a slice of bark. There is then an important sociopolitical argument. Are we really destroying the ecology and the botanical environment of a nation? In most instances this is utter

YOU WASH THIS SHIRT LIKE A CHEMIST

I didn't know there were chemist's hands.
Do I touch you like a chemist?

Grip your wrist
The way I grip the necks of Erlenmeyer flasks?

Hold your buttocks
The way I hold round-bottomed vessels?

What else does this chemist do?

Suck your tongue
The way I suck pipettes?

Carry your night scent in the morning
The way I carry my day's lab odor in the evening?

Wear your bath robe
The way I wear my lab coat?

Are chemist lovers different from others?

If anyone knows,
You do.

From: Carl Djerassi, *The Clock Runs Backward*. Story Line Press, Brownsville, Oregon, 1991, p. 26. Reproduced with permission.

nonsense in my opinion, unless you want to extract the material in tons. I know what I'm talking about because the last 20 years of my activities, up to about the early 1990s I was focusing on marine natural products which in many respects are even more sensitive in that context because people expect their national borders to extend sometimes to 200 miles out in the ocean. We, in America, recognize only 12 miles or something like that, but definitely not 200 miles. There are, however, a number of countries that are very sensitive to your entering these waters and diving in them and doing any collecting, and claim that we affect their sovereignty and exterminate their natural resources. In reality what we are looking for is one sponge or one fraction of a sponge because we can do all the work with that. This is significant too because the pharmaceutical companies have again become interested in screening because of the availability of mass screening approaches. Thousands and thousands of compounds can be screened at a time rather than doing one at a time. At the same time, that's not chemistry anymore.

How and when did you turn to marine products chemistry?

It was just another case of serendipity. In my case, serendipity was always the common denominator. I started with steroids in graduate school. Even though I did all kinds of departures, I always returned to steroids. In the 1960s, when we were working in mass spectrometry at Stanford, we were almost unique, and people were sending us materials from all over the world. We knew about the mass spectrometry of steroids more than anyone else in the world. At one point Paul Scheuer sent me a small sample of a new sterol from Hawaii that he'd isolated. By then mass spectrometry was so sensitive that we found at least eight compounds in that sample. One of them was a C30 sterol, something that nobody had seen before. We also had access to higher field NMR because we were next door to Varian. Mostly, these were the days of the 60 MHz NMR, and although we only had a milligram of the sample we found immediately that it was a totally different sterol. It was a sterol which had the cholesterol nucleus but a very different side chain with a cyclopropane ring. This was extraordinary for a sterol chemist because sterols were the oldest components of steroid chemistry and it was dull chemistry in which all the great stuff had already been done in the 1920s and 1930s. So this new sterol turned out to be biosynthetically unprecedented and that got me interested in marine sterol chemistry. Subsequently, we have isolated hundreds of new marine sterols, and spent over a decade on their biosynthesis and that led me into biological function. The question was, why do they exist in nature? Many of these sterols didn't have any counterpart in terrestrial organisms. We thought about cell membrane function and became interested in phospholipids. We discovered that marine phospholipids were again unprecedented, with different fatty acids, different chain lengths, and we initiated even some biophysical experiments. We made some synthetic membranes with these unusual steroids and phospholipids. But I want to stress that all this started with this one mass spectrum of the sample from Paul Scheuer. We published that first paper together, and published several papers together in subsequent years. Paul Scheuer is the father of American marine natural products chemistry, at least among our contemporaries. The very first was Werner Bergmann. Paul Scheuer deserves much more recognition than he has received.

You had a more difficult beginning of your career than most of your peers. How do you look back on it?

I certainly had a different path from most, and it appeared that it was harder. I made professional decisions which didn't seem to make any sense at the time but worked out at the end very well. One drawback that I had was very similar to that of Derek Barton, and that is that we both are very good organic chemists but neither of us had a real mentor, someone who would be helping us subsequently. You usually need that in the American academic system, and in the British even more so. Initially, we both did significant work in minor universities, Barton at Birkbeck College in London and I at Wayne State University in Detroit. In a way then, of course, it also became more noticeable that we did significant work in such minor places.

I always wanted to go into academia but I didn't want to do it in the traditional way. First I wanted to get experience in industry. Furthermore, I went to Syntex in Mexico in 1949. This seemed absurd. Everyone told me so. All the means of communication available today were absent at that time. Nonetheless, the two years I spent in Mexico were the most productive years in my scientific life. The first synthesis of cortisone from plant products, which was then one of the most competitive fields in organic chemistry, and then the first oral contraceptive, a lot of really important steroid stuff. It really put Mexico on the map scientifically. That's rather spectacular, for one research group (Syntex), to put an entire country on the scientific map where it didn't exist before. Within ten years, Mexico became the center of worldwide production and even of research in the steroid hormone field. This is the only time when a third-world country became the center of a sophisticated area of organic chemistry.

Press conference announcing the first synthesis of cortisone from a plant source at Syntex in Mexico City, 1951. Standing, from left to right: A. L. Nussbaum (subsequently, one of Djerassi's first Ph.D. students at Wayne State University), Mercedes Velasco, Gilbert Stork (then at Harvard and serving as consultant to Syntex), Juan Berlin, and Octavio Mancera. Seated, left to right: Juan Pataki, George Rosenkranz, Enrique Batres, Carl Djerassi, Rosa Yashin, and Jesus Romo (courtesy of Carl Djerassi).

Of course, I hadn't dreamed when I went there at the age of 26 that I would be able to earn an academic reputation in two years that also got me my first academic job at Wayne State University. At Wayne State I very quickly became a tenured Full Professor, and this is how I got to Stanford.

Somebody wrote somewhere that you'd produced enough results for three Nobel prizes. How do you feel about it?

I feel exactly the same way as do most scientists. It'd be great to get it. But paraphrasing a Swedish scientist, involved with the Nobel Committee at one time, the Nobel prize is marvelous for science and terrible for scientists.

I have been present at the Nobel prize ceremonies and was even present once at the deliberations of the Committee. I'm a foreign member of the Swedish Academy so I have the right to nominate each year. So I've had quite some experience with them but most of it, unfortunately, after I'd written my first novel, *Cantor's Dilemma*, although I wouldn't have changed much of the plot.

In *Cantor's Dilemma*, I write in great detail about the Nobel ceremony. I've been to some, and nominated many friends, including Don Cram. I wanted to know the menu of the dinner of the Nobel ceremony so I wrote to my friends whom I had nominated and who had got then elected. Don was one and Herb Brown was another one, and I asked them to send me the menu and the seating plan, and I used the Don Cram version in the novel. I also made up a conversation the Queen would have with her neighbor. Usually the person who sits next to the Queen at the dinner is the oldest Nobel laureate. In my novel it was the youngest one, the postdoc. In that conversation I had the Queen comment on the fact how Americans eat: they cut with knife and fork, and then they put the knife down, switch the fork, and use the fork only. I myself eat like a Central European and find it ridiculous to waste all this time by switching the fork. Then she goes on commenting on how people eat peas. Now I know that peas are usually not given at the Nobel dinner but it's in the novel anyway. The Americans chase the peas with their forks and they haven't got anything to help them getting on their forks while the Central Europeans just shovel them with their knives onto their forks. But then there are the English who hold their forks inverted, so that no peas could be kept on them. So you can divide the world between the British and ex-British colonials who eat the peas in this silly way and then the Americans who eat them in this counterproductive way, and the Central Europeans.

A few years later when I gave a lecture at the Swedish Academy of Sciences, they had a dinner for me where the President of the Academy offered a formal toast, and presented me a Peas Nobel Prize which was a can of peas wrapped in the Swedish flag. This showed me, too, that people read my novel. *Cantor's Dilemma* has been translated into many languages, most recently into Chinese.

I also give a lot of lectures on the culture and behavior of scientists, and I'm now returning to your question, how I feel about the Nobel Prize. I call these lectures "Noble Science and Nobel Lust: Disclosing Tribal Secrets." The desire is always there to win recognition and to win prizes, and the Nobel is just one of them although it is the ultimate one. If you ask the Nobel laureates quite honestly, do you want to win another one, they say, yes! This desire is a fuel but it's also a poison. How to learn to maintain a balance and not let the poison overcome the fuel part, is something that a number of scientists have not been able to solve. And this is getting worse these days with the competition getting increasingly brutal, and detracting from the elegance and even from the pleasure of scientific research.

I decided to write about that because earlier I had not really been very self-reflective which is probably true of most active research scientists. If you read autobiographies of scientists, and quite a few appear these days, you find very few which are self-reflective. Chemists are especially bad in this regard and most wouldn't even know how to do this. Chemistry is a very conservative science and chemists like to build high walls around themselves. Chemists tend to behave in a more traditional way than many others. I dare say, I suspect, that the divorce rate among chemists is lower than among biologists. I'm now making a completely unsubstantiated generalization.

I know that you'd published two autobiographies but, of course, not everybody has read them. Could you tell us about your background, and, in particular, what attracted you to chemistry?

I wasn't turned on to chemistry when I was a kid. I was a child of two practicing physicians. That's what medicine was to me, not research but the treatment of sick people. I would've become a physician.

Even when I left Vienna during Hitler's days and, via Bulgaria, went to the United States I took a premedical curriculum which also meant taking a lot of chemistry and biology. That was my first exposure to chemistry, and I had a marvelous chemistry teacher who turned me on to chemistry. This was in 1940 at Newark Junior College which doesn't exist anymore. This was a two-year community college; I just didn't have the money to go elsewhere. The chemistry lab was terrible but it was taught very well by a teacher named Nathan Washton, who is still alive.

Then after two years I went for one semester to the Midwest, to Tarkio College, which doesn't exist anymore either. Wallace Carothers, the inventor of nylon, had graduated from there. Then I went for my last year to Kenyon College in Ohio. At that time it was a very small school, but there were two absolutely outstanding chemistry professors, a man named Walter Coolidge, who taught me organic chemistry, and another named Bayes Norton, who taught me physical chemistry. I did my first research with Norton on the quantum yield of ethyl iodide. I was 17½ years old. It had a real impact on me. The classes were so small that it was more like a tutorial. These three teachers, Washton, Coolidge, and Norton converted me into a chemist. I still might have gone on to medical school but I couldn't afford it when I graduated and had to go to work. I tried to find a job with a pharmaceutical company. I was looking at the advertisements of these companies in a doctor's office where my mother worked. She emigrated with me but didn't have a license to practice medicine, and worked for a physician as an assistant. CIBA hired me as a junior chemist, and I worked for them for one year before I went to graduate school.

There was also a book that influenced me. Paul de Kruif's *Microbe Hunters*. Originally that book turned me to medicine. I read it when I came to the United States. That book was a romanticized sweet book on medical research. Now I would pooh-pooh it, but it had an enormous impact on budding scientists of my generation.

Did your mother eventually get a license?

Eventually she did, and so did my father who came much later to this country. They were never reunited though. Not only did I become estranged from my mother, but during the last few years of her life, she didn't even recognize me because she was suffering from Alzheimer's.

With my father it was the reverse. He didn't see me for ten years following my departure from Bulgaria. He was in Bulgaria during the War; then when he arrived in New York, I had just moved to Mexico. But later he moved to San Francisco with my stepmother and had always tickets in the Opera behind me. He died in an accident at the age of 96.

It's not very well known that most of the Jews in Bulgaria survived World War II.

Bulgaria is an interesting country. To Viennese, like my mother, it was judged the most primitive European country, other than Albania, at that time. This may have contributed to the divorce of my parents because my mother was a typical arrogant Viennese Jew and my father was a typical Balkan Jew; even though he got his MD in Vienna, but he was very Bulgarian. The Bulgarian Jews had lived there for 400 years — they were originally from Spain — and theirs was a coherent community; they spoke Ladino

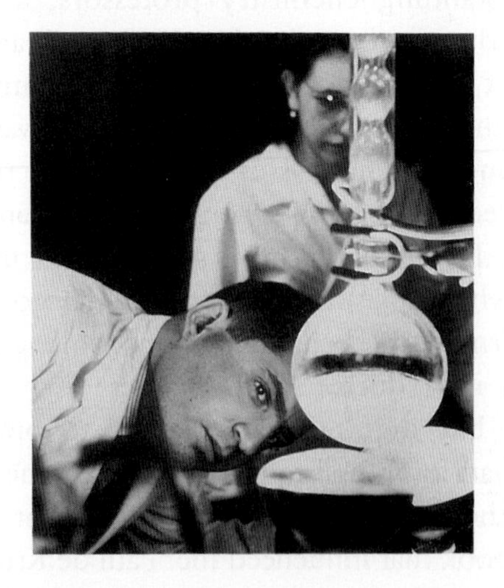

Carl Djerassi at Syntex in Mexico City, 1951, working on the synthesis of the progestational component of the Pill, with assistant Arelina Gonzales (courtesy of Carl Djerassi).

in addition to Bulgarian, didn't intermarry very much with Bulgarians, but were very much accepted, although were still considered partly outsiders and you could tell them by their names. Djerassi, for instance, is a very Sephardic name, perhaps Spanish or even Moroccan. The Bulgarians voluntarily joined the Axis, which consisted of Germany, Italy, and Bulgaria. But there was no real traditional anti-Semitism in Bulgaria, and the King was a very decent person. There were about 50 000 Jews out of a population of five million. The Germans seemed to have been much too busy in Yugoslavia and Greece, so the Bulgarian Jews survived. Then, immediately after the War, there was a window of less than one year when they could emigrate to Israel, and 90% of them did so.

At the beginning of our conversation you mentioned the exact day when your small team produced the ingredient of what eventually became the Pill. At which point did you realize that your work would have such a tremendous impact?

In some respects, immediately, and in some other respects, only gradually, over the years. Immediately, because we set out to synthesize a progestational hormone, a compound that would retain and even surpass the biological activity of the natural female sex hormone progesterone and be active by mouth. Progesterone is not orally active, and there were important therapeutic applications of progesterone already at that time. The most important was the treatment of menstrual disorders. That is not a life threatening condition but it affects, in a very painful way, millions of women. Another application was, and it sounds ironic today, the treatment of infertility of women who can become pregnant but cannot maintain the embryo or fetus because they don't produce enough progesterone, and therefore, suffer a spontaneous abortion. Progesterone was applied in the form of injection. The third potential application was a hoped-for treatment against cervical cancer. Again, this was done by injection but the direct injection into the cervix was a very painful procedure.

So we were working on chemical modifications of progesterone to develop an orally consumable synthetic progestational compound. This was not a trivial exercise because, at that time, it was assumed that anything one did to the progesterone molecule would destroy biological activity, or at least diminish it. In the estrogen field, the other female sex hormone, it was very well known by that time that lots of different compounds had estrogenic properties and you could make all kinds of modifications. In the progesterone field, it was assumed that this was not possible. The fact

DIARY ENTRY (11 AUGUST 1983)

We sit in Copenhagen:
Chemists from a dozen countries.
The talk is heavy; the words are long.

Male contraception,
Cures for cancer,
Morphine substitutes,
Drugs from the sea,
Medicines for the year 2000.

We've mouthed these words for many years,
Formulae hiding the chemists.
Who are these colleagues, students, strangers?
What do they do besides chemistry?

If this were the Holiday Inn,
Not the Royal Danish Academy,
Would I guess who they are?

A convention of grocers? Too serious.
Car salesmen? Too little polyester.
Bankers? Lawyers? No vests.
Clergymen? Wrong collars.
Poets? Nobody smokes.

How did they come to chemistry?
What do they do besides chemistry?
What do I do besides what I do
Besides chemistry?

From: Carl Djerassi, *The Clock Runs Backward*. Story Line Press, Brownsville, Oregon, 1991, pp. 13–14. Reproduced with permission.

that we were able to show that we could get a powerful analog of the progestational hormone had a great impact. Within weeks, we sent it out for bioassay because we didn't have biological laboratories at Syntex. We sent it to a commercial testing laboratory in Madison, Wisconsin; to Lipschutz in Santiago, Chile, another European refugee who was well known for his progesterone work; to Greenblatt at the University of Georgia; to Roy Hertz at the National Cancer Institute; and to Gregory Pincus at the Worcester Foundation who is generally considered, and with justification, the biological father of the Pill. He was interested, at that time, in studying the ovulation inhibiting properties of progesterone. Our product, called "Norethindrone," first earned FDA approval in 1957 for menstrual disorders. By that time it had also become obvious from the work of two groups, the group of Pincus and Tyler's group in Los Angeles, that Norethindrone had ovulation inhibiting properties, and within three years, from 1957 to 1960, there was a real jump as we gradually became aware that the oral contraceptive properties would be the important thing. No one, however, anticipated the speed by which it was then accepted, not the scientists, not the pharmaceutical companies, and not even the physicians. We didn't have any pharmaceutical marketing outlets at Syntex, so we licensed Parke-Davis to bring it to the market, when it was originally approved by the FDA in 1957 against menstrual disorders. But as a contraceptive, Parke-Davis at first didn't want to touch it. They were afraid of a Catholic boycott and we had to start looking for somebody else who would be willing to market it as an oral contraceptive.

How did you adjust to all the fame and everything else that came with it?

Easily. First of all, there wasn't much fame. Besides, I'm neither modest, nor conceited. I don't overestimate nor do I undervalue my achievements. At the same time, I've felt most of the time as an outsider.

In what sense are you an outsider?

I've always been an outsider within the American academic community, which at that time, was also understandable. I established my scientific reputation in exactly the naive way I'd planned to, namely, in industry. When I won the American Chemical Society Award for Pure Chemistry in 1957, I'd already been in academia; I was a Professor at Wayne State University, and this is probably why I got it, because no industrial chemist had been awarded that prize, but there's no question that the reason why I got it was the science I did in industry, at Syntex. In that context, I was an outsider and have remained as such.

When I came to Stanford in 1959, Stanford permitted me to also be a research executive at Syntex (when that company moved from Mexico to the Stanford Industrial Park), which was unheard of in an American university at that time. There was also strong jealousy and people assumed that I made a lot of money from the Pill. I did become an affluent person but only indirectly through the pill. My name was on the patent but I got exactly one dollar for it, because in 1951 I had been an employee of Syntex which had all the rights to it by law, and I didn't think I was exploited or that they cheated me. On the other hand, at Syntex, I believed

Carl Djerassi with graduate student Barbara Grant at Stanford University, early 1970s (courtesy of Carl Djerassi).

in the company and I acquired stock at a very low price at a time when other people could have done so too.

I'd like to stress that I never mixed my university research with what I did in industry. The outcome of one was a lot of publications and of the other, the patents. I was even audited once by the NIH and questioned whether I used NIH support (received in the United States) to carry out work in Mexico and patent the results. Stanford retained a lawyer who said, it may take years for you to disprove such allegation unless you find someone who's willing to read through all of your papers, which was about 400 at that time. Finally, Bill Johnson, the Chairman of our Department, volunteered to do that, and demonstrated that there was absolutely no connection between my published academic research and industrial patented work.

Then 15 years later, a Berkeley newspaper (*The Berkeley Barb*) described university scientists getting involved in biotechnology, using me as an example. Following my protest, they had a full page retraction.

In most cases I shrug these things off, except, for the *Berkeley Barb* article, which was simply too outrageous. That people still say "oh, he got all this money from the Pill", has gone by me. Such jealousy appears even now in a different way when I write fiction and people ask, "Why does Djerassi now have to write fiction?"

Any other reason of being an outsider?

I mean an outsider in terms of the American Chemical Society.

Let's take as a trivial example, the National Organic Chemistry Symposia which the American Chemical Society organizes every two years. I was only once invited to one, in 1957. Since then, never, whereas I'm flooded by invitations of all kinds from other scientific organizations worldwide.

Why?

Perhaps because it is an inside group of academic organic chemists. Take another example: since I came to Stanford in 1959, I haven't won a single ACS award for my academic chemistry, of which I have done a lot: all of our work in mass spectrometry; all of our studies on optical and magnetic circular dichroism; on computer-aided structure elucidation; marine natural products, etc., on the order of 800 papers or more. All my ACS awards since 1960 were for my industrial work. The academic community in the States has not really accepted me as an academician. I'm just an industrial chemist who pushed his way into academia.

Do you care?

In a way I do. If they ultimately ask me, what are you, an industrialist or an academic chemist, if it's up to me, I'll say, I'm an academic chemist. Intellectually, the most important things that I've done, with the exception of my work at Syntex, have all been in academia. If we talk about the impact of chiroptical methods, mass spectrometry, marine natural products, artificial intelligence, you name it, all have been done in academia. This may have been a component to my decision, late in life, but earlier than I would have done so otherwise, that I'd stop doing laboratory research and embark on one other totally different intellectual life: to write fiction. But to write fiction about my own life, my own scientific discipline, my own world.

In the end, like many of the important decisions in my life, it came from misery and turned out to be marvelous. When I had to leave Europe, that emigration really made me a chemist; I went to Mexico and it made me wealthy; the death of my daughter caused me to get involved in art support and start the Djerassi Foundation, so I'd say that in each case it ended well.

Let me also add to the question of being an outsider that when I came to Stanford, I was the first Jewish Chemistry Professor there, not because Stanford was anti-Semitic, not at all, it was just not yet done. Now, nearly half of my Department colleagues are Jewish. Of course, thinking about this was always on my mind as a Jewish refugee, a sort of paranoia that took a long time to overcome, nearly 30 years.

Another component of my perceived outsider status is that American organic chemistry has always been very fashion oriented. The fashion was physical organic chemistry during the fifties, starting already in the forties and extending into the sixties. It was the mechanistic approach to organic chemistry and that is why the Winstein-Brown argument got so much more

I HAVE NOTHING LEFT TO SAY

Five years after your death,
My only daughter,
I find this note:

"I have nothing left to say,
So I don't talk.
I have nothing left to do,
So I close up shop."

No date
No address
No signature
Your handwriting.

Written for whom?
Yourself?
To whom it may concern?

Written when?
Days,
Weeks,
Perhaps months
Before you walked into the woods?

If only you'd said these words to me.

From: Carl Djerassi, *The Clock Runs Backward*. Story Line Press, Brownsville, Oregon, 1991, p. 20. Reproduced with permission.

publicity than it deserved. Next to physical organic chemistry was synthetic chemistry, but never structure elucidation, which was never important in the United States compared to Europe or Japan. Or take natural products chemistry. That was hardly recognized in the United States during the postwar era. Of all the top natural products chemists in the USA, I am probably the only exception in being a member of the National Academy of Sciences. Paul Scheuer, for example, would have long deserved to get elected. He didn't even win a single award from the American Chemical Society until two years ago, when he was 75. It's shocking. And there are other examples.

The top fashion is no longer physical organic chemistry. It's now total synthesis of molecules with preferably 120 or more asymmetric carbon atoms. It's also moving into the next one which will be bioorganic chemistry. The cutting edge, high prestige field for which the new Assistant Professors are being hired, is along the borderline between molecular biology and chemistry, but from a chemical standpoint. I'm not saying that it's bad, I'm just saying that this is also fashion and therefore the power is there. The power of academic chemistry in the United States is really controlled by 12 institutions of which Stanford is one. Since I am in that institution, I should feel accepted in that context, yet for a psychologically curious reason, I do not.

In your book The Politics of Contraception, *published around 1980, you anticipated only minor adjustments in this area for the coming decades. You seem to have been correct.*

To my regret, I was 120% correct. Take the 10 biggest pharmaceutical companies in the world: not one of them is working on new methods of birth control, none of them is even selling any birth control agent. Birth control is just not on the high priority list of the pharmaceutical industry. The pharmaceutical industry is interested in making money, and I'm not blaming them for that, but they focus their activities on the wealthy parts of the world, that is, Japan, North America, and Western Europe. This northern tier of countries are also the geriatric countries, where nearly 20% of the people will soon be above the age of 60. I call the southern tier of countries the pediatric countries where about 45% of the people are below the age of 15. Their health priorities are totally different. The health problems of the geriatric societies are not birth control. Most of these countries no longer even reproduce themselves. They don't even have two children per family. Then you go to Nigeria where you find seven

Carl Djerassi and his image (photograph by I. Hargittai).

children per family on average. You need 2.1 just for replacement, but most European countries are way below. The lowest is 1.2 in Italy and Spain. They are on the steepest slope of economic development. Italy has very low pill consumption, practically no sterilization, but they can cope. So they don't need new birth control methods.

What then do the pharmaceutical companies focus on? They focus on inflammation because that is the most common disease of an aging population. Anti-inflammatory drugs are the most important drugs for millions of people, they are older people, and they have to take them all the time. That is ideal for a drug company. Cardiovascular drugs are another one, next is cancer, Alzheimer's disease; these are very important diseases, and they are high-profit ones, and for them people are willing to accept risks. For birth control, people aren't willing to take any risk. They say, I'm healthy, you've got to provide me with a birth control agent that's 100% safe, and there isn't such a thing. This is why they end up with litigation, which is very important in the United States, and the pharmaceutical companies are not interested in that type of headache. The World Health Organization has tried to do something in the birth control field for years, but its budget for research support is less than 20 million dollars and has barely changed for the last 20 years. To develop a fundamentally new birth control agent from the beginning to the end may cost 400 million dollars.

I really believe that the real problems are not scientific ones, but rather of a socio-cultural-economic-religious nature and can be solved only on the political and policy level.

You have now become a writer in addition to being a scientist. Do you have a special agenda that distinguishes you from the nonscientist writers?

If you'd interviewed me five years ago, I would have said that I was a chemistry professor who was also writing fiction. Today I'm a novelist who is still a professor of chemistry, and that is a big difference. During the summer we live in London, and it's then and there that I write my fiction. There I have no contact with academia, I work every day, seven days a week, seven to eight hours a day. I am a full time fiction writer in the summer.

Now I give more lectures than I ever did in my life — you should just see my travel schedule. Although people mostly invite me as a scientist, I prefer to speak about scientific behavior and ethics and culture rather than about specific scientific topics, and to do so using my fiction and my autobiography as important components. It has worked well. I think I'm having an effect in what I wanted to accomplish.

EXHORTATION

How does a chemist
Transmute himself into a poet?

Synthesize a poem?
Distill its essence?
Filter the impurities?
Evaporate it to dryness?

Stop the sophistry!
Write the poem.

From: Carl Djerassi, *The Clock Runs Backward.* Story Line Press, Brownsville, Oregon, 1991, p. 9. Reproduced with permission.

Do you envision a mission for yourself in bridging the gulf between the two cultures?

This is <u>the</u> mission. This gulf is one of the most important social problems today, the gulf between the scientifically literate constituency, which is a very small portion of the population, and the intelligent literate community, which is scientifically totally illiterate. This is also part of the reason for chemophobia in contemporary society. The important factor, of course, is the readership, and this is why I decided to use fiction. I call it "science-in-fiction" because I'd like to smuggle concepts of scientific culture of behavior into the conscience of people who are not interested in science.

THE CLOCK RUNS BACKWARD

At his sixtieth birthday party,
Surrounded by wife, children and friends,
The man who has everything
Opens his gifts.

Among paperweights, cigars,
Books, silver cases,
Cut glass vases,
Appears a clock
Made by Accutec Designs
In a limited edition.
A clock running backward. Cetucca.

Amusing.
Just the gift
For the man who has everything.

How Faustian, thought the friend,
Soon to turn sixty himself.
What if it really measured time?

As the hands reached fifty,
He stopped them.
Books, hundreds of papers, dozens of honors.
Not bad, he thought: I like this clock.

But fifty was also the time
His marriage had turned sour.
He let the clock run on:

Forty-eight years, forty-five years,
Then forty-one.
Ah yes, the years of collecting:
Paintings, sculptures, and women.
Especially women.

But wasn't that the time
His loneliness had first begun?
Or was it earlier?
Why else would one collect,
Except to fill a void?

Don't hold the hands.
The thirties were best:
Bursts of work, success, recognition,
Professor in first-rank university,
Birth of his son — now the only survivor.

What about twenty-eight?
Ah yes — he nearly forgot;
The year of THE PILL.
The pill that changed the world.
No — too pretentious, too self-important.
But he did change the life of millions,
Millions of women taking his pill, he thought.

The clock still regresses.
Twenty-seven years:
First-time father, of a daughter,
Eventually his only confessor.
Now dead. Killed herself.
The beginning of his second marriage.
The first undone.

Early stigmata of success to come:
The doctorate at twenty-one,
The eighteen-year-old Bachelor of Arts.
Backward: Europe. War.
Hitler. Vienna.
Childhood.
Stop. Stop.

The pater familias,
Surrounded by wife, children, friends,
The man who has everything
Is still opening presents.
More paperweights, more silver,
More books, ten pounds of Stilton cheese,
And one more clock.

Thank God it is moving forward,
Thought the friend,
The lonely one,
Who will soon turn sixty.

And smiled at the woman at his side,
The one he had met yesterday.
Who yesterday had said,
"Yes, I'll come with you to Oslo."

From: Carl Djerassi, *The Clock Runs Backward*. Story Line Press, Brownsville, Oregon, 1991, pp. 15–18. Reproduced with permission.

Paul Scheuer (courtesy of Paul Scheuer).

7

Paul J. Scheuer

Paul J. Scheuer (b. 1915, Heilbronn, Germany) emigrated to the United States in 1938 and graduated from Northeastern University, Boston, with a B.S. in Chemistry in 1943. His graduate studies at Harvard, interrupted by service in the U.S. Army, led to a Ph.D. in organic chemistry in 1950. In the same year he was appointed Assistant Professor at the University of Hawaii and has stayed in Hawaii ever since, from 1985 as Professor Emeritus. He continues to direct a research group in natural products — originally of terrestrial plants, but now exclusively of marine organisms. He has directed the research of 33 M.S. and 25 Ph.D. students, of over a hundred postdoctoral associates, and numerous undergraduates. More than 260 publications, including one authored and twelve edited books, have resulted from this work.

In 1992, his former students initiated a Paul J. Scheuer award in Marine Natural Products, of which he was the first recipient. In 1994, he received the Ernest Guenther Award of the American Chemical Society and the Research Achievement Award of the American Society of Pharmacognosy.

My wife and I spent the fall semester in 1988 and the spring semester in 1993 at the Department of Chemistry of the University of Hawaii at Manoa. We learned a lot about Paul during those stays, but were not aware of his Emeritus status. He was as hard driving in his work as a tenure track assistant professor. We shared interest in Hawaiian flowers. At the time of our stays in Hawaii I was not recording conversations with famous chemists, but when Carl Djerassi mentioned Paul's activities in the interview with him (see previous chapter), I decided to invite Paul to tell us about his life and work. Here is Paul's narrative:*

Your invitation to write "about your career and your pioneering work in marine natural products" was flattering and challenging; it proved to be rewarding. For the first time in my life, I have taken the time to examine the unlikely and meandering route which led me from a middle-class upbringing in southern Germany to an exciting academic adventure in the Hawaiian Islands.

Coral Reefs USA stamp displaying Finger Coral: Hawaii. The original title of Professor Scheuer's personal account was "Across Two Oceans onto the Reef — The Genesis of Marine Natural Products."

The journey which you will share had a slow and uncertain beginning, taking as it did nine years from high school graduation to a B.S. degree in chemistry and another seven to a Ph.D. But an assistant professorship at the University of Hawaii revealed a new world with unexpected challenges and rewards. Previously unexplored rain forests and coral reef organisms offered unlimited opportunities for natural product research. For a person who had never studied botany or zoology and whose natural product research in graduate school began with a pure crystalline solid, it was a true beginning.

*The complete version of this article originally appeared in *The Chemical Intelligencer* **1997**, *3*(2), 46–57 © 1997, Springer-Verlag, New York, Inc.

It was a beginning in another respect as well. The Chemistry Department at the University of Hawaii was not equipped for research, physically or intellectually. Rarely has any beginning assistant professor been able to experience the excitement and frustrations of starting with a tabula rasa. But the end result, as you will see, has been gratifying.

Education and Wanderjahre

Primary public education in the Weimar Republic (1919–1933) embraced a democratic element that had been absent in pre-WWI Germany. All six-year olds were enrolled in identical elementary schools (Volkschule) for four years, at the end of which a mandatory qualifying examination determined whether a student was eligible for a nine-year high school or a vocationally oriented middle school. Another decision point for high school students came after six years, when only those who aimed at a university education would continue for the last three years, which culminated in another comprehensive examination. Those who had left high school after six years would begin careers in industry or commerce. Presumably, that mid-point career decision was heavily influenced by the student's parents.

When I graduated from the Heilbronn Realgymnasium in the spring of 1934, the Weimar Republic had died. The national elections of 1933 made the National Socialists (Nazis) the majority party and Adolf Hitler the Chancellor, who rapidly became the dictator. Although I graduated first in a tiny class of 11 students, the racial laws of the Third Reich had eliminated the option to attend a university.

If university attendance had been allowed, I had no idea what I would have studied. It decidedly would not have been chemistry. Although I had two years of high school chemistry, my classmates and I were more fascinated by the antics of the instructor than by the subject which he taught (badly). There were few role models in a family of merchants. An uncle who was an attorney would occasionally invite me to attend civil trials, which I found utterly boring. In a way then, it was fortunate that I gained time before deciding on a career. The subsequent events, which eventually culminated in a professorship in the field of natural products, were serendipitous and remarkably devoid of deliberate planning.

During my teens I belonged to a youth group which combined elements of scouting with discussions of current events. One of the group leaders, a Ph.D. in economics, was working in his family's leather tannery. He asked me, whether I would like to be a tannery apprentice, an offer which I accepted. Following an old German tradition of apprenticeship, I started my career, not with sweeping the floor of the workshop, but with unloading raw cowhides onto trucks from freight cars parked on a factory siding. Since the factory specialized in making leather for shoe soles, the hides were big and heavy. It was the hardest physical labor I had ever done!

Over the course of twenty months I gradually worked my way through the entire operation from raw hides to finished leather, but leaving me with only a vague idea of the actual process that turns a cowhide into leather. My mentor believed that I should also learn how fine leather is made. He suggested a second apprenticeship in a tannery in Hungary, which was owned by a family friend. In December, 1935, I took the train to the provincial town of Pécs in southwest Hungary, leaving Germany as it turned out for good except for a brief visit in 1937 for my mother's funeral and a longer visit, courtesy of the U.S. Army, from May 1945 to June 1946.

I stayed in Pécs for a year before moving on to another tannery in Simontornya, a very small Hungarian town. It was there that I became interested in chemistry. The technical director of the tannery, a Ph.D. chemist, decided that I should learn some basic theory that would teach me something about the tanning process, historically one of the earliest and highly successful examples of biotechnology. Almost daily, after working hours — what else was there to do? — he tutored me in what might be called simplified classical biochemistry. The chemistry of the tanning process — whether accomplished by chromium salts or by the complex phenols of tree bark, remained a mystery, but I became fascinated with chemistry as an intellectual challenge. It was then and there that I decided to become a chemist.

Short stints in tanneries in Yugoslavia (now Slovenia) and England lasted through the fall of 1938. During spring and summer of 1938, a war in Europe became increasingly probable, and thoughts of my leaving the old country became more frequent. A cousin of my mother's, a New York attorney, was willing to sponsor me for an immigration visa to the United States. I arrived in New York harbor in mid-October, 1938, and with the help of my cousin a job of sorting and packaging calf and sheep leather materialized. By January 1939, I was a foreman in a small tannery in Ayer, Massachusetts. Once I was settled I looked for an opportunity to continue my study of chemistry. Northeastern University in Boston, a small engineering-oriented college, had the most extensive night school program. By fall of 1939, I was enrolled in freshman chemistry. In the course of the year it became clear to me that commuting to night school was perhaps not the best way to a university education, so I decided to quit my job, move to Boston, go to school full time, and try to make a living from part time jobs, nights and weekends. In the fall of 1940 I became a day student in the College of Liberal Arts at Northeastern University, majoring in chemistry. By the time I took organic chemistry it was immediately obvious to me that this was the chemistry I would want to study. Even by 1941 standards, the course was old-fashioned, but carbon was the most interesting element in the periodic table.

In December of 1941, the United States entered the war and everybody's life changed. As a German national, I was an enemy alien and had a few travel restrictions. Like everybody else, I registered for the draft, but was rejected, when I was first called.

In April 1943, I graduated from Northeastern with a B.S. in Chemistry. I applied to Harvard graduate school and was accepted; I even received a tuition scholarship. The Harvard chemistry faculty and the number of graduate students were small. G.P. Baxter, the Chairman, was everybody's informal and willing advisor. Of the three organic faculty members; Louis Fieser, Paul Bartlett, and Bob Woodward, only the latter two represented viable choices as thesis advisors since Fieser was rarely in residence as he was heavily engaged in war-related research. I chose Woodward (RB). He had been elevated from Junior Fellow to Instructor when Patrick Linstead returned to his native England for war-related activities. Group meetings were held in a small conference room; they were informal and lacked the sharply competitive atmosphere of the postwar years.

My research during the fall semester of 1943 consisted of many unsuccessful attempts to add ketene to alpha-vinylpyridine, probably in connection with the synthesis of quinine. In the spring of 1944 the Department needed an instructor for a short introductory chemistry course to be taught to about 30 GI's. Professor Baxter handed me a text, a class list, and a course schedule. There I was in front of a class in an ancient classroom in Sever Hall. I thought that I had prepared a good first lecture. I managed to say it all in ten minutes. Class dismissed.

Although I was assigned to the Chemical Warfare Service, the two years and four months in the U.S. Army were yet another hiatus in my chemistry education. My most spectacular chemical experience was the firing of WP (white phosphorus) shells from 4.2" chemical mortars in the Alabama hills. A two-month course in quantitative analysis at Edgewood Arsenal, Maryland, rounded out the chemistry. While at Edgewood I became a U.S. citizen in the U.S. District Court in Baltimore. By January of 1945 I had left the Chemical Warfare Service and was assigned for training in military intelligence to Camp Ritchie, Maryland. In early May 1945, a few days before VE Day, we were flown to Paris and from there traveled to Third Army Headquarters in Bavaria. Aside from temporary duty at the Nürnberg war crimes trials my fourteen months as a Special Agent were uneventful. After returning home I resumed my education in September, 1946, now to be financed under the G.I. Bill.

Postwar Harvard was a far cry from its skeletal existence during the war years. As was true all over the country, returned veterans together with the normal enrollment swelled the student ranks. The new organic instructors included Gilbert Stork and Morris Kupchan.

Discovering Natural Products

The Woodward research group had become large. Despite the group meetings, which now were held in the evening and rarely ended before midnight, it was difficult to know everybody. Dick Eastman was the only hold-over from the war years. He had graduated but stayed on as postdoc to manage the group while RB drove his old jalopy to Reno to be divorced from his first wife. Structure and synthesis of strychnine and steroid synthesis were the dominant projects. RB handed me a 500-gram bottle of crystalline strychnine (**1**) and asked me to follow a published procedure to prepare *neo*-strychnine (**2**). It was to be an intermediate that might resolve the principal difference between the Woodward (**1**) and Robinson (**3**) expressions,

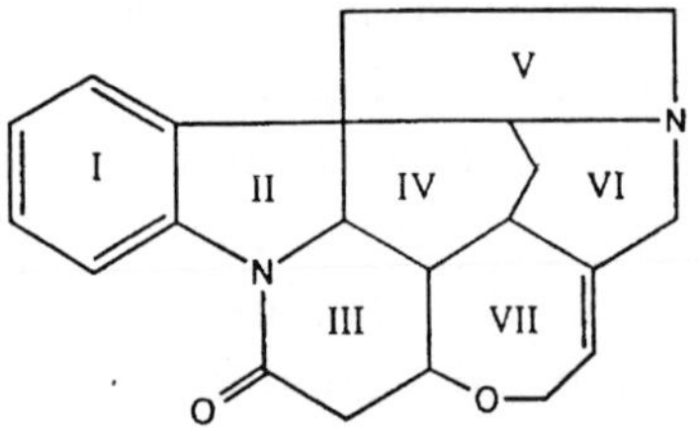

1

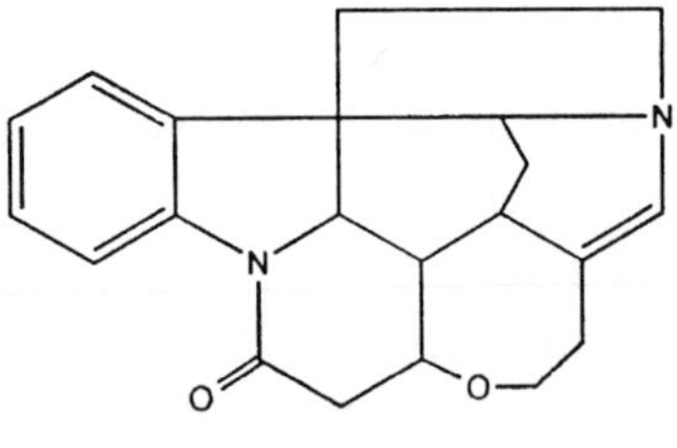

2

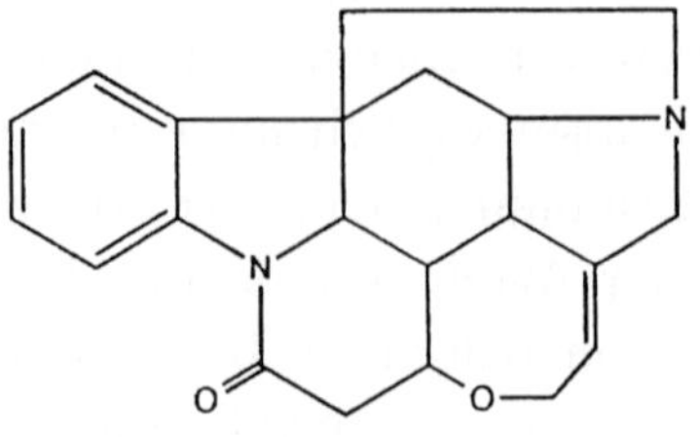

3

a 6-membered versus a 5-membered ring VI. In *neo*-strychnine, the double bond of ring VII had isomerized to ring VI, thereby setting the stage for cleavage and reclosure of ring VI, which would provide direct evidence for the size of that ring. In time, the *neo*-strychine route was abandoned. Warren Brehm's research[1] had provided conclusive evidence for the Woodward structure (**1**). Synthesis plans were in full swing and useful relay compounds were in demand. So it became my task to establish an unassailable structure for the hitherto elusive oxostrychnine.[2] I accomplished the project, but RB maintained only a marginal interest. When I needed advice, I consulted Gilbert Stork as did most Woodward students, who were not working on the hot project of the hour.

The course entitled "The Chemistry of Natural Products" was RB's teaching assignment, also inherited from R.P. Linstead — complete with lecture notes. (I can still hear RB's chuckle at the mention of it.) According to the University Catalog, it met Tuesday, Thursday, and at the pleasure of the instructor, on Saturday at 11. In his first lecture, RB made it clear that it was rarely the pleasure of the instructor to lecture on Saturday. In fact, there never were any Saturday lectures. But every single lecture was an intellectual summit and an aesthetic pleasure. Woodward had of course no notes. Colored chalk was the single visual aid. At the end of the hour, the single large blackboard would be covered with beautifully drawn structural formulas, each in a place that would relate it perfectly to its antecedents, contemporaries, and descendents. No eraser was ever used. Each lecture strengthened my desire to become involved in the study of natural products.

At the impressive and exciting commencement exercises in Harvard Yard I (with hundreds of others) was welcomed into the society of scholars; this event, however, could not obscure the fact that I did not have a job in early June of 1950. Then, Leonora Bilger, the chairwoman of the Chemistry Department from the University of Hawaii, who was on a faculty recruiting trip, visited us on June 24. The interview resulted on July 13 in an offer of an assistant professorship at an annual salary of $4140, plus a monthly territorial "bonus" of $48. My ignorance of Hawaii and of its university was profound. I had not even seen a catalog. Consulting fellow graduate students only revealed that my lack of knowledge was not unique. Only one of my contemporaries had visited Honolulu during his WW II naval service. But for me, somehow, the absence of facts only heightened the sense of adventure. Not to mention my resolve in the 1943 winter, when I dragged chemical mortars through muddy Alabama hills, never to be cold again once the war was over. I was ready to accept the offer and my fiancee, Alice Dash, who had been in my organic lab section during her senior year at Radcliffe, was willing to share this somewhat nebulous future. Alice, though born and raised on the eastern seaboard, knew more about Hawaii than anyone else around, but even she knew nothing of the university.

Destination: Honolulu

On September 5, Alice and I were married in the Harvard Chapel, followed by a reception in the Mallinckrodt conference room. On the following day we flew to San Francisco. On September 8, we boarded on the S.S. *Lurline* for the trip to Hawaii. At sunrise on September 13, the *Lurline* slowed near Makapu'u Lighthouse to take on the harbor pilot and scores of women laden with fragrant plumeria (*Plumeria acuminata*) leis. We slowly steamed toward a green mountainous island in a deep blue sea. It was an unforgettable beginning of a new life in a part of the world that I had never even read about.

We were met at the pier by two members of the Chemistry Department, who drove us to Faculty Housing on the University campus. A studio apartment, one of four in a converted army barracks, had been reserved for us. It rented for $35/month plus $6 for utilities. It was furnished with a refrigerator, a stove, two army cots, a table, a chest of drawers, and four chairs. The University farm, cows, chickens, and pigs were our nearest neighbors. If we wanted adventure, we surely were having it!

In addition to Mrs. Bilger and her husband there were two veteran faculty members. One of these was on leave of absence. The remaining six faculty members were a sophomore (who would leave at the end of his second year), a recently retired visiting professor, three new Ph.D.s, and a part- time instructor, who doubled as the Department's sole Ph.D. candidate. This situation apparently was not unique. Faculty turnover was rapid with Mrs. B. in firm control.

One of the attractions of the University of Hawaii (UH) position was a new chemistry building under construction. Before it was ready to be occupied, my quarters were in — guess what? — a reconfigured army barracks. This precluded starting of my own research, but I rapidly had two M.S. students to supervise. Shortly before the start of the semester, Mrs. B. informed me that I was to supervise a second year M.S. student. My joy over this unexpected good fortune was dampened when I was informed that all M.S. candidates must complete a research project and receive their degrees after two years' residency.

I knew that I wanted to do research in natural products, but I had not formulated a single specific project. And clearly, time was of the essence. My first two M.S. students, Seiji Sakata ('51) and Kiichi Ohinata ('52) did their research on projects that had occurred to me in the course of my strychnine studies. This gave me much needed breathing space to learn about teaching and about UH, about Hawaii, its people, its culture, and its natural resources. I rapidly became acquainted with scientists in the plant, marine and agricultural sciences. I soon discovered that Hawaii would be an ideal place for natural product research. In fact, the two major agricultural industries, sugar and pineapple, maintained excellent institutes, where applied as well as some fundamental research was conducted. Because of the isolation of the Hawaiian islands, 3,800 km from the nearest substantial landmass, Hawaii has a rich endemic flora which no chemist had ever studied. Early on, I audited Harold St. John's (Harvard '14, Ph.D. '17) course in Hawaiian botany. It included weekly field trips, which taught me some Polynesian botany and opened my eyes to the incredible natural beauty and biological diversity of the island of O'ahu.

Based on archeological findings, the earliest Polynesian settlers arrived in Hawaii from the Marquesas in their voyaging canoes around 650 A.D. Several hundred years later, a second group reached Hawaii probably from Tahiti. Captain James Cook is credited with the first western contact in 1778, which is well documented.[3] Although no evidence of any earlier European visitors to Hawaii has ever been discovered, a 1579 Dutch map shows a group of islands, *Los Volcanoes*, at the correct latitude but 1,600 km east of the correct location. Spaniards en route from Mexico to the Philippines may well have known of the Hawaiian islands in the 16th century, but apparently made no contact. The error in longitude is not surprising, since only after ships' chronometers were invented in the early part of the 18th century was it possible for sailors to log their east-west progress accurately.

Terrestrial Natural Products

When the Polynesian settlers arrived in Hawaii, they brought with them 24 food and fiber plants as well as pigs, dogs, and fowl. The only endemic mammals in Hawaii were a monk seal and two species of bat. In addition to essential food staples, among them taro, breadfruit, sweet potato, coconut, banana, and sugarcane, there was *'awa* (or *kava*, *Piper methysticum*) which furnishes the ceremonial beverage of that name. *P. methysticum* is pan-Pacific and has been used by native peoples from Indonesia to Hawaii. It aroused

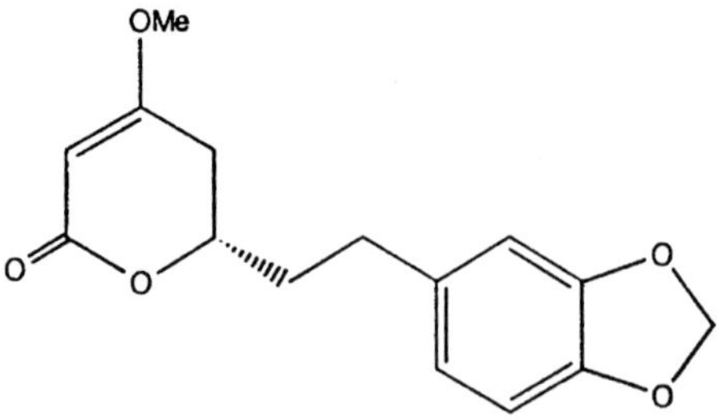

4

the interest of the early explorers and its chemical literature dates back to the 19th century. Early reports of its intoxicating and paralytic action proved to be unfounded, but like many firmly held errors die but slowly.[4] Perhaps these accounts reflect sensations experienced by Europeans unaccustomed to squatting or sitting on the ground for long periods of time during an 'awa ceremony. Rudolf Hänsel of the Freie Universität, Berlin, was first to demonstrate the soporific and non-addictive, properties of a major 'awa constituent, dihydromethysticin (**4**).[5] My own interest in "'awa chemistry"[6] brought Professor Hänsel to my laboratory in 1961, where he had an opportunity to collect 'awa on the island of O'ahu and to establish a connection with 'awa farmers on the island of Hawaii. Over the past 35 years, medicinal use of 'awa in Germany, resulting from Rudi Hänsel's research as a non-addictive sedative and soporific, has become well established.

Learning of the Ocean

An even more dramatic and new experience than Polynesian rainforests was the ocean — warm, blue, and rich in animals and plants of which I was totally ignorant. As I became acquainted with faculty members in marine biology and learned of their research interests, I soon discovered a vast and virtually untapped resource for natural product research.

Sea urchins, colored red to purple, many covered with calcareous spines, are conspicuous sessile invertebrates on Hawaii's rocky shores. I first saw members of this exclusively marine phylum of animals at Hanauma Bay, a submerged crater on O'ahu's southeast shore and one of the island's spectacular scenic attractions. Discussions with a faculty member in the Zoology Department suggested that little research had been done on the pigments of sea urchin shells and spines. A look into the chemical literature readily confirmed this. European and Japanese workers had studied sea urchin pigments, but only a single structure, that of echinochrome A (**5**)

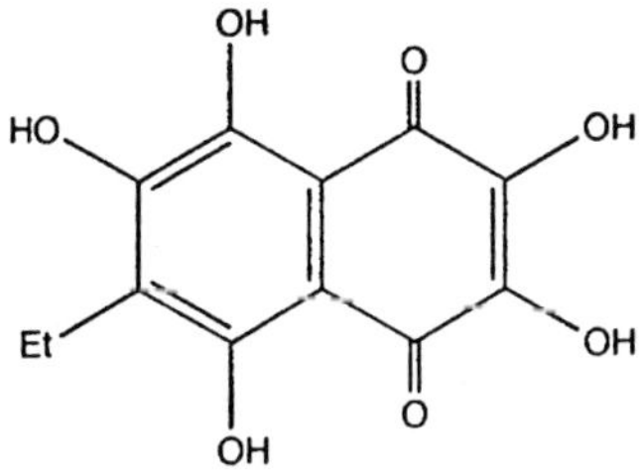

5

had been confirmed by synthesis.[7] A Friedel-Crafts reaction in an $AlCl_3$-NaCl melt at 180°C produced the desired compound in 1.5–2% yield. In an age when elemental analysis, melting points, and UV-visible spectra were the only accessible physical parameters, confirmation by synthesis was vital. I was ready to explore the void in our chemical knowledge of sea urchin pigments.

So it came about that a graduate student, Robert Amai, my wife, and I armed with screw drivers descended on Hanauma Bay to collect *Echinometra oblonga*. The research did not result in an unambiguous structure proof. It served to highlight the inadequacy of the separation method that the workers in the field practiced. The accepted technique was chromatography on a column of calcium carbonate. Colored bands did indeed develop, but — not surprisingly — nothing would move. After all, when the pigments are part of the live animal, they are tightly adsorbed on calcium carbonate of the shells and spines. A way had to be found to separate these closely related compounds, derivatives of naphthazarin or juglone differing only in minor structural features. A systematic and comprehensive investigation by Clifford Chang provided the answer: chromatography on severely deactivated (washed with 0.5N HCl, followed by air-drying) silica gel

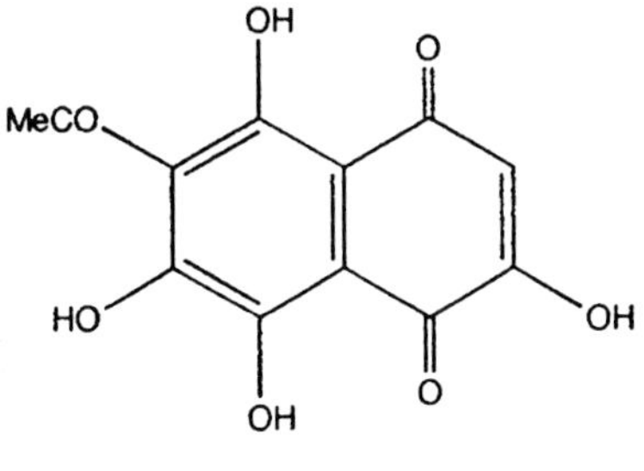

6

separated the individual pigments beautifully and finally opened up research on echinoderm pigments. Hand in hand with the discovery, how to separate effectively these acidic naphthaquinone derivatives, was the acquisition of a Varian A-60 NMR spectrometer. At last, there was a new and vital tool, which finally made possible a definitive structure of spinochrome M (**6**).[8]

A decisive influence on the direction of my future research came from my friendship with the late A.H. (Hank) Banner, a marine biology professor. While stationed in the Pacific during WW II with the U.S. Air Force, Banner became aware of ciguatera fish poisoning. In 1957 he assembled a multi-disciplinary research team to study ciguatera and invited me to participate. The major features of this involvement have been reported, as have some of the inadvertent byproducts of ciguatera research.[9] Palytoxin was perhaps the most significant and spectacular ciguatera offspring.[10]

Chemotaxonomic relationships have long been a valuable diagnostic feature of natural product research of terrestrial flowering plants. Our comparative study of echinoderm pigments led us to Bergmann's work, who since the early 1930's had used sterols as chemotaxonomic markers in his study of marine invertebrates. It was indeed a challenge if one considers the state of the art in chromatography and spectroscopy. In 1943 he isolated from a Caribbean gorgonian, *Plexaura flexuosa*, an unusually high-melting sterol, 180–182°C, with likely composition $C_{30}H_{50\text{-}52}O$, which he called gorgosterol.[11] Twenty-five years later Bergmann's student Ciereszko encountered gorgosterol in a number of coelenterates and secured the formula of $C_{30}H_{50}O$ by mass spectrometry. My student Kishan Gupta isolated gorgosterol from a *Palythoa* sp., where it was present in a mixture of five sterols. It had a long retention time on GLC, where it could be separated as its volatile TMS ether. An authentic sample from Ciereszko

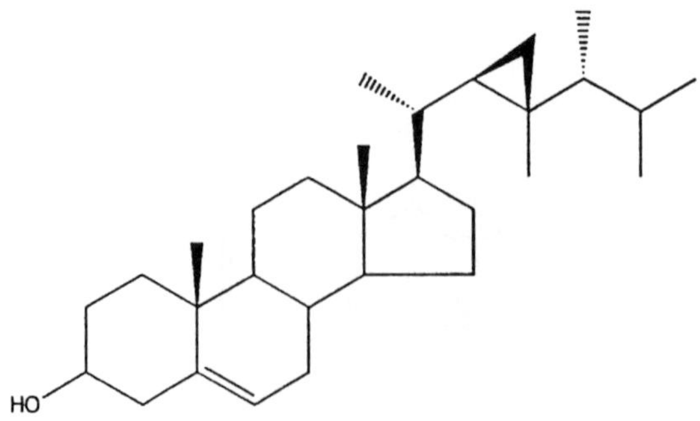

HO

7

confirmed the identity. Our Varian A60 NMR instrument was unable to probe its structural details. We turned to Carl Djerassi and his 100 MHz NMR instrument. He was able to solve the structure of gorgosterol (7)[12] and — as a veteran sterol chemist — became hooked on the study, particularly the biosynthesis of marine sterols. When Djerassi's coworkers helped him celebrate his publication millennium in 1982, they created a spoof communication disguised as a *Tetrahedron Letter*. I received a copy with the dedication: "To Paul Scheuer, whose original gorgosterol sample started it all."

Chemical Marine Ecology

My interest in ecology was piqued by my acquaintance with Bob Johannes, then a Zoology graduate student. He had added a nudibranch, *Phyllidia varicosa*, to his aquarium and discovered that all of his fish and shrimp soon died. He traced the cause to an unpleasant-smelling skin secretion of the nudibranch, which he discovered by stroking the animal. He showed by dialysis and heating to 100°C that the secretion was a small heat-stable molecule. Although we spread the word — including the local newsletter of the malacologists — that we would like to study the *P. varicosa* skin secretion, progress was slow. Once we received a call from Maui reporting a sighting of a school of mollusks in Ma'alaea Bay. Two coworkers promptly flew to Maui and returned with bags of Maui's famous potato chips, but without mollusks. At last, in the summer of 1973, Jay Burreson spotted

Nudibranch *Phyllidia varicosa* feeding on a sponge, *Ciocalypta* sp. (courtesy of Paul Scheuer).

the animal with its characteristic yellow and blue coloration at Pupukea Bay — on O'ahu's north shore. He rapidly verified Johannes's observations. More significantly, he discovered that within a few days in captivity the skin secretion of the mollusk would dwindle to zero. An inquiry to a noted malacologist brought no inkling of the nudibranch's diet. It would be another year, another dive season at Pupukea, which is inaccessible during the winter, when Burreson was in luck: he observed the animal feeding on an off-white sponge subsequently identified as *Ciocalypta* sp. The sponge indeed was the source of the *P. varicosa* secretion. Moreover, the chemistry was also rewarding: 9-isocyanopupukeanane (**8**) was a sesquiterpene with a new skeleton and a rare isocyano function. Isocyano natural products, most of them terpenoids of marine origin, have blossomed during the past twenty years into a fruitful area of research. The biological activities of isocyano compounds include antifouling and antimalarial activities.[13] The trivial name pupukeanane for this new sesquiterpene skeleton continues my practice, begun in 1961, to pay tribute to the rich flora and fauna of Hawaii by coining names that use the sonorous Hawaiian language.

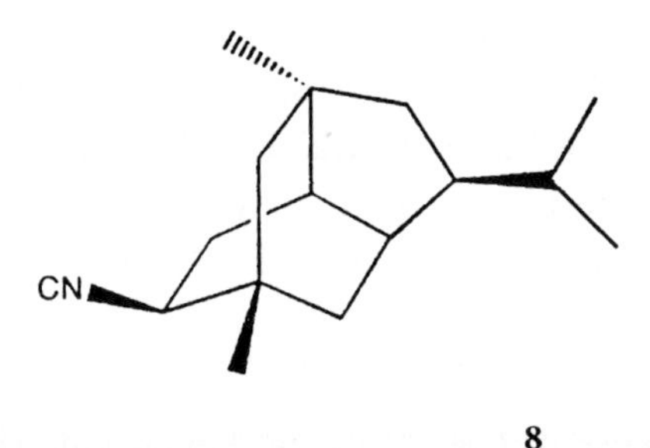

8

During the 1960's my research emphasis gradually gravitated from terrestrial to marine natural products. Although I published terrestrially oriented papers through the mid-1970's, the idea of writing a monograph on marine natural products began to germinate in the late 1960's. *Chemistry of Marine Natural Products* was published in 1973 (Academic Press, New York), the first book on the subject in any language.

Pupukea Bay with its underwater caves and lava tubes has remained an attractive dive spot. I appreciated it only after 1981, when on the persuasive urging of my graduate student Gary Schulte, I took scuba lessons and became a certified diver. Gary, while an undergraduate researcher in Bill Fenical's group at UC San Diego, had become interested in mollusks. On one of our night excursions to Pupukea in the late 1970's he collected some mollusks in the tidepools. They seemed to appear with the incoming tide to feed and mate. They were identified as *Philinopsis speciosa*, members of the gastropod order Cephalaspidea. According to conventional wisdom of the time, the anticipated metabolites were of polypropionate biogenesis. This prediction was borne out: we isolated two unexceptional polypropionate derivatives and an alkylpyridine reminiscent of an alarm pheromone constituent of a closely related mollusk, *Navanax inermis.* More significant than these results proved to be an observation by Steve Coval that the polar extract of *P. speciosa* appeared to contain a peptide. Isolation of this peptide and determination of its structure and stereochemistry proved to be an enduring experience. In the light of the long time lapse between our first collections and a definitive peptide structure we named the peptide kulolide (**9**) from the Hawaiian work *kulo* = taking a long time.

Two factors contributed prominently to the longevity of the kulolide project: instrumentation and ecology. Our 300 MHz NMR instrument in the early 1980's proved to be inadequate to resolve the complicated spectra of a $C_{43}H_{63}N_5O_9$ (794 Da) molecule, which existed in two conformations in all solvents over a wide temperature range. Acquisition of a 500 MHz instrument in 1989 resolved this difficulty. Attempts to determine the absolute

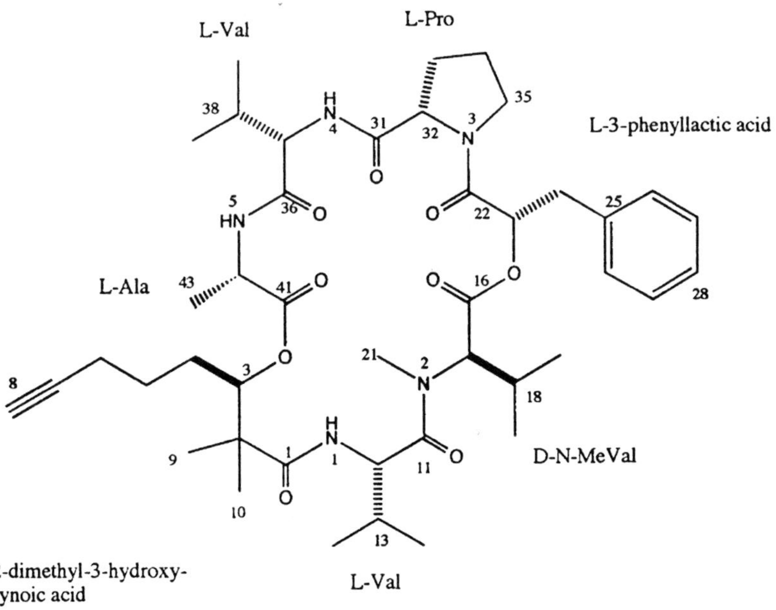

L-Val
L-Pro
L-3-phenyllactic acid
L-Ala
D-N-MeVal
R-2,2-dimethyl-3-hydroxy-7-octynoic acid
L-Val

9

stereochemistry of the two component acids, 3-phenyllactic and 2,2-dimethyl-3-hydroxy-7-octynoic acid, were frustrated by our dwindling supply of kulolide and by our failure to collect more animals.

Knowledge about the organismic biology and ecology of many marine invertebrates is rather limited. Alison Kay's authoritative treatise[14] lists no specific food source for *P. speciosa*; the Family Philinecea to which it belongs are carnivores, feeding on foramineferans, worms, and mollusks. A potential lead appeared in the work by Cimino and coworkers, who had linked the metabolites of the Mediterranean *Philinopsis depicta* (syn. *Aglaja depicta*) to its prey, the cephalaspidean mollusk *Bulla striata*.[15] According to Kay, the Hawaiian Bullidae are subject to dramatic fluctuations in population, ranging from virtually zero to a thousand shells at a single location.[14] During many years of collecting at Pupukea, we in fact never saw more than a few *Bulla* specimens at any one occasion.

Ecology came to our rescue and chemistry, in turn, solved the ecological puzzle. Both 1994 and 1995 seasons brought large populations of *P. speciosa* to Pupukea, which allowed us to reisolate kulolide (**9**) and resolve the remaining stereochemical lacunae, many years after our first collection.[16] The bountiful harvest of *P. speciosa*, followed by the tedious re-isolation of kulolide (**9**) provided a clue to the food source of the animal. During chromatography of the crude extract it became apparent that kulolide was the major, but by no means the only peptide in *P. speciosa*. In addition to several kulolide-related cyclic depsipeptides there was a linear tetrapeptide, which we called pupukeamide. It proved to be the key that would answer the question, "What do *P. speciosa* eat?" The structure of pupukeamide was instantly reminiscent of the majusculamides, which Moore and coworkers had isolated from the marine blue-green alga *Lyngbya majuscula*. But since *P. speciosa* is a carnivore, what is the missing link?

As luck would have it, the answer was found in my own research records. *L. majuscula* had been implicated as the source of a contact dermatitis, "swimmers' itch." We had briefly investigated it as a possible origin of ciguatera fish poisoning, with negative results. Watson's research into the toxins of sea hares (Notaspidean mollusks) was another pot-shot and unproductive approach to trace the ciguatera toxin.[17] While the original purpose failed, Watson's "ether-soluble toxin" seemed a worthwhile research objective. In time, we succeeded in determining the structure of aplysiatoxin (**10**) and debromoaplysiatoxin (**11**) from *Stylocheilus longicaudus*, a small but relatively abundant sea hare.[18] It was the last major piece of research in my laboratory carried out without the benefit of HPLC or high-field

NMR instrumentation on 12 g of crude toxin. Several years later, Moore and coworkers isolated debromoaplysiatoxin (**11**) from the same blue-green alga *Lyngbya majuscula*, which had yielded the majusculamides. Now it was crystal-clear: *Stylocheilus*, who feed on *Lyngbya*, are in turn eaten by *Philinopsis*. In fact, we had noted *Stylocheilus* in the Pupukea area, but

10 R = Br

11 R = H

Sea hare *Stylocheilus longicaudus* (courtesy of Paul Scheuer).

disregarded the animals, which had been studied long ago. But now we have a new puzzle: Why are there no aplysiatoxin-like compounds in *Philinopsis*? Like old soldiers, good research projects never die!

Drug Discovery

Dramatic biological activity and structures of natural products have formed an interwoven fabric since the science began, although its prominence has fluctuated. After a few decades in which the study of natural products took a back seat to laboratory synthesis and theory, natural products once again are looked upon as important drug leads. Not only is this an obvious liaison, but in an age when most funding agencies look for societal benefits even while advancing basic knowledge, it has become a *sine qua non*. New

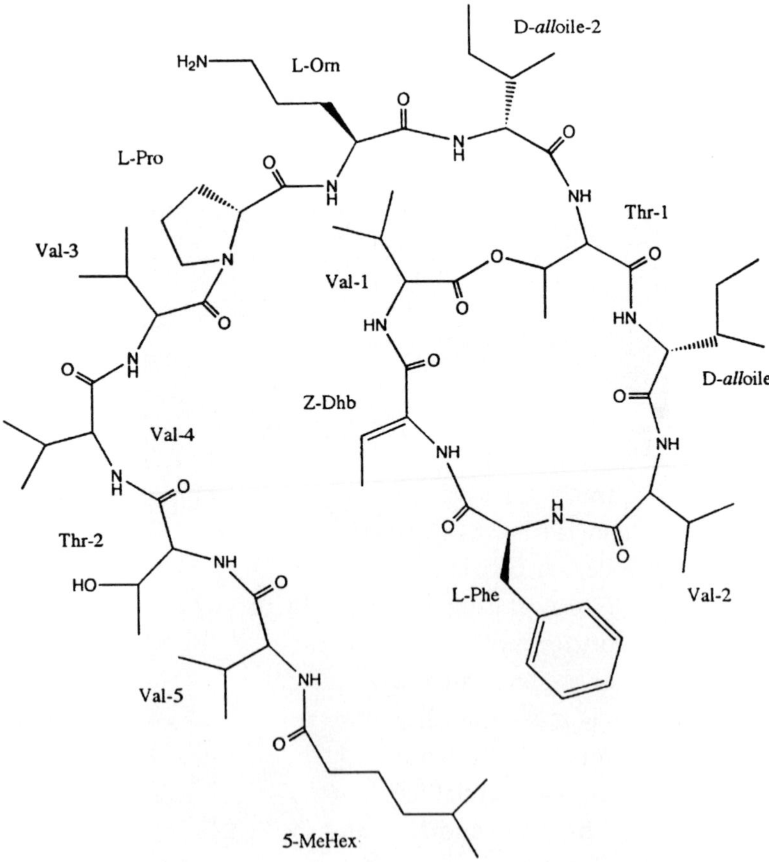

12

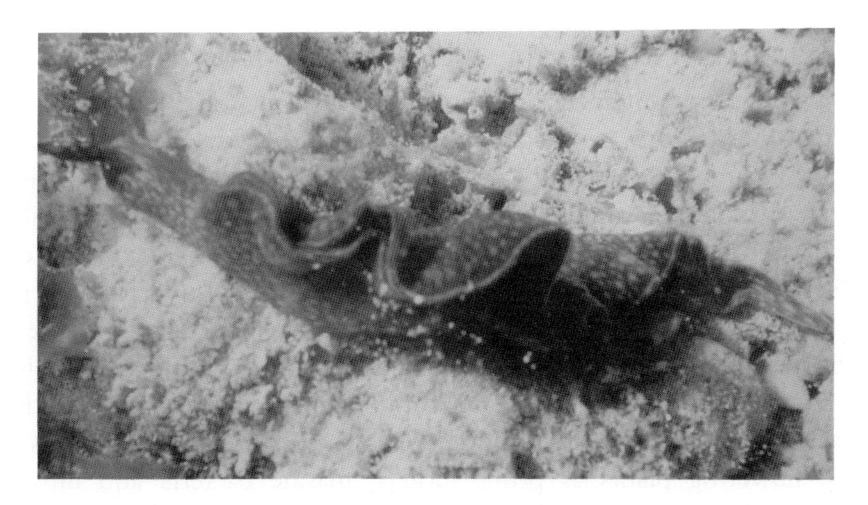

Sacoglossan mollusk *Elysia rufescens* (courtesy of Paul Scheuer).

13

anticancer agents have been the targets of choice ever since the establishment of the National Cancer Institute as a visible symbol of Richard Nixon's "War on Cancer." One of our candidates currently in preclinical trials against lung and colon cancers is kahalalide F (**12**).[19] It was an unexpected isolate from an herbivorous sacoglossan mollusk, *Elysia rufescens*, as the chemical and biological literature had led us to anticipate terpenoid constituents. The animal feeds on a green alga, *Bryopsis* sp., from which we were able to isolate kahalalide F as well as an acyclic analog as the major constituent.

Another significant product from my laboratory, while not a potential drug, has become a commercial chemical for use in molecular biology. Okadaic acid (**13**), first isolated from a sponge, *Halichondria okadai*,[20] selectively inhibits phospatases 1 and 2A. It is widely used for the study of cellular mechanisms.

What of the Future

What with genetic engineering, combinatorial drug design, detailed mapping of receptor sites, are we not witnessing the last hurrah of natural product research? I think not. There will never be a substitute for discovering new molecular architecture and for learning in greater detail the molecular composition of living organisms. As we all have experienced in our lifetime, technological advances insure that we continue to learn more from ever smaller samples. What about marine research? Is it not true that with increasing frequency we isolate known compounds? Indeed, this is the case, but two developments promise continued success. In parallel with the history of terrestrial natural product research, when microorganisms succeeded flowering plants as prime targets, marine bacteria and fungi are increasingly being scrutinized for new natural products. And their numbers are vast. A second exciting event is the use of rebreathing technology. This will allow ocean exploration at greater depths and for longer periods of time. Although mini-submersibles are wonderful inventions — and I cherish the memories of my first sight of bioluminescent gorgonian corals or suddenly viewing the wreck of a WW II divebomber (ours) off Makapuu — they are not well suited for collecting and they are expensive to operate. Rebreathing apparatus will allow a pair of human rather than mechanical hands to collect, say, bryozoans or encrusting sponges, which are beyond the capacity of scuba. Rebreathers will also provide access to a depth zone which up to now was out of reach. And finally, as we begin to study and understand the biosynthesis and genetics of marine organisms, exciting new natural products will be produced in quantity by fermentation technology.

Acknowledgment

I have indeed been fortunate to have travelled from the tanning vats in a German leather factory to the coral reefs of the "loveliest fleet of islands that lies anchored in any ocean."[21] This journey has been made possible and rewarding by the constant support of my wife Alice, my children Elizabeth, Deborah, David, and Jonathan. Without the talent and devotion of my coworkers there would be little to write about. Among the funding agencies which have supported my work I must mention Research Corporation, which gave me a start in 1951 with a $2,800 grant. Continued funding over the years has come from many sources, but most steadfastly from the National Science Foundation, the Sea Grant College Program, and PharmaMar, S.A.

References

1. Woodward, R.B.; Brehm, W.J. *J. Am. Chem. Soc.* **1947**, *70*, 2107.
2. Scheuer, P.J. *J. Am. Chem. Soc.* **1960**, *82*, 193.
3. Kippis, A. *A Narrative of the Voyages Round the World Performed by Captain James Cook*; Leavitt and Allen; New York, 1855, Vol. II, p. 111.
4. *The Merck Index*, 12th Ed. p. 1047 (Merck & Co., Whitehouse Station, N.J., 1996) still refers to 'awa as an intoxicating beverage.
5. a. Hänsel, R.; Beiersdorf, H.U. *Naturwissenschaften* **1958**, *45*, 573, b. Hänsel, R.; *Z. Phytotherapie* **1996**, *17*, 180.
6. Scheuer, P.J.; Horigan, T.J. *Nature* **1959**, *184*, 979.
7. Wallenfels, K.; Gauhe, A. *Chem. Ber.* **1943**, *76*, 325.
8. Chang, C.W.J.; Moore, R.E.; Scheuer, P.J. *J. Am. Chem. Soc.* **1964**, *86*, 2959.
9. Scheuer, P.J. *Tetrahedron* **1994**, *50*, 1.
10. Moore, R.E.; Scheuer, P.J. *Science* **1971**, *172*, 495.
11. Bergmann, W.; McLean, M.L.; Lester, D. *J. Org. Chem.* **1943**, *8*, 271.
12. Hale, R.L.; Leclercq, J.; Tursch, B.; Djerassi, C.; Gross, R.A., Jr.; Weinheimer, A.J.; Gupta, K.; Scheuer, P.J. *J. Am. Chem. Soc.* **1970**, *92*, 2179.
13. Chang, C.W.J. "Naturally Occurring Isocyano/Isothiocyanato and Related Compounds;" In *Prog. Chem. Nat. Prod.*, in press.
14. Kay, E.A. *Hawaiian Marine Shells*; Bishop Museum Special Publication 64(4). Bishop Museum Press, Honolulu, Hawaii, 1979, pp 430-431.
15. Cimino, G.; Sodano, G.; Spinella, A. *J. Org. Chem.* **1987**, *52*, 5326.
16. Reese, M.T.; Gulavita, N.K.; Nakao, Y.; Hamann, M.T.; Yoshida, W.Y.; Coval, S.J.; Scheuer, P.J. *J. Am. Chem. Soc.* **1996**, *118*, 11081.
17. Watson, M. *Toxicon* **1973**, *11*, 259.
18. Kato, Y.; Scheuer, P.J. *J. Am. Chem. Soc.* **1974**, *96*, 2245.
19. Hamann, M.T.; Scheuer, P.J. *J. Am. Chem. Soc.* **1993**, *115*, 5825.
20. Tachibana, K.; Scheuer, P.J.; Tsukitani, Y.; Kikuchi, D.; Van Engen, D.; Clardy, J.; Gopichand, Y.; Schmitz, F.J. *J. Am. Chem. Soc.* **1981**, *103*, 2469.
21. Clemens, S.L. (Mark Twain). On his visit to Hawai'i in 1866 as a reporter for the Sacramento *Weekly Union.*

Ayhan Ulubelen, 1995 (photograph by I. Hargittai).

8

Ayhan Ulubelen

Ayhan Ulubelen (b. 1931 in Turkey) is a member of the Turkish Academy of Sciences and its committee of governing body and she is Professor of General Chemistry of the Faculty of Pharmacy, Istanbul University. She was educated in Turkey and completed her Ph.D. in analytical chemistry at Istanbul University in 1956. Then she did two years of postdoctoral research at the University of Minnesota and, following a one-year stay in Turkey, spent four years at the University of Arizona on an NIH project in cancer research. Since 1967 she has been full professor at Istanbul University and also spent a few months at a time on several occasions in Japan supported by JSPS, in Germany supported by DAAD, and at the University of Texas at Austin under the auspices of a NATO project. Our conversation was recorded in Dr. Ulubelen's office on September 20, 1995 and it appeared in *The Chemical Intelligencer.**

How did it all start?

My father was an army officer and my mother was a housewife. Originally, I was thinking of becoming a newspaperwoman. However, when I was in high school, I saw a movie about Madame Curie, and I decided to become a chemist. All the girls in my class did too, but only I went through with it.

---

*This is an augmented version of the interview originally published in *The Chemical Intelligencer* 1997, *3*(1), 31–33 © 1997, Springer-Verlag, New York, Inc.

When I graduated, I couldn't find a job in the chemical industry. There were very few jobs there, and they may have had a bias against girls at that time. There was one opening in analytical chemistry in the Faculty of Pharmacy. A boy and I applied for it, but he failed the language examination so I got the job.

Has being a woman caused any difficulties since then?

None whatsoever. At that time, though, there were very few universities in Turkey; ours was the first. Today there are 57 universities for a population of over 60 million. The proportion of women is probably higher at Istanbul University, but there are many women at the others as well.

Do many of them dress traditionally?

Not really, because the creation of the Republic in 1923 changed everything and because we are a secular country and we don't want to go back to the religious ways.

But dressing traditionally, is it necessarily religious?

It is the radical Moslems who dress in a uniformlike outfit that is not the traditional dress. In the old days, that is, before the Turkish Republic, Turkish women used to have "carsaf" in the cities and baggy pants and a head scarf in the villages. Today, the women in the cities have modern clothes, and the outfits in the villages haven't changed much, except that the younger generation is switching toward modern outfits. During the last 10 years, increasing numbers of women started to wear "carsaf," and many young girls wear uniformlike outfits, very long and loose coats and very long, large scarves that come down to their waistlines. There is nothing traditional about this way of dressing. And in a profession one should have proper clothing; a medical doctor, a nurse, a lawyer, etc., should dress whatever way is appropriate for the profession.

Would any students appear in the lab in traditional dress?

Nothing like that has happened in my laboratory so far. It is dangerous because they have long skirts and the scarves are coming down almost to the waistline. When students come to the lab dressed this way, the University tries to close its eyes. Only in the medical school and especially for nursing students are the rules strictly enforced, especially the long skirts could carry all kinds of bacteria. Otherwise the students are free, of course, to wear what they like. There are more students dressed like this today than 10 years ago. This is because a lot of money is coming to the radical Moslem students from sources outside of Turkey. The radical Moslem countries don't like Turkey being a secular country. Having a secular country that is 98% Moslem seems very dangerous for them. Their government style is based on Moslem rules. That's why I believe that these countries give money to our students. Especially attracted are the poor people who come from Anatolia and migrate into the big cities. They are the most open to all kinds of influences. According to a highly accepted belief among educated people in Turkey, the students get a place to live and a monthly payment if the girls cover up and the boys grow a beard.

How about support for scientific research?

Research money is very tight. Here at Istanbul University we have a Research Fund. People may apply for support and may receive a few thousand dollars if their application is granted. This money they can spend for their research. Every university has such a research fund. The money in the fund is a portion of the university's earnings from outside contracts. The Government also contributes to these funds.

Another source is TÜBITAK (*T*ürkiye *B*ilimsel ve *T*eknik *A*rastirma *K*urumu; Scientific & Technical Research Council of Turkey), which in some ways corresponds to the National Science Foundation in the United States. TÜBITAK helps us buy some instruments, not any large-scale equipment though, not an NMR, for example.

You are a natural product chemist and obviously you need an NMR apparatus. Do you have one?

Not at Istanbul University but I'm also associated with a TÜBITAK research lab, and I go there twice weekly and take our samples with me to run them on their NMR. They have a 200 MHz instrument, a very good one. Now they are trying to get a 400 MHz instrument. They already have such an NMR in Ankara. My dream is, of course, to get an NMR machine for Istanbul University. If this happens, we could give service not only to everyone at this university but to many places outside too. We would be charging less than TÜBITAK charges.

Turkey being a NATO country, do you get research support from NATO sources?

We Turkish people don't know how to ask for money from outside of Turkey. There are joint projects, most of them just starting, with the European Community (EU). Turkey is not a member of EU, but we can participate if we have partners in the Community. Special topics are favored such as energy-related ones and environment-related ones. Scientifically they may not be very strong though.

Please, tell us about your research interests.

My main research is the structure determination of natural products and pharmacological investigation of the isolated compounds.

First, I started with plants that are found in great abundance in Turkey. We had worked with triterpenes and flavonoids at the beginning and then switched to diterpenoids. We are getting interesting new compounds and find their structures by UV, IR, NMR, mass spectrometry, and other techniques.

The sources are all Turkish plants. We work mostly with traditional medicinal plants. I mean the old Turkish folklore medicine. We would like to find out what is the actual ingredient of the plant that produces the desired effect.

For instance, the last two plants we have worked with were sent to us from eastern Turkey by a former student of mine, who works as a pharmacist. She said that one of these plants was very good for spontaneous abortion, and the other for wound healing due to its antimicrobial effects. The villagers use these plants extensively. We also have special stores, even in Istanbul, called "AKTAR," that sell plants and plant extracts used as traditional medicine.

Did you find the abortive agent?

Unfortunately, it was not very good. When we fed the mice with the compounds that we had isolated, we found that they developed cystic degeneration in the ovaries. The tests are not yet completed though, because we are not quite sure that the cysts developed due to this compound. They sometimes occur during ovulation. The abortive factor was obviously present, and we have to continue our tests.

Every single compound, and we isolated about 10 of them from this plant, is administrated to 10 mice each. Then another group, at the Faculty of Medicine, studies the ovaries, the livers, the kidneys, and the brains of these mice involved in the tests. This medical group is looking for any harmful effects that are possibly caused by these compounds.

At this point we can't recommend that women should be using these plants as an abortive agent. We have so informed the pharmacist who sent them to us, and I'm sure she's passing this information on to the villagers.

They may have been using it for hundreds of years.

Yes, and not only in Turkey; it is present also in Chinese folklore. We have even made a joint study with a Chinese group. They'd asked us to send them a sample.

Our birth rate is rather high. Not very long ago it was 2.3% and, more recently, according to the newspapers, it has dropped to 1.5%.

Of course, I'm aware of the fact that researchers in many other places are involved in finding such abortive agents, and the World Health Organization also supports such research. The goal is to find something from natural sources that women can just drink.

I happened to be in America when Carl Djerassi's birth control pill was being introduced, and it has prevented a lot of unwanted pregnancies. But it also contributed to the spread of free sex all over the world and that has seriously damaged family life. Just yesterday I read an article about England which said that people don't want that type of living anymore and they want to bring back the families. The pill may also cause some cardiac problems and may also be carcinogenic. The agent causing spontaneous abortion may be much more advantageous, and it would be used only when truly needed. The women in our villages and in the villages of India and Pakistan and many other countries could use it when really needed, and its use would be very easy. However, so far there is nothing like that has been found.

But you're working on it?

Yes, I am. But I'm also working on many other things such as cancer, HIV, and diabetes. All this work is with natural products. And lots of other people are working on these problems all around the world. My samples all come from Turkey, but the methods I employ are the same as anywhere else. I go to international meetings and exchange our experience with my colleagues.

How do you compare your work with that of others, and your conditions and possibilities?

I continuously ask myself this question, and I know I am not one of these big scientists, but I sincerely believe that if we had the possibilities and conditions of so-and-so in the United States or in Germany, and I don't want to mention names, we could do much better than they do. Today our papers are being published in distinguished international journals.

If you were given a large amount of money, what would you spend it on?

The first would be good instrumentation, then I would improve the library, and I would pay my students much better than I can now and would have more students too. Now we publish about 10 papers annually, and we could easily double that if only we had a better equipped laboratory.

I've heard that Turkish scientists receive a premium for each article they publish in refereed international journals. That means that the authorities appreciate and encourage scientific research in a very conspicuous way. Why don't they understand then your need for an NMR spectrometer?

NMR is needed by the chemists, and in this country the medical faculties are the leaders. They have the largest number of publications. Of course, they are very important because they are related to human health. They also get the best instruments, and most of the support is going to them.

Why is then the premium for publications?

It comes from the desire to improve the level of Turkish science. On the basis of the number of the publications in periodicals covered by the *Citation Index* of ISI, Turkey is ranked 34th in the whole world and the Turkish scientific community would like Turkey to move up to somewhere around no. 20 to 25. Some years ago Turkey was down to no. 45 or so.

The premium for publications comes in two parts. One part goes to the authors' pockets in proportion to the number of papers in journals covered by the *Citation Index*, and this premium is awarded by TÜBITAK. The other part goes to the authors' research group for research expenses, and this comes from the university research funds. For determining this part, it is not only the number of papers that counts but also the impact factors of the journal where the article appeared. The higher the impact factor, the larger the premium is. Just this week a medical colleague received TL 270,000,000 (1 USD = TL 48,000 at the time of the conversation) for his three publications to support his research.

My research group has also bought various small pieces of equipment using premium money. This year we are planning to buy a personal computer with this money, for writing papers, etc.

You don't have a PC yet?

Yes, we do. It is in the Dean's office, and there is another one in the library. But my group would like to have a PC of its own.

Addendum: *Professor Ulubelen acquired her own PC soon after our conversation. Then in August 1998, she added the following information:*

Recently we found antituberculous compounds from *Salvia multicaulis* and we published our findings in the *Journal of Natural Products* **1997**, *60*, 1275–1280. An archeologist friend of mine mentioned a plant from archaic times that was used in birth control in Northern Africa. We checked the literature for Africa and found out that it is a plant of the umbelliferae family, and we are planning to work with it. We are also checking diterpenoid alkaloids for their antifeedant, insecticidal, and insect repellent activities and are getting some promising results.

John and Rita Cornforth 1997 (photograph by I. Hargittai).

9

JOHN W. CORNFORTH

Sir John Warcup Cornforth (b. 1917 in Sydney, Australia) is Professor Emeritus at the University of Sussex at Brighton, U.K. He shared the Nobel Prize in chemistry in 1975 (with Vladimir Prelog) "for his work on the stereochemistry of enzyme-catalyzed reactions."

Professor Cornforth is Fellow of the Royal Society of London, Foreign Associate of the National Academy of Sciences, Corresponding Member of the Australian Academy of Science, and member of many other learned societies. His many distinctions include the following Medals of the Royal Society: Davy (1968, with George Popják), Royal (1976), and Copley (1982). Professor Cornforth was knighted in 1977 and became a Companion of the Order of Australia in 1991.

We recorded our conversation at the Cornforths' home in Lewes, U.K., on September 13, 1997. Rita Cornforth (née Harradence) is also a D.Phil. of Oxford where she did her doctoral work with Professor Sir Robert Robinson. She was a great and most kind help in our conversation. The interview appeared in *The Chemical Intelligencer*.*

Could we please start with your family background and education?

I was the second of four children. My father was a schoolmaster. My mother, before she married, was a maternity nurse. I was born in Sydney but spent five years in the northern part of rural New South Wales, at Armidale.

*This interview was originally published in *The Chemical Intelligencer* **1998**, 4(3), 27–32

The Cornforth Family in 1949 on Trafalgar Square, London (courtesy of John Cornforth).

My father was an Englishman but my mother's family had been in Australia for a considerable time. She got her name, Eipper, from a German missionary who came out in 1852, settled there, and married an Irish woman. So I am a mixture of German, Irish, and English. There was no tradition of science in my family. My father taught classics, Latin and Greek.

Did you study Latin and Greek too?

Yes, and I have never regretted that. My choice of science was conditioned by deafness. This was noticed first when I was about ten. It is a condition which might or might not go on to total deafness. In my case it did. But one does not know beforehand. By the time I was at high school, it was obvious that I wasn't going to be able to teach or to become a barrister which had been another possibility. I quite liked the idea of going into law. I'm glad now that I didn't. But I had to find something in which my hearing would not be a fatal drawback. I had a good chemistry teacher at the time, his name was Leonard Basser. I began to be interested in the subject.

Were you also interested in other subjects?

I was never much interested in mathematics so I probably couldn't have become a physicist. I was interested in astronomy. I also had the idea of studying biochemistry, but as it was then it was largely taught as an analytical service to the medical schools. I didn't like that and went in for chemistry.

I entered Sydney University at 16, and there was this Exhibition, a scholarship that exempted me from paying fees.

I was getting deafer all the time. I could not use the hearing aids that were available then because the sound came distorted. I did not use lip-reading very much at the time. Even if I had been an expert in lip-reading, I could not have used it for lectures because lip-reading is a guessing game. You always have to interpret what you think is the content. It is not good for learning new ideas. I have to tell that to deaf children who sometimes ask for my advice about how to get through university. They have more help now.

I read a lot, especially the original literature. At that time, in chemistry, you had to know German. I had been taught French at school but not German. So I bought a German chemistry text and a German dictionary, and I looked up every word until I knew them all. Now I read German very easily, but my grammar is weak.

Looking back, would you advise deaf children today to get into chemistry?

Yes, if they are interested.

You were selected in 1939 to go to Oxford.

Rita and I were both selected. Rita was one year ahead of me at university. There were only two of these scholarships every year between the six Australian universities for any science subjects. We were selected independently, we were not yet attached to each other at that time. We both wanted to go to Oxford and work with Sir Robert Robinson. Robert was on the selection committee and he must have been very persuasive because he got both scholarships for us.

What was it like to work for Robert Robinson?

I found it very stimulating. I could never learn to lip-read him; he was an extremely difficult subject for lip-reading, but he was very good at hand-writing, and would write things down for me. We started arguing from the first day.

What did you argue about?

Structures, strategies, and miscellaneous science topics. You would put forward an idea and he would immediately find an argument against it. Unless you had a very good case it was extremely hard to win an argument against him.

Would you care to say anything about the Robinson–Ingold controversy?

Once I terribly upset Robinson by saying that Ingold deserved a Nobel Prize. That was, of course, much later. About the controversy I would just say that the two men came into conflict because both were very strong characters and their approaches were different.

You did some work on penicillin at the beginning. Would you, please, tell us something about it?

We arrived in Oxford in 1939. Just about that time people were already working there on extending Fleming's work on penicillin. The trouble was that penicillin is very unstable in its crude form. Fleming had asked a chemist named Raistrick to see whether he could get anything out from the broth, but this attempt was unsuccessful. However, Chain, another chemist, knew about the sensitivities of penicillin, and he with Florey and others took up this work. They devised a beautiful assay to measure the quantity of penicillin in arbitrary units. They were then able to find out what conditions wouldn't inactivate the stuff and then gradually to concentrate it. Fleming's original culture was not a good producer by modern standards. So production had to be on a large scale because large volumes were needed to scrape together enough penicillin for a treatment. The first time they challenged the staphylococcus *in vivo*, it was a spectacular success. The infection killed all the controls in a few hours, and all the treated animals survived. That was in 1940.

When was the first human case?

It was an Oxford policeman, in 1941. He had developed septicaemia, blood poisoning from an untreated sore. They did not have much penicillin. If they had given him all that they had in one dose, they might well have cured him. However, they decided to divide it up. Upon the first injection his temperature went down and his blood cleared. But then there was a relapse. They gave him another injection and the same thing happened. This went on until none was left and the man died.

Rita and I entered the penicillin project in 1943, when the structure of the molecule was being determined. We were able to synthesize D-penicillamine, a fragment representing nearly half the molecule. The chemistry occupied hundreds of chemists here and in the U.S. during the rest of the war. There was a monograph published on the combined efforts, *The Chemistry of Penicillin*, Princeton University Press, 1949. I wrote one of the chapters.

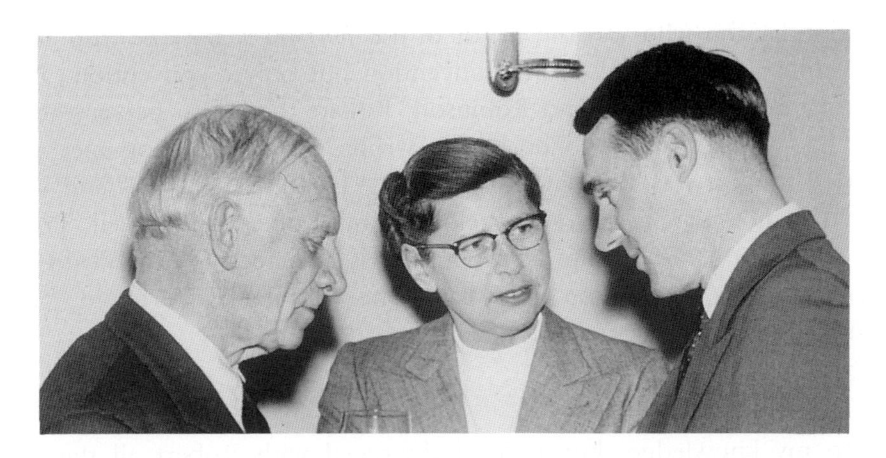

John Cornforth with Robert Robinson and coworker in 1959 (courtesy of John Cornforth).

How long did your association with Robinson last?

As a scientific collaboration, 14 years. In the meantime, we left Oxford in 1946 and joined the Medical Research Council and worked at its National Institute. Rita and I were married in 1941 and we had our first child in 1943. Rita never stopped working except for short periods of time. She kept working until 1975. We always worked together.

When I went to the National Institute, I continued my earlier work, the synthesis of cholesterol. Robinson persuaded Charles Harington, the Director of the Institute to let me do this. I finished the synthesis in 1951. It was the first total synthesis of a nonaromatic steroid. Robert Woodward was also working on the cholesterol synthesis, and we came to completion at about the same time.

Was there communication between you?

No.

Was there competition?

Yes.

Would you tell us about your interactions with Vladimir Prelog?

There was no collaboration but we did publish one paper together. I was always interested in stereochemistry. For the designation of stereochemistry about an asymmetric center there was the system, or convention, by Ingold,

Prelog and Cahn who was the editor of the *Journal of the Chemical Society* at the time. It was a very useful convention but only a convention. There is a story between Prelog and Robinson. Robert tended to be very abrupt when he was discovered to be in the wrong. He and Prelog met at an airport, I think it was Tel Aviv. Robert said, "Hello, Katchalsky," and Vlado said, "I'm not Katchalsky, I'm Prelog." And Robert said "Well, Prelog, your designation of absolute configuration is all wrong." And Vlado said, "But it's only a convention, how can it be wrong?" Robert said, "Well, it's unnecessary."

Did he ever use it?

Not to my knowledge, but I do. I disagreed with Robert all the time.

Vlado's work on stereochemical problems goes back quite a long time. It was some of his best work too. One of the things he did was a contribution to the problem of absolute configurations. At the beginning of stereochemistry it was impossible to tell, when you had a molecule that could exist in two mirror image forms, which isomer was which. That is, it was impossible to know the absolute configuration, the absolute arrangement in space of the groups around the asymmetric center. Emil Fischer guessed the stereochemistry of the simplest possible sugar. He got it right. But he didn't really know, and the whole of this science could have been rewritten without much effect had he proved to be wrong. There was no way of discriminating between the two possibilities until the early 1950s. Then a Dutch crystallographer, Bijvoet, examined the X-ray diffraction of a rubidium salt of tartaric acid. The rubidium atom causes anomalous dispersion of X-rays of a certain range of wavelengths. This dispersion makes it possible to correlate the phase differences with the absolute configuration. For a long time after that, tartaric acid was the only substance the absolute configuration of which had been shown. If you wanted to deduce the absolute configuration of something else, you had to establish a connection by chemistry with tartaric acid. There were long sequences which brought in all sorts of substances with tartaric acid at the center. Vlado made an extremely useful rule by which he could deduce absolute configurations about a chiral center. He established the rule by starting from alcohols which had been correlated with tartaric acid, things like menthol. He was then able to get at the unknown absolute configuration of an alcohol simply by making an ester from it and reacting this with an organometallic reagent, then hydrolyzing off the alcohol and measuring the sign of optical rotation of what was left. If it was positive you could make one deduction about the absolute configuration, and if it was negative you made the opposite deduction. He tested this with dozens of examples. It is a very good rule and I know of no exception.

Well, Rita and I wanted to synthesize the natural form of mevalonic acid, an important intermediate in sterol biosynthesis, and we thought of a route starting from linalool. That is a terpene alcohol found in many perfumes. Vlado had just published a paper assigning absolute configurations to the two mirror-image forms of linalool. This told us that if we started from commercially available linalool we would get the natural form of mevalonic acid. So we did this synthesis and got the wrong, unnatural form! It was not that Vlado's rule was wrong, but in this one paper he or his collaborator had misapplied it and drawn the wrong conclusion. We wrote to him and the three of us published a correction, our only paper together. And we have been friends since that time.

How do you describe yourself in chemistry?

I describe myself as an organic chemist who is interested in the biological aspects of the subject. This interest strengthened when George Popják joined the National Institute for Medical Research shortly after we did, in 1947. Popják came from Szeged, Hungary, where he had studied medicine and got his M.D. He came to London some years before the war. He was not a student of Szent-Györgyi in Szeged and later Szent-Györgyi regretted that very much. Popják is now in Los Angeles; he is emeritus at UCLA. In London he worked at St. Thomas's and used phosphorus as a radioactive tracer. George had an interest in lipids and it was a time when there were some exciting developments in lipid chemistry. There was the acetate hypothesis of fatty acids, that they are built up from acetate units, and that acetate is also the favored starting material for cholesterol synthesis. Already in the late 1930s, there were radioisotope experiments with the acetates, using deuterium instead of hydrogen. That arose out of the works of Rudolf Schoenheimer. I remember reading his little book, *The Dynamic State of Body Constituents.* It was a revolutionary viewpoint for me, the sense that we have the same kind of stability as flame. What I mean is that all the constituents of our body are being torn down and built up again all the time. Just like flames. I had never had that perspective.

Looking back, what would you single out as your most important result?

It's the realization that you could use stereochemical methods to understand what was happening at the enzymic level.

John Cornforth with George Popják in 1963 (courtesy of John Cornforth).

Was this research planned?

It grew. Popják and I had become expert in studying the course of sterol biosynthesis by making and feeding isotopically labeled precursors and tracking down the isotopes in the products by chemical degradations. As knowledge of the pathway advanced, we were able to answer quite complicated questions; for example, the course of an intramolecular rearrangement. Well, we were looking at the step — we know now that it is two steps — in which two molecules of a fifteen-carbon precursor come together to make a thirty-carbon product called squalene. It looks like a perfectly symmetrical process but it isn't: we found that one hydrogen atom from one of the two precursor molecules had been exchanged. That started me thinking about asymmetric treatment of symmetrical substrates by enzymes and I worked out that the pathway from mevalonate to squalene via the intermediates then known might proceed in 16 384 different ways, all stereospecific. And then ideas began to form about how to find out which way was actually used. Soon it was clear that succinic acid, a simple four-carbon compound, was going to be a key degradation product, and that we would need to measure the extremely small optical rotations caused by replacing one hydrogen in a methylene group with deuterium. By great luck, an instrument capable of doing just that was under development at the National Physical Laboratory. We did our first trial with a coenzyme that supplies some of the hydrogen in cholesterol, the nicotinamide dinucleotide NAD. In its reduced form this transfers hydrogen to a substrate from one side or the other of its nicotinamide ring; which side depends on which enzyme it is assisting. We used deuterium label to mark one of the hydrogens and then degraded the reduced coenzyme to deuterosuccinic acid. It was optically active and the sign of rotation depended on which hydrogen we labeled. At the same time we synthesized a deuterosuccinic acid in such a way that we knew the absolute configuration. From the sign of its rotation we could deduce the configuration of any other specimen obtained by degradation.

This set the pattern for what followed. Mevalonic acid is a key precursor of steroids and it has six methylene hydrogens, all stereochemically distinct. We devised synthetic methods for labeling each of them with deuterium or tritium, in such a way that we knew where the label was. Then we fed these labeled acids to suitable enzyme preparations that would take them along the biosynthetic pathway, all the way to cholesterol or part way to an intermediate like squalene. Sometimes it was then just a matter of finding out if the label had been lost or retained; sometimes we had to degrade to deuterosuccinic acid to find out the configuration at the methylene group. We had to work out our methods of synthesis, enzymic transformation, degradation and analysis as we went along. It was enormous fun. And in the end we reduced 16 384 possibilities to 2. We knew that we could go no further without solving the problem of the chiral methyl group. In one step of the biosynthesis, a methyl group is formed by adding hydrogen to one side or the other of a methylene group — but which side? A methyl group to be chiral must contain all three isotopes of hydrogen. We could generate such a group by biosynthesis and we would solve our problem by finding out its absolute configuration, but how to analyse it? The key was provided by Hermann Eggerer, then working at Munich. Malate synthase, an enzyme of the glyoxylate cycle, uses a substrate with a methyl group (the Coenzyme A ester of acetic acid) and turns it into a substance with a methylene group (malic acid). There was good reason to believe that it would do this stereospecifically and that it would pull off the lightest hydrogen isotope preferentially from the methyl group. We already knew of an enzyme that would discriminate between the two remaining hydrogens on the resulting methylene group by removing one of them. So by chemistry John Redmond and I made two specimens of acetic acid. Each of them contained chiral methyl groups; the absolute configurations were known from the synthesis and were opposite in the two specimens. At Munich each specimen was added to a cocktail of enzymes that converted it first to acetyl-Coenzyme A and then to malic acid. The malic acid was treated with the second enzyme and the loss of the easiest hydrogen isotope to measure, radioactive tritium, was monitored. Sure enough the results from the two specimens were very different. Now we had an analytical method that would tell us the absolute configuration of any chiral methyl group that could be brought into the form of acetic acid. We applied it to the appropriate stage in squalene biosynthesis and at last reduced our 16 384 alternatives to 1. Of course, the result had many other applications to a large number of enzymes in which methyl groups are generated or transformed. With a few interesting exceptions, stereospecificity has been the rule.

May the asymmetry of substances in living matter have arisen as a result of chemical processes? This is a question that is difficult not to ask, yet I know we can only speculate about it.

And I'm not immune to speculating about it either. The way I see this is, the only life we know operates in a medium of water. The big molecules of life, the proteins and the nucleic acids and polysaccharides, are built up from simple units by what amounts to the elimination of water. But these big molecules are unstable in water and they hydrolyse if left long enough. In a situation like that, the way to preserve a molecular pattern is by self-replication of molecules. The individual molecules decay but the pattern survives, just like what Schoenheimer found with modern organisms. And because the simple units, sugars and amino acids, are mostly chiral molecules, the first self-replicating molecule had to be chiral even if both mirror-image forms of the simple molecules were available. For replication, you can't take randomly right-handed and left-handed molecules as building blocks. That won't work. So the real question is, why don't we have a mirror-image life side by side with the one we have? I suspect it is because the appearance of a self-replicating molecule is an extremely improbable event. The molecule to replicate itself had to have a particular chirality and it passed this on more or less exactly to its successors, and so life is asymmetric. It happens like a baby's smile.

???

Spontaneously. And to maintain the pattern of life you have to keep providing energy. This happens to be sunlight for most of the life on earth but sunlight is not the only possible energy source, especially at the beginning. So I see the origin of life as a minor victory against decay.

John and Rita Cornforth (courtesy of John Cornforth).

In addition to chemistry, what else has interested you?

I read a lot, and the art of writing has always interested me. I know a lot of poetry, which I use as a substitute for music. I learned to play chess as a schoolboy, and became an expert, but never spent enough time on it to get hooked. I played blindfolded against 12 players simultaneously, won 8, drew 2, lost 2. This is still an Australian record, after 60 years. I used to play chess with Robert Robinson, first at Oxford over the board, later by post. After his sight failed we went on playing blind postal chess. He couldn't see the positions and I didn't look until the game was over. The quality of the chess was quite good. I used to play golf and I still play tennis. Being deaf I could play them but team games were difficult. Later in life I have spent more time at gardening. I especially like growing fruits, the garden here has more than 20 different varieties.

Is there anything you would like to add?

If you have the predicament of deafness, you cannot get anything much from meetings or from conversations involving more than two people. This is why I tend to go for the printed record. I disbelieve in abstracts and reviews automatically. An abstract is what the abstractor thought was important of what an editor and his referees allowed to appear of what the author thought was important! By the time you read either a review or an abstract it has been filtered through too many brains. The best you can do is to go for the original paper, but what I get from a paper is almost never what was intended. So I am worried about the tendency to avoid full publication of results and to rely on databases, while access to the whole of the literature is becoming more difficult because libraries cannot afford full coverage. The quantity of information being produced is stupendous but the quality is nobody's business. The usual "preliminary" publication is almost all interpretation, not the actual experiments which are the only things of lasting value. The problem can be solved only on a multi-national scale and I should like to see a European central library whose readers can call up any paper, however old or obscure the source, on their computer screen and decide if they wish to copy it. For that reason among others, I take a lot of care with my papers and I am content that they will be my memorial. "What you did, why you did it, and what were your results" is as valid an ideal as when Rutherford enunciated it.

Quotable Cornforth

John W. Cornforth gave a thought provoking public lecture in September, 1992, to the Royal Australian Chemical Institute as part of the celebration of its 75th anniversary. The lecture was printed in the Australian Journal of Chemistry [**1993**, *46*, 265–275]. Its title is "Scientists as Citizens." It is about the position of scientists in society, their three dilemmas: secrecy, history, and truth, and about the weaknesses and strength of science. The following compilation of quotations is not a summary of the paper and is no substitute for reading the paper in its entirety. It only highlights some of its ideas. A few of the quotations were edited slightly as they were taken out of longer trains of thought. I am grateful to Professor Cornforth for checking this compilation and for his kind permission to use it in this volume.

It may once have been possible for one human brain to grasp the essentials of all sciences, but not now, not ever again. Even in my own discipline, organic chemistry, I keep having to make assumptions that I have not tested.

What one scientist assumes in a statement by another is that evidence about it has been recorded and can be checked, and that facts incompatible with it have been looked for and not found.

Science is the art of the probable; and I am using that word not just in its modern sense of "likely", but in its older and more exact meaning: "testable".

Scientists do not believe; they check.

If we are citizens of anything, we are citizens of the earth.

The closer one gets to the chemical and biochemical essentials, the greater is the unity of life.

For the past 150 years most of the discoveries have been made by people who earned their living as teachers.

Most scientists spend their careers in applying the results of earlier work, not in extending the frontier.

The business of chemists is matter.

The pressure of industry to generate short-term profit is always tending to reduce industrial scientists to the level of parasites on the body of scientific knowledge.

Scientists have these dilemmas, that their role as loyal citizens and employees, and respectable members of society, can be in conflict with their science, based as it is on free exchange and recording of information.

For most of its existence the human species has occupied its biological niche as a parasite on the ability of green plants to collect solar energy, and as a predator of other animals that do the same.

Humanity has exploited the resource of fossil fuels with all the restraint of a fox in a chicken house.

Most of the fossil fuel is spent on uses that are totally frivolous when measured against the basic needs for survival. And as a chemist I just hate to see all that lovely irreplaceable raw material going up in smoke.

If you are a scientist, you realize before long that if the future is in anyone's hands it is in yours.

If you judge that overpopulation, poverty and mass starvation are inevitable for the great majority of the species, you may concentrate on the survival of a wealthy minority that can monopolize the world's limited resources.

The supreme irony is that although there are several hundred religions now, and there have been many more, each of them claims to be true and exacts belief from its members and, most unfortunately, their children.

The discipline of science generates a special relationship with truth. There is public truth: the obligation to record what you have done as accurately as you can, never fabricating, never distorting, and never suppressing findings unfavorable to your conclusions. Private truth is even more important. As a scientist interacting with your experiments, you receive an education in the implacability of truth and in your own capacity to be deceived by your expectations, your hopes, or just your stupidity, that is unlike any other experience I know.

All new discoveries have to be introduced gradually and with circumspection.

Scientists are apt to overvalue the importance of new discoveries, and to underrate the extent of their own ignorance. … As citizens they cannot afford to be humble, but as scientists they must be.

Scientists of different disciplines are understanding each other less and less, the search for information outside their own speciality becomes more and more laborious, and cross-fertilization of ideas becomes rarer.

Scientists need to try harder to show their strength, which is that they thrive on being doubted. In this, they differ from all others among the shapers of society.

The sciences are different from nearly all other subjects taught. … They deal with the natural world, which men did not make at all; and the honest way to teach them is "these things are probably so because of this, and this, and this."

Scientists have in the schools the opportunity to start the sceptical revolution.

Vladimir Prelog, 1995 (photograph by I. Hargittai).

10

Vladimir Prelog

Vladimir Prelog (1906–1998) was born in Sarajevo, Bosnia-Herzegovina. He was Professor Emeritus at the ETH in Zurich till his death. He was co-recipient (with John W. Cornforth) of the Nobel Prize in Chemistry in 1975 "for his research into the stereochemistry of organic molecules and reactions."

He attended high school in Osijek, Croatia, and wrote his first paper when he was 15.[1] From 1924 he studied at the Prague Institute of Technology in Czechoslovakia and received his Diploma in 1928. His mentor was Rudolf Lukes. He stayed on in Prague after graduation and got his doctorate, Dr. Ing., in 1929. From 1935 to 1941 he taught at the University of Zagreb, Croatia. Then he moved to Zurich and spent there the rest of his life, rising to full professor at the ETH (Eidgenossische Technische Hochschule) Zurich in 1952. For some time he served as Head of the Laboratory of Organic Chemistry, as Leopold Ruzicka's successor. He retired in 1976, but stayed on, officially as a special student, in fact as a Professor Emeritus.

Vladimir Prelog died in January, 1998.

Professor Cornforth remembered: "When we shared the Nobel Prize in 1975 we were already friends. He told me that he wanted me to be the one to return thanks for both of us at the banquet after the ceremony. He gave as a pretext that he really had no native language except Serbo-Croat which nobody would understand. But he was a fine linguist and could easily have done it in several other languages.[2]

This is then what Cornforth said in Stockholm, on behalf of the two of them: "Our backgrounds, and the experience that has shaped us as scientists, are very different. We were born, and we grew up, on opposite

Leopold Ruzicka (1887–1976) and Vladimir Prelog in October 1975, on the occasion of the news about Professor Prelog's Nobel Prize (courtesy of Vladimir Prelog).

sides of the globe. What we have in common is a lifelong curiosity about the shapes, and changes in shape, of entities that we shall never see; and a lifelong conviction that this curiosity will lead us closer to the truth of chemical processes, including the processes of life."

I spent a lively and very interesting morning with Professor Prelog in his office, decorated with chiral objects, among others, in March, 1995. What got into the interview reproduced below was only a small part of the conversation. Professor Prelog, insisting to be called Vlado, was in a reflective mood, and was very much willing to communicate.

His birth place, Sarajevo, came up naturally in the conversation, and he was gravely bothered by the tragic events there. He told me that he used to think very seldom about Sarajevo, but lately his memories came back and came back greatly focused: "I see the origin of all what has happened and it is so cruel, and I see the origin of it in the different groups not speaking to each other, in making groups that are marked by religion, by language, by whatever. Of course, all this is in addition to 500 years of the Ottoman oppression. People have learned that if they wanted to have justice they had to do it by themselves. This is often done by revenge. You kill my brother, I kill your brother. This is another important factor."

His own childhood was a model of animosity: "I spent my first years in Sarajevo, and later a few more years, but I never played with a Moslem child, I never played with an Orthodox (Serbian) child, and I did not even play with Croatian children whose families were originally Bosnian. We were in a special group of civil servants, professors, advocates, and so on, and they made an isolated group, a kind of club of their own. The Croats were Catholic and the Serbs were Orthodox and they did not speak to each other." *Then as if an explanation, he added a story about a behavioral experiment*: "Margaret Mead, the anthropologist, made the following experiment. There was a class of boys in a school, a normal group of children, some of them were friends and others were not. Then they put half of these pupils in green uniform and the other half in red uniform. Very soon they started to sit in different sides of their class-room and they started to fight. There seems to be an important characteristics of the human brain and soul related to this."

Another topic that he told me occupied much his mind recently, was his relationship to his mother: "My mother came from a family of Italian masons who came from Northern Italy to a small town in Croatia, Petrinya. The peasants of that region built their houses by themselves but they invited Italian masons to build their churches. These masons became rich, but my mother's family was big. She was one of 12 children. Four of them were girls, and without a dowry. They had little hope for marriage, so they all had to go to study to become teachers. My father had to teach these girls and fell in love with my mother. But she did not fall in love with him. Nevertheless when he proposed to marry her, she could not resist as she was told by her family that she had to grab this unique opportunity. My parents were not ripe for the marriage and divorced within four years. Later they came back together for a few years, but finally they separated forever.

"For 17 years I did not know anything about my mother and I think this was psychologically very hard on me. There was no contact between us from when I was about 10 to 27. Then some friends arranged a meeting for us, and since then we had a very good relationship. She died very old, at 92 years of age. But there were 17 years when my mother did not ask for me and I did not know anything about her. That in some way formed me. On the other hand, when I was growing up without a mother, everybody was trying to replace her. So I got more love than normal children usually get from their families. During my active life I never thought about this but you just came at the moment when my mind is very occupied by looking back to these times.

"My mother eventually found the right man for her and they got married, and we had a very good relationship with both as long as they lived. I remained her only child."

In connection with Professor Ruzicka's love of Dutch paintings we talked about the unlikely situation of an art historian being interested in chemistry. Vlado Prelog was bothered by the separation of the cultures and told me about a noted historian, Carl Jacob Burckhardt (1891–1974) who was also a Swiss diplomat, served the League of Nations as High Commissioner in Danzig (1937–39), and was President of the International Red Cross in the critical years of 1944–48. He wrote biographies of historical personalities, such as Richelieu in four volumes (1935–67). Prelog and Burckhard, both members of the Board of Ciba-Geigy in Basel, used to travel together in the train from Basel to Zurich. Burckhardt found it difficult to believe that astronauts would not be falling down during their space walk or that a feather and a lead ball would be falling with the same speed in vacuum. His physics was not even pre-Galilean.

The following interview appeared in *The Chemical Intelligencer*:*

You were born in Sarajevo, and it is impossible to avoid this question: How do you look back to your birth place?

I was born in Sarajevo, capital of Bosnia, in 1906 when it was a province of Austria-Hungaria, which occupied it in 1878 in a war with Turkey. It was still under military government. In 1904, my father, a Croat from Zagreb, took up a position there because, just married, he needed a job, but for political reasons could not get one in Croatia. He had started to study at the Croatian University in Zagreb. As a Hungarian, you may know that Croats and Hungarians, who had lived in the same state for centuries, became bitter political enemies in the 19th century. In 1898 Croatian students burnt a Hungarian flag in the court of Zagreb University, whereupon the university was closed and many students, among them my father, were relegated. He was able to finish his studies only because the Charles University in Prague offered hospitality to some of these students. That enabled him to get a Ph.D. degree there but did not qualify him for a teaching job in Croatia. Anyway, my father's pleasant memories motivated me to study in Prague. I spent altogether 11 years of my life in Prague

*The interview was originally published in *The Chemical Intelligencer* **1996**, *2*(2), 16–19

and married my Czech wife there. All that happened because of a burnt Hungarian flag.

Let us now switch to science. Would you characterize your field as natural products chemistry?

I have always been interested in natural compounds chemistry. As there are many stereochemical problems, I became interested in stereochemistry. I lacked clear concepts and unambiguous words and symbols to deal with these problems.

Is natural products chemistry as important today as it was before?

It will always remain important. Natural products which survived screening by evolution carry a message. To decipher this message we have to know their structure and reactivity. Of the many millions of known organic compounds, natural products are a small fraction but they are the basis of life. When we think or talk, sleep or walk, reactions of natural compounds take place to make it possible. Life depends on them and therefore they are and will always remain the most fascinating area of chemistry.

I would like to ask you about your teachers. You had Ivan Kuria in Osijek, Rudolf Lukes in Prague, and Leopold Ruzicka in Zurich. Whom did I miss?

Vladimir Prelog in the company of (from left to right) Jack Dunitz, André Dreiding, and Albert Eschenmoser (photograph by I. Hargittai).

Vladimir Prelog holding a gold-plated sculpture of the backbone of the transfer ribonucleic acid molecule (photograph by I. Hargittai). The sculpture was given to him by Alex Rich (MIT) through Jack Dunitz (ETH-Z). When Prelog saw the sculpture, he exclaimed, "God's Signature!" (private communication from Alex Rich and Jack Dunitz, 1998).

Sir Robert Robinson in Oxford. He was not my teacher in the usual meaning of the word, but I was very much influenced by his papers. Later, when I met him, I found that he was also an unusual person, not easy to deal with. In spite of that, I admired him. For many years, until about two weeks ago, his photograph has decorated this room, but I replaced it by a portrait of Ruzicka because I realized that in fact he influenced me much more than Sir Robert did.

After my arrival in Zurich in December of the fateful year 1941, Ruzicka assigned to me the investigation of the lipophilic extract from 5000 kg of boar testes, which he ordered from Wilson laborataries (next to the famous slaughter houses) in Chicago. He hoped it would be possible to isolate from it biologically interesting compounds. The results of my work were rather disappointing. The only exception was the isolation of small quantities of a compound with a musk-like scent. I determined its structure as 3-hydroxyandrosten-16. Ruzicka was interested in this result because the polycyclic steroid resembles structurally the monocyclic musk, civetone, which he had investigated at the beginning of his scientific career. Later, the compound became interesting from different other reasons. The same compound was isolated by other investigators from truffles. Since antiquity, sows are used to detect this delicacy under the ground, where it grows, by smell. Finally, it was also detected in the sweat of human males and is considered to be the mammalian male pheronome. It is used today in pig breeding as well as in perfumery.

Shortly after the war, Ruzicka changed considerably. For several years he neglected chemistry and spent most of his time collecting Dutch paintings. He had become a modest collector already during his short stay as professor of chemistry at the University of Utrecht, but after the war he started collecting on a larger scale due to chemistry. During the war GIBA Ltd. produced quantities of testosterone in the USA by Ruzicka's method. A few million Swiss francs accumulated there as royalties which the Americans did not want to transfer to Switzerland, as they feared that it would be occupied by the German army. After the war, when the money was transferred, a considerable part of it became subject to tax. To avoid this, Ruzicka donated it tax-free to a Foundation which enabled him to create a collection of Dutch paintings exhibited at the Kunsthaus in Zurich. He concentrated his attention on buying paintings and spent most of his time visiting auctions in Switzerland and abroad, mainly, in England. By all that Ruzicka was not any more the shining example I wanted to imitate.

There are many stories from this period of his life, of which I would like to tell what is perhaps the most fascinating one. The most valuable

Vladimir Prelog's *ex libris* by the Swiss artist Hans Erni (courtesy of Vladimir Prelog).

painting Ruzicka purchased in England was a portrait by Rubens of King Philip IV of Spain. In order to export such a painting from England, a permit of the National Gallery of London was necessary, because this institution had the privilege to purchase unique objects of art to prevent their loss for England. After receiving such a permit, Ruzicka traveled to London to make final arrangements. During his stay there he was invited by his friend, Sir Ian Heilbron, for dinner, where he met the Director of the National Gallery. They spoke about the permit which had just been issued. The Director mentioned that the painting was in a very bad condition and should be thoroughly cleaned. He also offered the service of his experts, an offer that Ruzicka gratefully accepted. They even agreed about the price. After cleaning it, the experts found that the painting was of remarkable beauty and the National Gallery withdrew its export permit. Ruzicka was, of course, furious but he knew that, even if he could not export the painting, he could own it as long as it stayed in England. So he offered the portrait, then worth 750,000 Swiss francs, on loan to the Swiss Embassy in London.

The Rubens painting Philip IV from the Ruzicka collection (courtesy of Vladimir Prelog).

There it stayed until the diplomats arranged for its transfer to Switzerland, where it was exhibited as part of the Ruzicka Collection in Zurich. One day in 1985, an evidently frustrated and also insane young man from Munich went to the Kunsthaus, poured an inflammable liquid on the painting and burnt it to ashes. If Ruzicka had not been so successful in everything he undertook, this Rubens masterpiece could still be admired in England.

You may conclude from our conversation that I am always strongly impressed by fateful consequences of often inconceivable sequences of events.

References

1. Prelog, V. *Chemiker Zeitung* **1921**, *45*, 736.
2. Cornforth, J. *The Chemical Intelligencer* **1998**, *4*(2), 50.

Derek Barton in 1988 (photograph by I. Hargittai).

11

Derek H. R. Barton

Derek Harold Richard Barton (1918–1998, born in Gravesend, Kent, U.K.) did his studies at Imperial College, London, got his B.S. and Ph.D. in 1940 and 1942, respectively. His first teaching positions were also at Imperial College. Then he held professorships at Birkbeck College of the University of London and at Glasgow University before returning to a professorship at Imperial College. While there, he and Professor Odd Hassel of the University of Oslo received the Nobel Prize in Chemistry of 1969 for their contributions to the development of the concept of conformation and its application in chemistry. After 20 years at Imperial College, in 1977, Dr. Barton left for directorship of research at the CNRS (Centre National de la Recherche Scientifique), Institut de Chimie des Substances Naturelles (ICSN) at Gif-sur-Yvette in France. When retirement age caught up with him in France too, he left for College Station, Texas.

Derek Barton was elected Fellow of the Royal Society (London) in 1954, received three medals of the Royal Society: Davy (1961), Royal (1972), and Copley (1980). He was elected Foreign Associate of the National Academy of Sciences in 1970. The American Chemical Society awarded him the Priestley Medal in 1995. He was made a Knight Bachelor in the United Kingdom (1972); an Officier de la Légion d'Honneur in France (1985); and given the Order of the Rising Sun in Japan (1972).

In 1996, Professor Barton published a comprehensive volume of his works, titled *Reason and Imagination: Reflections on Research in Organic*

Chemistry. Selected Papers of Derek H. R. Barton.[1] It contains 137 reprints of his "most important and most original" papers published between 1943 and 1993. He himself made the selection and wrote the commentaries. Some quotes from his Preface: "...originality is the most important quality in Academic Research. ... the invention of new reactions has been the major activity in my life from about 1960 onwards. ... This book will be of value if it teaches that if you know, in the academic world, how to do a reaction, you should not do it. You should work on reactions that are potentially important and that you do not know how to do! Also, experimental observations that you do not at first understand are much more interesting than those that you do."

At the time of the interview Dr. Barton was Dow Distinguished Professor of Chemical Invention at the Department of Chemistry of Texas A&M University. We had met in person before, but in the Spring of 1996, when I was organizing this interview, although we visited each other's Universities within weeks, our mutual visits did not overlap with our presence, hence this interview was conducted by correspondence. It has appeared in *The Chemical Intelligencer.**

The 1995 ACS Directory of Graduate Research *lists your current research interest as "Invention of new chemical reactions of relevance to the synthesis and partial synthesis of natural products." What new reactions are you presently working on, and what natural products would you like to make?*

We are working on new ways to make hindered peptides under mild, room temperature conditions. Hindered peptides, like the medicinally important Cyclosporin A, contain a number of hindered residues. The synthesis of peptides with hindered backbones continues to be of current importance.

We are also interested in the synthesis of selectively functionalized saturated hydrocarbons under mild conditions such as room temperature, neutral pH and atmospheric pressure. We practice what we call Gif Chemistry. This has a remarkable mechanism of action which is probably used by Nature in the enzyme methane mono-oxygenase (M.M.O.). Our intention is to provide new reactions which will be utilized by others.

How large is your research group and how is your work supported?

---

*This interview was originally published in *The Chemical Intelligencer* **1997**, *3*(1), 31–33

At present, I have 16 collaborators from a dozen different countries at Texas A&M. Two other collaborators work in Hungary, and another in France at Gif-sur-Yvette. The group is supported by the NIH, the NSF, the Welch Foundation, and an important number of industrial friends.

What has been the most important reaction you invented and the most important natural product you synthesized during your career?

The most important reaction for practical use was the application of the thiocarbonyl function for the generation and synthetic application of carbon and other radicals. The thiocarbonyl function exerts a disciplinary effect over the carbon radicals, and very high yields of desired products can be obtained under mild conditions.

The most spectacular synthesis accomplished was the preparation of 60 g of the essential steroidal hormone, aldosterone, when the World's supply was only in milligrams. This short, three-step synthesis was attained by the invention of a new reaction (nitrite photolysis) for the intramolecular functionalization of an inert neopentyl group.

How did you get involved with conformational problems?

The concept that *chemical reactivity* could be correlated with preferred conformation was, in 1950, an original conception. The short theoretical paper that I published changed organic chemistry as well as biochemistry and eventually molecular biology. Most molecules have a preferred conformation. When this is established, then the stereochemical consequences enable one to predict the selectivity of chemical reactions and their relative speed. My involvement was simple. I read the literature of chemical physics and of complex natural products like steroids. It was just needed to see that what happened in these apparently different fields could be correlated in a harmonious theory. The importance of this theory was obvious to preparative chemists. So within a year, everyone was using the concept of conformational analysis. The original article[2] was reprinted in the same journal in a celebration of the journal's 50th anniversary.[3] I hope to see the article reprinted on its own 50th birthday in the year 2000. I shall be 82 years old.

How much interaction did you have with the Norwegians who were active in conformational analysis, Odd Hassel and Otto Bastiansen?

Derek Barton in 1952 (photograph courtesy of John D. Roberts).

When I wrote my 1950 key paper, I had never met Odd Hassel or Otto Bastiansen. I had just read their publications. I later met Hassel at a Chemical Society Meeting in London. He was rather aloof, and the quality of his lecture gave no indication of the importance of the subject. I met Bastiansen much later, after the death of Hassel, when I gave a Hassel lecture in Oslo. He was charming and urbane. It was not at all difficult to talk to him.

You have spent considerable portions of your life in Britain, France, and Texas. Would you care to make a comparison between the academic atmospheres of the schools you have been involved with?

British chemistry departments have only a limited number of professors, who have power over their sections of the department. When the professor with the power is very able, this can produce very good science. However, not all professors who showed ability when young continue to do so in later years. Without the appropriate leadership, the structure can become chaotic or simply nonproductive. There is no mechanism to change this by retirement. In fact, in some Scottish universities, in the not so distant past, there was no retirement at all. The late professor John Read at St. Andrews University used to boast that the only way that he would retire from the department was feet first and dead. Essentially, that was what happened about the time when I returned to London from Scotland.

The American system is much fairer. All members of the academic staff have responsibilities in teaching, research, and administration. If you can raise much government money from grant applications, then you can enlarge your research group. Since government and contract money have "overhead" added, an outstanding research man or woman can bring in the money needed to pay for the president and other university dignitaries. He (or she) thus has status because of his (or her) pecuniary contribution to the administration. Thus he (or she) can have the space that is needed for their work.

The "overhead" is calculated as a percentage which is added to the normal grant application. It can vary from 45% as in a reasonable university like Texas A&M, up to 100% for MIT. Those who do not bring in "overhead" can spend more time teaching. In the end, it is the "overhead" money which determines the age of retirement. It is flexible and related to the "overhead."

In France, the research that is carried out is dominated by the CNRS (Centre National de la Recherche Scientifique). This government body has established numerous institutes of research, which specialize in a given subject. In my own case my institute was a natural products institute. Salaries in the CNRS are sacrosanct. So one's salary arrives whatever one does and whether one is present or absent. Only about half the money to carry out the work is provided by the CNRS, the rest has to come from industry. This certainly fosters CNRS–industry relationships!

In the institutes, the work, as in universities, is not done by those adequately paid to do it (the professors), but by the students. In the institutes of the CNRS there are, theoretically, no students. However, it is still the theoretically absent students who do the work. They are registered in a neighboring university.

The relationship between the universities and the CNRS is interesting. Naturally, professors would like to abolish the CNRS and redistribute all the money to themselves. In reality, it is the extra money provided by the CNRS that finances groups in universities to work hard and publish in reputable journals. The constant threat of withdrawal of the money is a strong, but disliked, stimulus to the professors to do something really interesting. France owes a lot to the CNRS.

A few years ago, the American Chemical Society published your autobiography, entitled Some Recollections of Gap Jumping. *As I understand, gap jumping meant to connect, perhaps intuitively, data and observations and experience where there was a gap in the chain of reasoning. Can you give any example of this? Is this something others can learn from you?*

Well, the concept of conformational analysis was, in its time, a perfect example of gap jumping. Anyone could have written the article, but only I did! I mentioned earlier the synthesis of aldosterone using a new reaction. This was gap jumping because you had to believe that alkoxy radicals, generated by photolysis of a suitable nitrite, would indeed participate in radical exchange of NO.

To be a good gap jumper, you have first to recognize the gap that you want to jump. Most people never learn to perceive the difference between the important and the trivial. In the case of chemical reactions, you have only to recognize a reaction that would be important if it existed. For example, in the deoxygenation of hydroxyl groups in complex antibiotics, there was no precedent for radical chemistry. All the chemistry attempted, often with disappointing results, was ionic chemistry. We thought how we might convert hindered secondary alcohols into thiocarbonyl derivatives which might undergo radical fission and then deoxygenation with a tin hydride. This beautiful reaction is still of major importance. The gaps were between ionic (good yield) and radical (bad mixtures). Then the concept of thiocarbonyl activation was another gap between traditional alcohol derivatives and radical inducing nontraditional functions. Tin radical chemistry was already well-known, but not for the rupture of the carbon-oxygen bond, only for weaker bonds. But it all fell into a harmonious whole.

What was your background originally, and what attracted you to chemistry?

My father was a carpenter who eventually built up a small business in the wood industry. He sent me to a good school, but I left early on his premature death when I was 16 years old. I was already a self-taught intellectual from my reading both in the school library and at home. As you will suspect, I am an only child. After two years in the wood industry, I felt that I must seek an intellectually more interesting life. This came with the university (Imperial College) in the war time years. I chose chemistry because I was naturally gifted for it. Have not later events proven this?

What were you knighted for? Also, to me it seems strange that even in a book published in America you, as an author, are listed with your title, Sir Derek H. R. Barton. Does such a title have relevance outside the country where it was issued? Do your students call you Sir Derek?

For Service to the Crown — the Nobel Prize. The use of my title in the book, which is international in scope, was chosen by the editor, not by me. Nobody here calls me Sir Derek — unless they want to annoy me.

British stamp honoring Barton's Nobel Prize. This stamp was member of the series issued for the 100th anniversary of the Royal Institute of Chemistry in celebration of British achievement in chemistry.

You are considered to be a tough person. Are you really? In fact, I remember the Michael Dewar celebrations, on the occasion of his 70th birthday, in Austin, Texas. At the beginning of your lecture, you posed an organic chemistry question to Dr. Dewar and gave him to the end of your lecture to solve it. Then, since he could not solve it, you gave the solution following your lecture. Some felt rather uncomfortable during this episode. I have forgotten the puzzle, but you must have been quite certain that he would solve it, I'm sure. Do you remember the occasion and the puzzle? Any comment?

No, I am not tough at all. I only expect my students to work as long as I do, not much longer than the professor, as in most parts of the world. Towards Michael Dewar, I was simply applying the Dewar approach to little problems. Think about them and solve them. The younger Michael, whom I knew, would have solved my problem easily. Clearly, too much calculation had had a bad effect on him. I remember the problem.

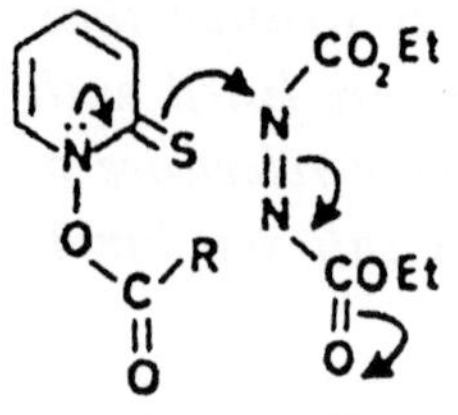

When **1** and **2** are left in CH_2Cl_2 at room temperature in the dark, there is a rapid reaction to afford an adduct quantitatively. What is the structure of this adduct?

The answer is:

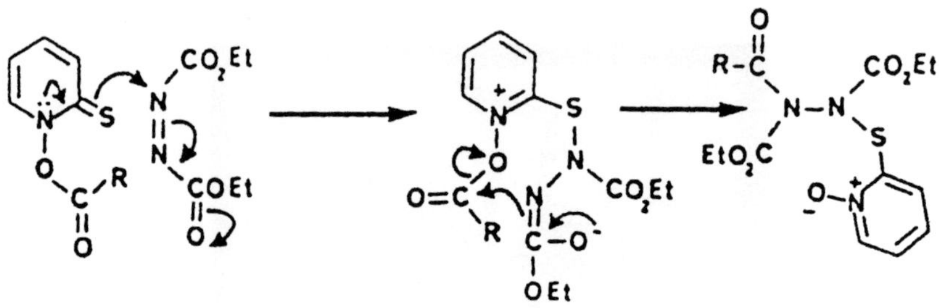

The question Derek Barton posed to Michael Dewar at the Dewar Symposium in February, 1988, in Austin, Texas. The reaction was subsequently published in a paper, "dedicated with appreciation to Professor M. J. S. Dewar on the occasion of his seventieth birthday." The reference is Barton, D. H. R.; Ozbalik, N.; Vacher, B. *Tetrahedron* **1988**, *44*, 7385–7392.

Who was your most remarkable teacher, colleague, pupil?

The answer to this question is clearly myself.

Has natural products chemistry changed because of the success of computational chemistry?

Natural products chemistry has changed because of the effect of the computer on X-ray crystallography and because of the development of NMR techniques of remarkable power. Structures are solved with ease, and there is really no longer any challenge. Of course, computational chemistry helps to understand structure. It helps more in consideration of mechanism.

Would you care to make a comment about the retirement systems of different countries?

I have already replied to this question. My own decision is that I will retire when I can no longer persuade anyone to give me any more money for my research. Of course, I will spend some of my own money too.

References

1. Barton, D. H. R. *Reason and Imagination: Reflections on Research in Organic Chemistry. Selected Papers of Derek H. R. Barton*, Imperial College Press, London and World Scientific, Singapore, 1996.
2. Barton, D. H. R. *Experientia*, **1950**, *6*, 316–320.
3. Barton, D. H. R. *Experientia*, **1994**, *50*, 390–394.

Odd Hassel (courtesy of Hans M. Seip of the University of Oslo).

12

Odd Hassel

Odd Hassel (1897–1981) shared the Nobel Prize in Chemistry with Derek Barton in 1969, "for their contributions to the development of the concept of conformation and its application in chemistry." Hassel and his graduate student, Otto Bastiansen, investigated the structure and conformation of cyclohexane and its derivatives in the gas phase by electron diffraction. When different conformers were possible, and none of them could account individually for the experimental data, they had the simple, in hindsight, notion that a model of their mixture should be tried. Thus, for example, neither the "ee" nor the "aa" conformer of trans-1,2-dibromocyclohexane would account for the experimental data, but their 60% + 40% mixture would. Hence, the discovery of conformational equilibria, which is part of any introductory text today in organic chemistry.

Odd Hassel was born in 1897 in Kristiania (at that time the name of Oslo). His interest in chemistry developed in high school and there is no information as to what may have excited his interest in this subject. Hassel excelled in mathematics and the sciences but was mediocre in other fields. Later in his life though, he developed an interest and knowledge in literature, languages, music, and archeology. Hassel entered the University of Oslo in 1915 and majored in chemistry in 1920 (cand. real. degree), with minors in mathematics, physics, and mechanics. He then spent some time in France, Italy, and Germany. He attended the physics lectures of Paul Langevin in Paris. In 1922, he went to Munich and worked some time with Kasimir Fajans in analytical chemistry. Then he moved to the Kaiser Wilhelm Institute in Berlin-Dahlem. There he got his introduction to X-ray crystallography and published a

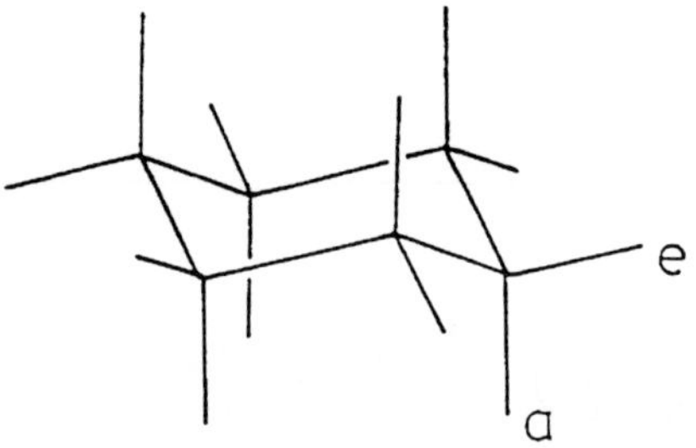

Model of the cyclohexane molecule with labels of an axial (a) and an equatorial (e) position.

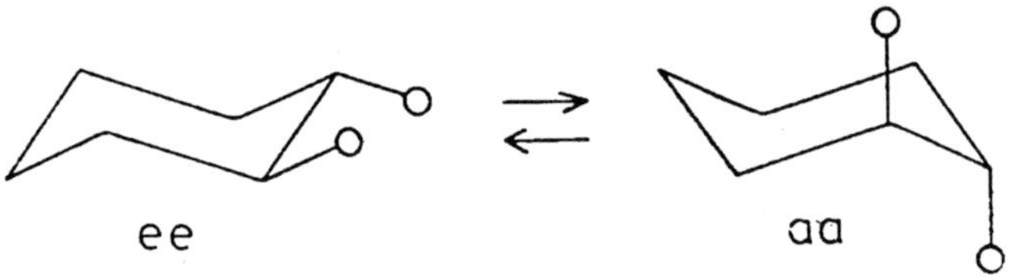

The conformational equilibrium of trans-1,2-dibromocyclohexane.

pioneering paper on the structure of graphite with Herman Mark. For his doctorate, he was examined in physics by Max von Laue; in chemical technology by Fritz Haber; and in physical chemistry by Max Bodenstein. He got his doctorate in 1924. A few years later, Herman Mark with Raimund Wierl conceived gas-phase electron diffraction, the technique Hassel was to use in his conformational studies.

Following one more year in Germany on a Rockefeller Fellowship, Hassel returned to Norway in 1925 and started working at the University of Oslo. He did some research in inorganic crystallography with Victor Moritz Goldschmidt, and established the Norwegian gas-phase electron diffraction group. Molecular structure and conformational properties of cyclohexane and its derivatives remained Hassel's main research area until 1950. Then he developed another line of research on the structure of charge-transfer complexes. Although he received his Nobel Prize for his conformational studies, he gave his Nobel lecture on the charge-transfer complexes.

The scientific world learned about the discovery of conformational equilibria in 1946, when Hassel and Bastiansen published a paper in *Nature*.[1] However, the original report by Hassel already appeared in 1943, in Norwegian. Norway was under German occupation at the time and Hassel was not allowed to publish his paper in English and he refused to publish it in German. It appeared in the recently established Norwegian journal

CYKLOHEKSANPROBLEMET

Av O. Hassel.

Summary.

In the highsymmetrical form of the cyclohexane molecule (symmetry D_{3d}) shown to exist both in the hydrocarbon itself and in a series of derivatives two geometrically different types of *C—H* bonds are present. The six bonds of the first type (ε bonds) are directed parallel to the principal axis of the cyclohexane molecule, the six other bonds (ϰ bonds) form an angle of 109°28′ with this direction. The conversion of the molecule between the two possible models of symmetry D_{3d} transforms ϰ bonds into ε bonds and vice versa. Contrary to the mother substance itself the molecular configuration of its derivatives will in general be different before and after this conversion has taken place. The corresponding difference in internal energy is supposed to be high enough in most cases to make the concentration of the less stable form considerably smaller than the concentration of the other form in an equilibrium mixture. The energy of activation of this transformation, however, seems in general to be small enough to make a separation of the two forms impossible under ordinary conditions. In the case of monochloro-cyclohexane an electron diffraction investigation of the vapor reveals the presence of the ϰ form but gives no indication of the existence of the ε form. The significance of the experimental results given in this and in earlier papers and concerning the stereochemistry of cyclohexane and its substitution products is discussed.

Facsimile of the English language Summary of Hassel's seminal paper, which was published in Norwegian in 1943 (Hassel, O. "Cyklohexanproblemet." *Tiddskrift for kjemi, bergvesen og metallurgi* **1943**, Nr. 5, pp. 32–34).

Tidsskrift for kjemi, bergvesen og metallurgi. This paper was "often quoted but seldom read," until it was republished in English in the same journal, following Hassel's Nobel Prize.

I met Hassel in 1968 on my first visit to Oslo. He was by then retired but still had an office and a personal assistant. To me he was a legend, but in spite, or because, of his dry and laconic way of communicating, he quickly put me at ease. He seemed to be interested in Hungarian scientists, he had met quite a few of them in his Berlin days, Michael Polanyi, Eugene Wigner, Leo Szilard, and others. Upon my departure, Hassel gave me two

Odd Hassel, Lars Onsager, and Leo Szilard (from left to right) in 1924 (photograph by physics professor Johan P. Holtsmark, courtesy of Otto Bastiansen).

books, both edited by Peter Debye and published by Hirzel in Leipzig, *Elektroneninterferenzen* (1930) and *Molekülstruktur* (1931).

When Hassel turned 70, in 1967, the University of Oslo established "The Hassel Lecture" to be given annually, later biennially, by a foreign scientist. A committee was established and every year, months before the event, a list of names would be presented to Hassel who chose the next lecturer. For the 1981 lecture, he chose me. Sadly, this was to be his last contribution to the selection process. At dawn of the day of the Hassel

István Hargittai and Otto Bastiansen with Odd Hassel's portrait at the Department of Chemistry, University of Oslo, in 1988.

Lecture, on May 11, 1981, Hassel died. For a moment the organizers were at loss, not knowing what to do. Then Bastiansen remembered an episode from Hassel's life. For underground resistance activities, Odd Hassel was arrested by the Norwegian nazis and turned over to the German Gestapo during the War. People were upset and some contemplated shutting down the University of Oslo in protest. However, a message came from the prison, from Hassel, "The lectures must go on." And this is what was decided on May 11, 1981, too, "The Lecture must go on." The 1981 Hassel Lecture was to be not only a scientific presentation but also a memorial to a great scientist.

Acknowledgment

I thank Professor Hans M. Seip for his kind assistance in collecting references on Odd Hassel's works.

Bibliography

Bastiansen, O. "Minnetale over Professor Dr. philos. Odd Hassel." *Yearbook of the Norwegian Academy of Science and Letters*, 1981.

Andersen, P.; Bastiansen, O. "Odd Hassel." In *Selected Topics in Structural Chemistry.* Eds.: Andersen, P.; Bastiansen, O.; Furberg, S. Universitetsforlaget, 1967, pp. 9–12.

Lund, E. W.; Rømming, C. "Forty-five Years of Achievement." In *Selected Topics in Structural Chemistry.* Eds.: Andersen, P.; Bastiansen, O.; Furberg, S. Universitetsforlaget, 1967, pp. 13–23.

References

1. Bastiansen, O.; Hassel, O. *Nature* **1946**, *157*, 765.

Michael Dewar in 1988 (photograph by I. Hargittai).

13

MICHAEL J. S. DEWAR

Michael J. S. Dewar (1918–1997) was one of the most remarkable chemists of his time. Shortly after his death, John W. Cornforth remembered him with the following words: "Michael had the best brain of anyone I ever knew. Unlike many theoretical chemists, he was also a chemist; and his critical contribution to the subject has still to be appreciated."[1]

John Cornforth and Michael Dewar (courtesy of John Cornforth; the photograph was taken by Mary Dewar in the 1960s).

Dewar published an account on his life and work in 1992.[2] Then, in 1996, the Canadian chemist, John T. Edwards produced a brief comparative analysis of Derek Barton and Michael Dewar,[3] titled "Both Entrepreneurial Instinct and Superior Intelligence Are Required for a Nobel Prize: D. H. R. Barton and M. J. S. Dewar Have Exceptional Intelligence and Careers with Many Parallels, But Only One Got the Nobel Prize." Since a Barton interview had already been scheduled for publication, in July, 1996 I contacted Michael Dewar and asked him some questions; I had known him for a long time.

Michael Dewar and István Hargittai in 1988 (in Austin, Texas, by unknown photographer).

I met him in person for the first time in 1969 when I was a research associate at the University of Texas at Austin for a whole year. Once he almost banned me from the Experimental Science Building where his office had been before the more spacious Welch Hall was completed. My office was in the old physics building but I had to use a hood to transfer a sample of ethane-1,2-dithiol. Although the hood was quite efficient, some smell managed to escape. On another occasion we were discussing the structure of adamantane. I had just completed its electron diffraction analysis and Michael had done some calculations on it and our results differed somewhat. He said graciously, "Don't believe the calculations."

I was back in Austin from time to time, mostly to attend the Symposia on Gas-Phase Molecular Structure, and witnessed his growing disappointment in the University and then his enthusiastic preparations for his move to Florida. In the preparation of his response to my question, we exchanged a lot of e-mail messages. He kept rewriting and refining his text and consulted at least one other person about what he was writing and how he was writing it. His response appeared in full in *The Chemical Intelligencer*, "Some Comments on 'A Semiempirical Life.'"[4] Here I edited out two sections; one is about one of his sons' experience with the Texas Draft Board and the other is a brief outburst about the late C. K. Ingold, which generated considerable criticism.[5]

First, my question to Michael Dewar, and then his response:*

Please, write some comments on your own work, present interest, activities, your views on the progress of organic chemistry, science. I would also appreciate your looking back and singling out your most important contribution to chemistry from a present-day perspective. I was happy to be at the Symposium in Austin, Texas, honoring you, at the beginning of March 1988. I remember the story told by Nat Bauld [Professor of Chemistry at the University of Texas at Austin] according to which the UT football team was No. 1 nationally the same year when they brought you to UT. The Chemistry Department was very proud of the success of the football team. Bringing Michael Dewar to UT, Nat continued, was the first step in building up a Chemistry Department of which the UT football team could be proud."

I will start by explaining my present position so far as chemistry is concerned.

My last appointment before moving to Gainesville was at the University of Texas in Austin (UT) where I had the Welch Chair of Chemistry. My wife, Mary, who was a Tudor (English sixteenth-century) historian, and I had decided that when I retired from UT we would move to Gainesville; I would give up chemistry, and we would collaborate in writing detective stories. However, when I did retire, in 1986, the University of Florida (UF) offered me a Graduate Research Professorship, and since I had some work I wanted to finish and research funds to support it, and since some members of my research group were willing to come with me to Florida to help to finish it, I accepted. However, three years ago, having completed

*The complete version of this article originally appeared in *The Chemical Intelligencer* **1997**, *3*(1), 34–39 © 1997, Springer-Verlag, New York, Inc.

the work I wanted to do and made arrangements for that line of research to be continued elsewhere by an ex-member of my research group, I did finally retire, and although Mary died before we could write anything together, I have kept my promise to abandon chemistry. Therefore, so far as chemistry is concerned, I will here be looking only backward because my career in chemistry is now closed.

What has been my most important contribution to chemistry? That is an easy question to answer. I was the first organic chemist to really find out what quantum theory is about and to use it to interpret chemical behavior. When I began my work, organic chemists interpreted their results in terms of resonance theory, a simplified version of the valence-bond approximation which had been introduced by Pauling. I showed that Mulliken's molecular orbital approximation provides a far better picture, and I developed simple ways to use it. These, in turn, led to the prediction of a whole series of new concepts which have now been generally accepted by organic chemists, for example, antiaromaticity, sigma conjugation, aromaticity and antiaromaticity in nonplanar cyclic systems, the idea that cyclic transition states may be aromatic or antiaromatic, and aromaticity in rings where one of the contributing atomic orbitals is a *d* atomic orbital instead of a *p* atomic orbital. I also used these ideas myself in interpretations of a number of chemical reactions.

I also pioneered the use of quantitative quantum-mechanical calculations as a practical chemical tool. In principle, the proper approach here is to solve the corresponding Schrödinger equation, by numerical integration if necessary. This, however, is possible only for a few extremely simple systems. Otherwise, simplifying assumptions have to be applied to make the calculations feasible. However, the resulting errors in the energies we are trying to calculate soon become larger than the quantities we are trying to calculate. This, indeed, is the case for organic molecules in general, even the very simplest ones. The use of the term ab initio to describe such procedures has, therefore, been unfortunate because it conveys a wholly misleading impression of accuracy, a situation that has been shamelessly exploited by ab initio theoreticians.

This kind of situation has occurred before in chemistry in other connections, and chemists have found that practically useful results can often be obtained in such cases by introducing parameters whose values are determined empirically, by fitting the calculations to experiment. This approach had, in fact, been tried in the present connection but abandoned because it could not be made to work. My group showed that the failure was due simply to failure to optimize the parameters properly, and the so-called *semiempirical* procedures we developed have proved very effective. We ourselves have used them very effectively in the studies of a wide range of chemical reactions and other chemical problems.

We have, moreover, done this in the face of opposition, obstruction, and abuse from the "official" quantum chemists who have, quite rightly, felt themselves threatened by our work. Those using our procedures in their work can still run into problems with referees when they try to publish their results or when they apply for support for their research. However, since this is all discussed in detail in the book which forms the basis of this article, I need not go into it further.

One of my most practically significant contributions arose in a curious way. I was giving a lecture at a conference at Montpellier in France soon after I had developed the concept of π complexes. At the end of the lecture, in the discussion period, someone asked if olefin-metal complexes could be so interpreted. Being evidently in a rather inventive mood that day, I said that of course they could and that they would moreover be further stabilized by back-coordination involving a filled *d*-MO of the metal and the empty antibonding π-MO of the olefin. Chatt had, in fact, recognized that back-coordination must be invoked to account for the stability of such complexes, but he had explained it differently. He assumed that the olefin isomerizes to a carbene — ethylidene in the case of ethylene — so the complexes are really carbene complexes, not olefin complexes. When he got home after the conference, he and Duncanson set out to distinguish between the two possibilities by using the then-new technique of IR spectroscopy, and they found I was right. Since I did not publish my work independently, the π-complex theory of metal-olefin complexes came to be referred to as the Dewar-Chatt-Duncanson theory, and many indeed have come to believe that Chatt was in fact responsible for it. So Chatt got credit for the idea by showing that I was right and he was wrong!

My work on theoretical chemistry may also have drawn attention to one of the major problems of science today, namely, its subdivision into an increasing number of progressively smaller areas, each of which sets itself up as an independent discipline with its own rules and list of members. Since an area of little real importance is unlikely to attract good people, the members are then likely to be mediocre or worse. They will form a mutual admiration society, holding meetings at which they congratulate one another on their "achievements," with all the paraphernalia of really significant groups. If a paper describing work in such an area is submitted for publication, the journal in question sends it for review to official experts in the area, the so-called "peer review" system, so if the paper reaches any really novel conclusions, it is likely to be rejected, even if the evidence and logic used in reaching them are flawless. If the authors are not officially members of the area, the situation is even worse. Such a paper is almost certain to be rejected, regardless of whether it is right or wrong. The "peer review" system can thus act as a major barrier to progress by preventing the publication of new and significant material.

The trouble here is that people who work only in one narrow restricted area are likely to get fixed in their thinking, so they are unlikely to come up with anything really novel. New ideas are likely to come only from outside. Now my early work officially classified me as an organic chemist, not a theoretician. This is why I was able to bring new ideas into the theoretical area. I had not been indoctrinated with the "official" ones. But because I was not an official quantum chemist, I often had major problems getting my work published, even when it was really important, and I am sure that many others have found themselves in the same boat.

What can be done about this? Frankly, I don't know. If the rules about reviewing were relaxed, this would lead to journals being cluttered with garbage. What one really needs is a way to assess the ability of potential reviewers so that journals could choose them better. But how could one do this? And since most of those with good reviewing qualifications are likely to be doing good work themselves, would it be helpful to have them waste time that they could spend more profitably on their own research by putting them to review garbage?

My thinking about most things tends to be unorthodox. This is because I have had a very strange career. Since this is reported in the book that was the incentive for this article, I need only mention a few essential points here.

I was born in India, my father being in the Indian Civil Service, the organization which governed India when it was part of the British Empire. Till I was eight, I lived in a community in which there were no other white children. I used to go to the club which was the social center for the other Britons in the place. Indeed, I learned to play bridge there when I was five. So I met children for the first time when, at the age of eight, I was sent back to a boarding school in England. The other boys there naturally thought me peculiar and bullied me unmercifully. Indeed, I formed no real friendships until many years later, when I went to Oxford University. So I have always been a bit of a loner, ready to believe that people were against me, and I have always tended to form my own ideas about everything by reading about it without consulting anyone else.

My next school, Winchester College, had the reputation of being the top one scholastically in England, and scholarships there were particularly sought after because they were exceptionally valuable. I not only got a scholarship there but got it a year early and was also at the top of the list. The scholars lived together in College. For many years, one of the mathematics teachers at Winchester ran intelligence tests on the students in his classes. With two exceptions, everyone from College was off the scale, and the two exceptions were only just on it. One of them was there when I was. We all thought him a bit dim. So College formed a rather strange society. Everyone there was a genius, so the place was an intellectual hothouse. Everyone discussed everything at a very high level. This, of course, was another reason why I was never prepared to accept anything on the grounds that everyone else accepted it. It also made it difficult for me to marry because I could not have stood living with someone less intelligent than I was, and there are very few who meet this requirement. By incredibly good luck, I met someone who was at least my equal, better still, I was able to persuade her to marry me. As an added bonus, she was also not merely good-looking but very beautiful. Whether she was wise to say "yes" is another matter!

Mary had had a normal childhood and so was not, like me, isolated from the world. So at least I had someone I felt was at least my equal, someone I really respected, to guide me. I certainly needed guiding! Mary used to say that I might be good at chemistry, or anything else that interested me, but I was hopeless at everything else. She took on the job of running me. Not that I was easy to run, and, looking back, I can see that I have been my own worst enemy, through doing things against her better judgment.

My worst failing has been a refusal to accept defeat when I have been dirtily treated by others. In many cases of this kind, the least bad alternative is just to surrender and accept defeat. I have never been prepared to do this. I have always fought back, even when I did myself harm by doing so. In such cases, when I dug my toes in, Mary would in the end go along even when my action hurt her too. Her situation was like that of the owner of a china shop with a pet bull. Most of the time, the owner is able to keep the bull under control. At intervals, however, somebody comes into the shop wearing something red. The bull then runs amok, and all the owner can do is try to minimize the damage.

These occasions are not mentioned in my book because in writing it I did follow Mary's advice. Perhaps it will help others if I point them out now and some conclusions that can be drawn from them.

The first was my move to Courtaulds in 1945. This did me nothing but harm. I was at the time all set for a distinguished academic career. I had a postdoctoral fellowship at Oxford which let me free to do anything I wanted; my work on stipitatic acid had already given me wide recognition, and I had the backing of one of the leading chemists in the world, Sir Robert Robinson. I would certainly have got a lectureship somewhere as soon as one became vacant. Instead, I went into industry. It is true that the job was not in a normal industrial laboratory. It was in a new laboratory that Courtaulds had recently set up in Maidenhead, a small town to the west of London, to do fundamental research, and the people in it were essentially free to carry out any research they liked so long as it had some connection with polymers and fibers. That, however, really made very little difference because, to the outside world, industry was industry, and getting back into academia from any position in it was then unheard of. The move also infuriated Sir Robert. Why do I do it? According to my book, I had begun studying the kinetics of some reactions, and when I did so, I found I did not know enough about physical chemistry. Since the position I had been offered at Courtaulds was as a physical chemist, this seemed an opportunity not only to learn about physical chemistry but also to be paid a good salary to do it. This is not what really happened. Let me now set the record straight.

Given that I wanted to study science, I should really have gone to Cambridge rather than Oxford. I went to Oxford, and to Balliol College in particular, because I was able to get two scholarships there. This was important because my father had died early, when I was nine, and my mother was left with a very small pension. By winning the two scholarships, I was able to support myself.

At that time there was just one Chemistry Fellow al Balliol, R. P. ("Ronnie") Bell, a physical chemist. The College had had plans to appoint a second Chemistry Fellow, this time in organic chemistry, but these had been shelved because of the war. Now that the war was over, the College started putting these plans in action.

I was the obvious choice for the new fellowship. Ex-members of colleges always had priority in such cases; I had an outstanding record, and I had even been made a member of the Senior Common Room. Indeed, the only alternative candidate was a very dim type called Waters (I used to refer to him in private as "old wetness"). However, at the meeting to make the appointment, there was one strong dissenter, Ronnie Bell. Since his opinion was clearly crucial, Waters was appointed.

When I heard about this, I was outraged. Mary said I came home foaming with rage, saying "I'll tear his testicles off." So my move to Courtaulds was really a case of shaking the dust of the place off my feet, even if it meant cutting off my nose to spite my face. It was also very bad for Mary. She was well on her way to getting her D.Phil. (as Oxford calls the Ph.D.), and the move stopped her.

The years in Maidenhead were grim, particularly for Mary. During the week I was working all day at Courtaulds, and I spent the evenings and weekends writing a book for Oxford University Press and developing my MO treatment of organic chemistry. I had of course to do this if I was to have any hope of getting back to academic life, but it was hard on Mary, particularly since she also had obstreperous young children to look after, no neighbors she could leave them with, and very little help. It was also several years before we were able to get a car; the nearest grocery store was a mile and a half away, and we had no refrigerator, so Mary had to go shopping there every day pushing a baby carriage, and I had to travel the six miles to the laboratory on a bicycle. When the end came, it did so in a quite unexpected way. I was offered not only an academic position but the Professorship of Chemistry at Queen Mary College (QMC), one of the component parts of the University of London. This was almost incredible. At that time there was just one professor in each department at most universities in Britain. The professor had absolute power over everything to do with the department; no nonsense about democracy! So getting a professorship was the goal of everyone in academia in Britain. The University of London was, moreover, one of the three top British universities, so professorships there were especially valuable. I was moreover the youngest person (apart from one or two mathematicians) to be appointed to a professorship in London — and this when I had had no other academic appointment!

The move let Mary get a belated Ph.D. in history, and her thesis was published by the Athlone (London University) Press, one of several books she wrote on English Tudor history. We also had a very attractive house right on the edge of Richmond Park. Indeed, deer used to come to the fence at the back to get food we put out for them. I did admittedly have problems with the Department at QMC. The main one was the Chemistry Building, a huge brick monstrosity on five floors with walls so thick that it had, unfortunately, survived the bombing of London almost unscathed. My predecessor, Partington, had rarely come into the place, so he had allowed himself to be banished to the top floor, where he had two offices, one for himself and one for his secretary, and a single huge open laboratory. The two next senior people on the staff (faculty), Jones and Hickinbottom, had cornered the better accommodation lower down. I could, of course, have turned them out, but I didn't want to upset them. Not that I got any thanks for it; they both hated me from the word "go" because each of them thought he should have got the professorship — and naturally they also hated each other. There were, of course, no elevators — elevators hadn't been invented when the place was built — and there were just two staircases, spiral stone ones, one at each end of the building, with no glass in what should have been windows, so in winter snow often drifted onto one while one was using them, which I had to do from time to time because the only rest rooms in the building were on the ground floor. I was, however, told at my interview for the professorship that Chemistry was next on the College's building program and that they were working hard on trying to find the necessary money.

There were other problems, for example, the fact that the Department had no equipment and no money to buy it. The unorthodox ways I solved these are described in my book. However, no amount of originality on my part could solve the problem of the building itself.

When I went to QMC, the chemistry department there was virtually unknown. Within six years, thanks to my efforts and those of the other members of the Faculty, some of whom I had myself appointed and nearly all of whom blossomed with the help and encouragement I gave them, it had become one of the best chemistry departments in Britain while the Physics Department was still at the bottom of the heap. At this point, the College decided that since Chemistry was doing so well and Physics so badly, the next building should go to Physics rather than Chemistry in the hope that that might make Physics do better. Needless to say, I was outraged. This was the real reason why I moved to America. I could not move to another professorship in Britain because the one I had was near the top of the heap and all the ones I might have moved to were unlikely to fall vacant because the current occupants were quite young.

The move really shocked the academic establishment. No professor from any reputable British university had ever before defected. Indeed, when news got out that Professor Sir Derek Barton, the Professor of Organic Chemistry at Imperial College, one of the other London colleges, had had a very attractive offer from another major American university, the Rector (President) of the College got a personal communication from the British Cabinet saying "Anything that needs to be done to keep Professor

Barton here must be done." Derek, who is one of my oldest friends, had in fact also been trying to get a new chemistry building but had been told that the College had no money. Shortly after I left, I got a note from him saying "you will be amused to hear that I am getting my new building after all!" Derek did, in fact, also end up on this side of the Atlantic because when he reached retirement age in London, Texas A&M offered him a Professorship.

I need not have left QMC. I had things set up there in such a way that I could easily have waited a bit longer for a new building. However, when I later saw the chaos that ensued when it was being built, I realized that I was really well out of it. And the move to America was a total success in every other way, for our two sons as well as for both of us. This was the one case when my refusal to put up with people doing me down paid off!

One essential point is, however, misstated in my book. I say there that when Kharasch died, the University of Chicago offered me the Chair of Chemistry he had held. This was not in fact the case. I was offered a Full Professorship at a salary corresponding to that of the Chair, but not the Chair. The Chair, as such, was left vacant for technical reasons concerned with the endowment. I was, however, assured that I would get it as soon as these problems had been cleared up. So when it was given, three years later, to someone else on the Faculty I was outraged. This is really the reason why I went to the University of Texas in Austin, which at the time was virtually unknown. Everybody thought I was crazy to go, and in the long run they were right. However, it took some time for things to go sour, and since this is all covered well in my book, I need not say more about it here. But life might have been much easier for all of us if I had stayed in Chicago, unless and until I got some better offer from some other major American university.

[...]

So my main conclusion is really quite simple. If someone hits you, don't hit back. At least, wait till you have decided whether or not hitting back may make things even worse than they would be otherwise. If it is, just put up with the situation. You may think it unfair for skunks to get away with being skunks. But then life is unfair, and nothing you can do is going to make it fair. This is my second conclusion. The third is even simpler. Keep your unfavorable opinions of others to yourself. I have never been good at doing this. Indeed, some of my comments about other chemists have been unprintable. Needless to say, all my chemical friends are not merely good chemists but outstandingly good. But the others have naturally tried to make life difficult for me in every way they could, and they have frequently succeeded.

My problems here were certainly made worse by Sir Robert Robinson. I had a tremendous admiration for him, and he too was not one to put up with abuse from people he despised. I learned from him the need for total integrity in science and never to allow one's work to be influenced by personal interests. But he was very ready to express himself very forcefully when other chemists failed to come up to his own high standards. His feud with Ingold was the classic example. [...]

Mary's problem might have been softened if her college in Oxford had not washed their hands of her after she left. The trouble was that she went to the wrong college, Lady Margaret Hall (LMH) instead of Somerville. LMH was a very snobbish place, full of girls trying to pretend they were higher in the social scale than they really were and very godly to boot, and of course the Fellows were the worst offenders, because the social superiority of the girls they were teaching rubbed off on them. Since Mary came from the industrial north of England, officially a very low class area, nobody at LMH really liked her being in the place. They couldn't refuse to accept her because her examination results had been so totally exceptional. Somerville, on the other hand, was concerned only with scholastic achievement. Nobody there was in the least interested in where the students came from, only in how good they were. Why then did Mary not go to Somerville? This was another low blow. She had, in fact, intended going to Manchester University, the major university in the north of England. However, when her examination results turned out to be so outstanding, the headmistress of her school suggested that she try for Oxford. Since she knew nothing about Oxford, she had to rely on the headmistress for advice. Now the headmistress had never had one of her girls go to LMH and was interested in finding out what the place was like. So she advised Mary to put LMH as her first choice.

So life has been a bit of a struggle for both of us. As long as we could face it together, we did well, but I have found carrying on alone hard.

References

1. Cornforth, J.W. *The Chemical Intelligencer* **1998**, *4*(3), p. 32.
2. Dewar, M. J. S. *A Semiempirical Life*. American Chemical Society, Washington, D.C., 1992.

3. Edwards, J. T. *The Chemical Intelligencer* **1997**, *3*(1), 25–30.
4. Dewar, M. J. S. *The Chemical Intelligencer* **1997**, *3*(1), 34–39.
5. See, Davenport, D. A. *The Chemical Intelligencer* **1997**, *3*(4), 53–55; Laidler, K. J.; Leffek, K. T.; Shorter, J. ibid. **1997**, *3*(4), 55–57.

John Pople, 1995 (photograph by I. Hargittai).

14

JOHN A. POPLE

John A. Pople (b. 1925 in the U.K.) is Board of Trustees Professor of Chemistry at Northwestern University in Evanston, Illinois. He is credited primarily with the computational revolution in chemistry. The 1998 Nobel Prize in chemistry was shared by John A. Pople "for his development of computational methods in quantum chemistry" and Walter Kohn of the University of California at Santa Barbara "for his development of the density-functional theory."

Dr. Pople received his degrees in mathematics from Cambridge University, including the Doctor of Philosophy in 1951. He was Carnegie Professor of Chemical Physics between 1964–74 and John Christian Warner Professor of Natural Sciences between 1974–91 at Carnegie-Mellon University in Pittsburgh, Pennsylvania. He has been at Northwestern University since 1991. Dr. Pople is Fellow of the Royal Society (London), Foreign Associate of the National Academy of Sciences, Corresponding Member of the Australian Academy of Sciences. His many distinctions include the Linus Pauling Award of the American Chemical Society (1977); the Davy Medal of the Royal Society (1988); and the Wolf Prize (Israel, 1992). Our conversation took place in his Evanston office on December 29, 1995 and it first appeared in *The Chemical Intelligencer.**

What is the essence of computational chemistry?

*This interview was originally published in *The Chemical Intelligencer* **1997**, *3*(3), 14–19 © 1997, Springer-Verlag, New York, Inc.

Computational chemistry is the implementation of the existing theory, in a practical sense, to studying particular chemical problems by means of computer programs. Some people draw a distinction between computational chemistry and the underlying theory. I really prefer not to and to think of computational chemistry as the implementation of theory for the understanding of chemical problems. The theory has preexisted, and the computers have enabled it to be implemented much more broadly than was possible before.

Does the widespread availability of computers stimulate the development of theory?

You are quite right, it is a two-way flow indeed. Many of the theories that are now used were developed initially in a much simpler form when people could do very simple calculations with hand calculators or log tables. In reverse, the need for efficient application in computational programs has some kind of back effect on the theory. Sometimes one can improve the theories in the sense of discovering a quicker, more efficient way of doing a given calculation. Most of the progress in recent years has been in the improvement of the efficiency of the algorithms that are used.

What is currently the most important problem to be solved by theory?

Theory applies to the whole of chemistry. Almost any problem in chemistry can be approached by theoretical methods at any level. Some old theories are being now quantitatively implemented. There are also some new theories.

The Møller-Plesset theory, which originated in the 1930s is being widely used to include electron correlation in ab initio molecular orbital calculations. May there be other old theories that might be implemented today?

Yes, indeed. We are looking for better methods in the old literature, all the time.

What are your current interests?

The theories I am currently developing include the density functional theory and are aimed to treat quantum mechanical problems in a more efficient way than has been done previously, by cutting out some difficult steps of the previous computational methods. The essence of the density functional theory is to try to use the density distribution of electrons, the same thing which you measure in X-ray crystallography. We would like to go directly from the electron density distribution to the energy. This is possible in principle but it is not obvious how one can do it in practice. There is some promise that it is possible to do this more efficiently than by the currently existing methods.

How about the nuclear positions?

In the electronic structure theory, you study the distribution of electrons for a particular nuclear arrangement in the molecule, work out the energy at that nuclear arrangement and then explore different nuclear arrangements. This is also how chemists think about molecules. At first the nuclei are in a fixed position, then they are moving around on a potential surface. The density functional theory says that if you knew exactly the electron density distribution of the molecule for a given nuclear position, then you should be able to deduce from that the total energy of the molecule.

What makes the application of the density functional theory easier to apply than the quantum chemical methods?

It is easier because the electron density is only a function of three dimensions. To get the full wave function of the electrons, that's a problem in $3n$ dimensions where n is the total number of electrons. The full solution of the quantum mechanical problem for a fixed nuclear position involves the Schrödinger equation in $3n$ dimensions. In principle, if you could only find how, to get from the density to the energy directly would be very important. The trouble is that people don't know how to do it precisely.

How far has it advanced so far?

It's already being used. The simplest form of the density functional theory is called the local density approximation. This approximation is essentially assuming that at every point in the molecule, if you know the density at that point, you assume that the properties in that region are the same as the uniform electron gas of the same density. It's a crude approximation but it has been used and attempts are being made to improve that. This is a very active area currently in the electronic structure theory.

What are the properties that can be produced?

When we are talking about the electronic structure of the molecule, one can compute virtually any property of the molecule. The structure is the location of the minimum on the potential energy surface. For a diatomic molecule, you just have a potential curve with a minimum which gives you the bond length. For a polyatomic molecule, the potential energy surface is a function of many coordinates. The structures are the locations of various minima on that surface. The various minima correspond to the various isomers of the same molecule. Once you have that, you can get other properties, for example, the vibrations of the molecule are determined by the local curvature of the minimum and they can give you the harmonic force constants. From the potential surface it is also possible to calculate the energy change of breaking the molecule in two pieces, that is the dissociation energy. It is possible to determine the bond strength, to calculate the activation energy of reactions which is the energy required to go from a local minimum up to a transition structure, that is a saddle point on the surface corresponding to the supermolecule. It is also possible to get other properties, such as the electric dipole moments, magnetic properties, nuclear magnetic shielding constants. Virtually any physical property of a molecule is determined from its wavefunction in the electronic structure theory.

How do you define the supermolecule in your investigation of reactivity?

Supermolecule is just two molecules considered as one large molecule.

What would correspond to the "experimental error" in computational work?

This is a good question. The way I like to do this is to set up a theoretical model. You apply one theoretical model essentially to all molecules. This model is one level of approximation. Then you apply this one level of calculation to a very large number of different molecules. In fact, one level of approximation is applied to all molecules, giving you an entire chemistry corresponding to that approximation. That chemistry, of course, would not be the same as real chemistry but it would approach that chemistry and if it is a good model, it will approach real chemistry well. What I try to do is to take a given model and then to use that model to try to reproduce a lot of well-known facts of experimental chemistry. For example you try to reproduce the bond lengths in a large number of simple organic molecules, or the heats of formation for that set of molecules, in a situation where the experiment is beyond question. Then you can actually do statistics and say that this theory reproduces all known heats of formation to the root-mean-square accuracy of 2 kcal/mol. When you've done that you build some confidence in the level of theory. If you then apply the same theory in a situation where experiment may not exist, you know the level of confidence of your calculations.

Is this a semiempirical approach?

Yes, this is semiempiricism in a limited sense. Semiempiricism in an extreme sense is when you have a theory and you feed in many parameters and then you determine new parameters for other systems. However, here I am talking about taking just one simple, well defined mathematical approximation and testing that approximation. When you've done that, then you have exactly what you're asking for. You have one level of uncertainty of theory. If, using this theory, you reproduce all the known heats of formation, say, within 2 kcal/mol, and then calculate the heat of formation for another molecule, you'll have some basis to believe that your result is good within 2 kcal/mol.

Why is it then that computational results usually appear without indicating any error?

I usually produce theories which are calibrated exactly this way. We have developed, for example, the so-called G-2 theory and it was deliberately set up this way. First we collected all those molecules for which the experimental heats of formation had been determined very precisely. They were not too many, about a hundred. Then we set ourselves the target to produce a theory which would reproduce all of these with a mean absolute error of 2 kilocalories, and that was achieved. So predictions of heats of formation by the G-2 theory are accurate to 2 kilocalories. Such an objective, concerning the errors, should always be there in computational work. I fully agree with that.

What should be the ideal relationship between experimental and computational work?

I view computations as a technique which any chemist can use, including experimentalists. These calculations should augment the experimental work.

Is there progress in this direction?

There is indeed, but not as much as I'd like to see. The older generation is especially suspicious of theory. They are not used to it and probably think that it's too difficult. In reality, the application of the theory is not difficult at all. The development of the theory and the development of the programs is difficult but the use of the theory is very straightforward, just like any other techniques, as crystallography or magnetic resonance.

Should computational chemistry be considered to be a separate discipline?

There are Professors of Theoretical Chemistry. However, I don't think computational chemistry should be a separate discipline. It's a technique that all chemists should use. So it should be in the general curriculum, it should be taught, but not even necessarily by a theoretical chemist. The programs should be considered a black box, just like a complicated spectrometer is, and chemists should learn how to use these programs, and to use them in a critical manner, to understand the limitations of what they get out, just like any other technique.

There seems to be a proliferation of computational publications from laboratories where buying a computer and the necessary software may be a shortcut to do research rather than developing experimental chemistry. A lot of tables of numbers emerge which are not necessarily useful.

I agree that the mass of publication of numbers is not very useful. What is useful is to study a particular class of compounds and to carry out a number of computational studies on these compounds, and carry them out at the best level feasible with current equipment and which then throw some light on the properties of those compounds and on how they react. That would then constitute good science. That would also fit in well with the corresponding experimental studies.

Is there a bias in good journals against submissions which contain only computational results?

There is such a bias in the *Journal of the American Chemical Society* which has a rather strong position against such papers.

You are considered to be the person who, more than anybody else, has transformed chemistry, making it also a computational science.

In the electronic structure aspect, yes.

Were you more interested in making the computational approach available for application rather than going for the highest level of sophistication for some selected simple systems?

That's correct.

How did you chart your strategy when you first started this work?

It was a long time ago, when I formulated this concept that one level of theory implies an entire chemistry. It was during my postdoctoral studies, in 1952. At that time nothing was really possible in practical terms. The first theory in this category was the so-called PPP theory, Pariser-Parr-Pople, which handled essentially only one electron per atom and was used for the p-electrons of aromatic hydrocarbons very successfully in the 1950s. It was a very simplified theory which you could handle without computers in those days. But you are right, my general objective has always been to produce theories and associated computational techniques which would be extensively applicable, and illuminate as many chemical properties as possible. This has proved to be a very successful approach and I was helped a great deal by the huge advances of electronic computation.

Did you anticipate this?

I didn't anticipate it would be as extensive as it has been, but, nonetheless, I thought it was going to happen. Electronic computers existed in the 1940s but they were very unsophisticated. To get two and two equals four you had to feed in two and two and then reconfigure the computer and press the button for the result and get the answer four.

Where do you feel that the computer is the limitation today?

There is always a limitation. If you have this objective, you're always pushing for more. Programs may run too long, use too much space and memory, but these limitations are changing rapidly.

When did you move to the U.S. and why?

My early research career was in England. I started working in chemistry in 1948 under Lennard-Jones as a graduate student, and held appointments in England until 1964. I was on the Faculty of Mathematics in Cambridge and taught mathematics rather than chemistry. I got my degree in mathematics. That's traditional in Cambridge. People do mathematics and then do theoretical science. This goes back to Isaac Newton. Many of the best known theoretical physicists in Cambridge, people like Dirac, were Professors of Mathematics. I was a Lecturer of Mathematics. Then from 1958 to 1964 I worked for the National Physical Laboratory which is the British equivalent of the National Bureau of Standards in the U.S. I moved to the Carnegie-Mellon University in 1964, as Professor of Chemistry. Part of the motivation to move here was the better audience in America for computational science than in England. The very first time I visited the United States in 1955, when I was 30, I had never given a lecture in a chemistry department in England. I had given seminars within the theoretical chemistry group but never a general lecture. Then I came to this country and made a tour of several universities, including UCLA and Chicago. I found it quite remarkable to get a hundred people to come to listen to what I had to say. The U.S. was the best place to develop theory at that time.

John Pople around 1970 (from the *Proceedings of The Robert A. Welch Foundation Conferences on Chemical Research*, Houston, Texas, 1973).

Did you need a lot of support?

No, just for basic computer equipment and a few students, and the National Science Foundation has provided that support. I have been most fortunate in the quality of the students that have come to work with me.

What is your involvement in the Gaussian programs?

Gaussian 70 was the name of the original ab initio program. There were semiempirical programs before that. Originally I started the Gaussian project (in 1968) which later on became a commercial product. However, I am not associated with the Company anymore; since 1992. I'd rather not make any further comment about this project at this point.

You have a lot of awards and distinctions. Would you care to single out any of them?

When I was a graduate student in Cambridge I won the Smith Prize. It's a very old award which goes back to the eighteenth century. It's a fairly well known prize for graduate students in mathematics. Then the latest was the Wolf Prize of Israel for 1992 which is a highest level award, and which was a pleasant experience. It's like the Nobel prizes, given in various fields. Actually, it's even wider than the Nobel prizes; they give them in the arts as well.

What was the incentive in your childhood that set you onto this career?

I suppose I was a mathematically curious child and started a research program when I was 12. I was interested in permutations, although I didn't know this term at the time. I started producing tables of factorial N, but I kept everything to myself; 12 is an age when one doesn't like to be seen as academic. Later on I had some very good teachers in high school.

What was your family background?

My family was nonacademic; my father ran a clothing store in the West of England, in a small town called Burnham, about thirty miles from Bristol where I went to high school. It was the best high school in the area and from there I went to Cambridge as an undergraduate.

Coming from a middle-class family, how did you feel fitting into Cambridge?

It's always been possible to get into Cambridge, or Oxford, from any background by a scholarship arrangement. I competed for a scholarship, sat for an examination at Trinity College, and became a Scholar of the College. This is a tradition that goes back a long way. People can become Scholars of the Cambridge Colleges, whatever their background, provided they make it through this examination. This gives you privileges when you get to the University. This way people of all sorts of background move into professional classes. It has been a way in which England is not really totally a class-ridden society. People who go to Oxford and Cambridge may go there because it's the thing to do or because their parents are upper class, but others come in as Scholars. Students are classified as Scholars because they had won a scholarship, and the rest of the students are called Commoners. The Scholars have some privileges. When you go to the College Chapel, for example, the Scholars sit in privileged places and the Commoners sit at the bottom.

Who were the people who had a strong influence on you?

My supervisor, Lennard-Jones, was a great inspiration and he was very helpful when I was entering chemistry.

How about your family?

My wife is a piano teacher, now retired, and we met when I was a graduate student in Cambridge. We have four children, none of them in an academic position. Three of my children were born in Britain and one was born here. One of them lives in Ireland and the others live in the U.S.

Why did you move to Evanston?

We moved in 1981 to Chicago because our daughter lives here. So, for a while, I was commuting to Pittsburgh. Then, when I retired at Carnegie-Mellon, I assumed a part-time professorship at Northwestern University and now we live nearby.

Any hobby?

Listening to classical music.

You set up a program in 1952 and now, some decades later, it has really been accomplished. Do you see people among today's young generation setting up such long-range programs?

I was very fortunate in being in the right place at the right time, and the emergence of electronic computers made it all possible. Other people have also set up bold strategies, such as those who have developed computer logic and the whole computer science. It's hard to say though whether anyone is starting anything of that magnitude now. But it is possible. People may have the ambition, for example, to understand human consciousness.

Roald Hoffmann, 1994 (photograph by I. Hargittai).

15

Roald Hoffmann

Roald Hoffmann (b. 1937 in Zloczow, Poland, now the Ukraine) was John A. Newman Professor of Physical Science at Cornell University at the time of the interview. Now he is Frank H. T. Rhodes Professor of Humane Letters and Professor of Chemistry there. He shared the Nobel Prize in Chemistry in 1981 with Kenichi Fukui "for their theories, developed independently, concerning the course of chemical reactions."

Having survived the Nazi occupation, Roald Hoffmann arrived in the United States in 1949. He graduated from Columbia University in 1958 and received his Ph.D. from Harvard University in 1962 under the supervision of W. N. Lipscomb (Nobel Prize in Chemistry, 1976). He stayed on at Harvard for three years as a Junior Fellow. A result of his cooperation with R. B. Woodward (Nobel Prize in chemistry, 1965) was the Woodward-Hoffmann rules of chemical reactions, based on the notion of the conservation of orbital symmetry. Dr. Hoffmann has been at Cornell University since 1965.

He has been a member of the National Academy of Sciences, the American Academy of Arts and Sciences, the American Philosophical Society, Foreign Member of the Royal Society (London), the Indian National Science Academy, the Russian Academy of Sciences, and many other learned societies. He has received the Priestley Medal, the A. C. Cope Award in Organic Chemistry, the Award in Inorganic Chemistry, and the George Pimental Award in Chemical Education of the American Chemical Society, among many other awards and distinctions. Dr. Hoffmann has also written poetry and published popular science books.

Roald Hoffmann and his wife Eva were spending a week in Budapest at the end of June, 1994. Because of Eva's interest in laces, we went to visit Kiskunhalas, some 130 kilometers south of Budapest, famous for laces. We taped our conversation during that car ride to Kiskunhalas. The interview appeared in *The Chemical Intelligencer*.* I am grateful to Dr. Hoffmann for letting me use his poem "Two fathers" to illustrate our conversation in this edition.

First of all I would like to ask you about your schooling and about your teachers. You write in one of your poems about a boy, presumably yourself, who did not own a book until he was 16.

My schooling was interferred with by the War. First, there were a few months in a Ukrainian school in Zloczow. Then there was the second and third grade in a Catholic school in Krakow in Polish. My fourth grade was taught in Yiddish in a Displaced Persons refugee camp in Austria. Then a little bit in German in Germany, and eventually in the fifth and sixth grade everything was taught in Hebrew in Munich. I learned algebra for the first time in Hebrew. All of this just to show my refugee background, with the mixture of languages that many children in the chaotic postwar period experienced. Then we succeeded in getting to the United States, of course, where all things went well. I attended New York City public schools, first a public school in Brooklyn, then a special science public school called Stuyvesant High School. There was a wonderful concentration of talent in that all-boys school in New York City. Then I went to college at Columbia. So I grew up in New York City. Eventually I went to graduate school at Harvard.

We were poor; it was not easy for my parents to begin in a new country. So in fact, we had no money to buy a book for me until I was 16. I can't remember any particularly inspiring teachers until high school. I remember though that I was not particularly interested in chemistry. At Stuyvesant High School there were advanced courses in every field. They would be called in the United States today advanced placement courses. This was a kind of second course in each subject. I took such courses in biology and physics but not in chemistry. I also took a lot of courses in mathematics. At the end of high school I intended to go into medical research. That would have been a compromise between my parents' desire that I should

---

*This interview was originally published in *The Chemical Intelligencer* **1995**, *1*(2), 14 and 18–25

Roald at the age of 8 in lederhosen, in front of their home in Krakow, Poland, 1945 (courtesy of Roald Hoffman).

become a doctor and what I wanted to do, which was some sort of scientific research. I started the university as a pre-medical student. But that did not last more than a year, and I somehow drifted into chemistry.

I had some unusually good teachers in high school and they were in mathematics and in biology. At Columbia University the teachers I had in humanities courses were just wonderful. The ones in science courses were okay, but I don't think I hit one that inspired me until the last year in college. I really think that if I had not encountered those teachers in my last year, George Fraenkel and Ralph Halford, I would have gone into the humanities.

The humanities were so seductive, especially history of art. I had fantastic teachers in Japanese literature, in the history of art, in English literature, poetry, and other literature courses. The world was just opening for me, and those were the most inspiring teachers I had.

What turned you to chemistry?

There were several factors. One was that in my last year in college I took some really good courses in theoretical chemistry. But I think that the main thing was summer research experiences. One was between high school and college, at the then National Bureau of Standards (NBS), then in Washington, D.C. I worked there two summers. The first summer was pretty boring, but it was my first introduction to research. I worked on the thermochemistry of cement with a man named Ed Newman. The very first scientific paper I published appeared in the research journal of NBS, on the heat of formation of hexacalcium aluminoferrite. This is a compound occurring in cement. After that, I knew I did not want to do cement chemistry, but research *was* exciting.

I came back the next summer and had something much more interesting to do. It was the low-temperature pyrolysis of hydrocarbons, low temperature meaning around 300°C. Then the next summer I worked at Brookhaven on some radiochemistry with Gerhard Friedlander and Jim Cummings. We constructed a low-level counting apparatus for carbon-11. That was very exciting. Imagine making a thousand atoms of carbon-11, and measuring quickly their amount while they decayed with a 20-minute half-life!

Although I did not subsequently do any of these kinds of chemistry, they constituted my introduction to real research. They were very important for me, and I still think that for many kids research is the way into science. It does not matter actually what one does in detail, but the social setting of a small group, the closeness to science and scientists, the turning into reality of what is taught in courses, all this is very important.

I could have done other things. However, at that time I thought that I was not good enough for physics, even though I got A's in all my physics courses. I saw kids, my friends, who were better at the subject. Now I know I was wrong and could have done physics just as well. For similar reasons I was pushed away from mathematics. I don't know why, but I was not that interested in biology at that moment in my life.

So you see I came to chemistry rather late and with no strong motivation, except that I got my introduction to research in chemistry. It's true that I had a chemistry kit when I was a kid, but I was not that much in love with the field. I really did not decide to become a chemist until I was 20 years old, or even older.

How would you characterize your relationship to your graduate advisers? What is important for a graduate student?

I worked for my Ph.D. with two people. I started out with Martin Gouterman and finished with William Lipscomb. For both of them I was their first graduate student at Harvard. Originally, I wanted to work for Bill Moffitt, a professor at Harvard, who was the world's leading young theoretician at the time. But he died in the year I came to Cambridge. Martin Gouterman was a postdoc of his who was appointed as an instructor, and then as an assistant professor. Gouterman was a student of Platt's at Chicago and was interested in porphyrin spectra. He is now at the University of Washington, and he continues to be active in spectroscopy and in theory. After about one year of work with Gouterman, I went to Russia for a year, to work with Davydov on exiton theory. Upon my return, I switched to working with Lipscomb. It took only one year of work with Lipscomb to finish my Ph.D. in 1962. Subsequently, during a three-year junior fellowship at Harvard, when I was independent, I developed the collaboration with R. B. Woodward. So I was not a postdoc or a student of Woodward's. There was a different kind of relationship there.

Both Gouterman and Lipscomb were very important to my scientific development, and my interaction with them was really day-to-day. This close contact might be understandable for Gouterman, who was just beginning his academic career. From the moment he started though, he had four graduate students. There was a great demand for doing theory by graduate students at Harvard, and no one on the faculty at the time wanted to do that. With Lipscomb, who was much further along in his career, there was really also daily interaction.

I remember clearly one important interaction with Gouterman after I had come back from one of Per-Olov Löwdin's great summer schools, where I first learned some group theory. There was someone on the Harvard Faculty who was trying to make cubane. I decided I was going to carry out a molecular orbital calculation on cubane. I was very proud of my newfound skills in group theory. So I set up the problem with eight orbitals, one on each carbon, I guess. What I had in mind was a Hückel calculation on eight hydrogen atoms at the corners of the cube, and it was all doable by group theory. I went to Gouterman, showed him the work. He said gently that it was very nice but you have not done cubane, you have done eight hydrogen atoms. And he slowly led me into the complexities of the problem, to setting up the matrix involving all the valence orbitals of cubane. This was essentially a forerunner of an extended Hückel calculation. He said you must consider $2s$ and $2p$ orbitals on the carbons and the $1s$ on the hydrogens, so that you have a total of 40 basis orbitals for C_8H_8. He said to go ahead and set it up, which I did. Then I did a Hückel-type approximation, and I had all kinds of s–s, s–p, and p–p σ and π interactions. I still have the matrix somewhere in my notebooks, with α's, β's, γ's, δ's, ε's, for the various resonance integrals. But I could not go any further, until we had programmed, two-to-three years later, an extended Hückel program in the Lipscomb group. At the outset I had no idea of taking the β's proportional to the overlap.

Anyway, the interaction with Gouterman was really quite close and with Lipscomb as well. There was also a research group in each case and a lot of learning from other people. I think that from both Lipscomb and

The stereochemistry of the ring-opening or cyclization reaction was studied theoretically by means of the extended Hückel theory[1]. The model

1. R. Hoffmann, J.Chem. Phys. 39, 1397 (1963), 40, 2480 (1964); identical parameters were used in this work.

geometries and the degree and direction of approach to a realistic transition state varied from molecule to molecule and are best described individually.

In the study of the cyclization of butadiene, the starting geometry was one of the cisoid form (VII) with 12 and 34=1.34 A, 23=1.48 A, all C-H 1.10 A, α=120°. The terminal CH_2 groups were then twisted in syn (disrotatory) and anti (conrotatory) modes through a range of angles, while retaining the trigonal conformation and other model parameters. This calculation actually indicated that in the ground state syn (disrotatory) twisting was very slightly favored; however in the syn (dis) mode the 14 overlap population (bond order) became more negative or antibonding, as the twisting increased, while for the anti (conrotatory) mode the 1,4 bond order became increasingly positive. The calculation was then repeated as a function of α, and it was found that syn (disrotatory) twisting led to a repulsion between atoms 1 and 4, and only anti (conrotatory) twisting had as a consequence 1-4 attraction and bond formation. At α < 117° anti twisting was favored. As α decreased and a 14 bond order developed, the 23 bond order increased and the 12 and 34 bond orders decreased correspondingly. In the first excited state ($\pi \rightarrow \pi^*$, spin multiplicity unspecified) the above these conclusions ~~apply only with syn and anti interchanged~~ are precisely reversed.

Approaching the transition state from the cyclic form, a model cyclobutene (VIII) geometry with 23 1.34A, 12,34,14 1.54 A (β=93.7°) was chosen. Hydrogens at 1 and 4 were ~~located~~ placed so that the C-H bonds formed tetrahedral angles with each other and 12 or 34. Syn (disrot) and anti (con) modes of twisting while retaining

A facsimile page of the original manuscript of the electrocyclic paper with hand-written corrections by both Roald Hoffmann and Robert Woodward (courtesy of Roald Hoffmann).

Gouterman I learned the importance of interaction with experiment. Lipscomb stressed that a lot.

From Woodward I also learned much. I was at a later stage then, it was now 1964 when I began to work with him. What I learned from Woodward was how to simplify explanations to an absolute essential. It is these simple explanations that make an impact on chemists. You don't have to cloud your explanations in mathematics.

I learned quickly. The first paper I wrote for Lipscomb, or the draft of a paper, was not very good. He revised it extensively. But I learned immediately and I remember that for the next paper he had only minor changes to make. With Woodward, writing was much more difficult. The language mattered to him, and he rewrote things in excruciating detail. I have a manuscript of the first orbital symmetry paper, with his comments on my draft somewhere. I have always had an easy time writing scientific papers, and that has been important. Since the collaboration with Woodward, my papers have been very pedagogic, taking the time to explain.

There is the question of how to choose research projects. With Woodward, one was looking clearly at important problems (though Woodward saw that more clearly than I). One could argue that part of talent is a kind of differential insight as to what is important and what is not — which anomalies to disregard and which to pay attention to.

Coming back to the teacher-student relationship, I feel that the role of the teacher is to awaken in the mind of the student those capacities which are already in some way there, and that need only that awakening. This idea I think actually goes back to St. Thomas Aquinas. I think it's not a bad summary of what a teacher does. In that sense, I think you need interaction on a continuous basis.

I think students learn in group meetings; and this is a great advantage of the American system. Students have to learn how to discern good science from routine work. In the beginning it's very hard to do that yourself. You need guidance.

Unfortunately, merely studying the chemical literature does not help you. Everything looks great there. You can publish any piece of junk somewhere. I think JACS accepts 60% of all full papers, and that's in one of the best journals. If you go to the next lower quality group of journals, they accept 95% or more. There is just no selectivity in publishing science. The discernment between routine and good occurs at the level of the practicing scientist, who often learns this from group meetings. I learned that from Gouterman, from Lipscomb, from Woodward. It's not important actually what group you're in. What is important is to be in a reasonably

Roald Hoffmann (left) and Robert Woodward (right) receiving the first Arthur C. Cope Award of the American Chemical Society in 1973. In the middle, Herman Bloch, chairman of the board of the ACS and Harriet Cope (courtesy of Roald Hoffmann).

international center where there are a lot of seminars, people coming and going, talking about their work. So I think someone in a provincial university in Hungary is in trouble. There the literature might serve as a substitute. To teach the students to go to every seminar in sight is very important.

What was your relationship with Kenichi Fukui?

I met Professor Fukui for the first time during one of his first visits to the United States. I think it was in 1964. At that time I was a Junior Fellow, two years past the Ph.D., and just about to begin the work with Woodward. Fukui came to Harvard to give a talk, and I was pleased to talk to him. I knew him primarily through his work on reactivity indices. Our relationship has always been very good. One reason is that the traditional model of competition in science does not apply to Fukui's and my work. We never actually competed on anything. In a curious and maybe unique way, our work intersected in such ways as to be of benefit to each other, and reinforced the work of the other, without us working directly together.

Fukui came up with his theory of orbital reactivity and the idea of frontier orbitals, the highest occupied and lowest unoccupied molecular orbitals (HOMO and LUMO). He derived these ideas from perturbation theory. With the formulas he had in place by 1960 (and the most important work was already done in 1955 or around then), he could have solved the problem of the stereospecificity of electrocyclic reactions. But he didn't, because the problem was never posed by experiment. Woodward and I did it first, and we did it essentially with a mix of extended Hückel calculations, and our own "discovery" of the role of frontier orbitals. This was a discovery for ourselves, not a real new finding. We also used perturbation arguments. I knew the Fukui and Dewar papers at the time. Dewar had written that absolutely inscrutable series of four papers or so in JACS in the fifties. It was clear, however, that Dewar was not interested in teaching or explaining specifics; he just liked the generality of his mathematical formulas. Interestingly, Dewar's failure (as a researcher) in these papers was a failure of teaching, I think. So his work had no impact, essentially, on the community he needed to reach.

Fukui was more interested in explaining but was still caught up a little bit with the mathematical formulas and the reactivity indices, very popular then. And he lacked a beautiful experimental case, such as that of the electrocyclic reactions, to demonstrate the power of his approach.

The moment that we did the orbital symmetry work, Fukui could do it. His formalism was certainly up to the task. A reasonable summary of the situation is that our orbital symmetry work with its focusing on HOMO and LUMO reopened interest in the community in Fukui's work. People saw that he had done similar kinds of things, that many reactions could be profitably studied with his formalism.

I don't think Fukui and I competed consciously or unconsciously. Our interaction just worked very well. Subsequently, I had no less than three of Fukui's talented coworkers as my postdocs, Fujimoto, Imamura, and Akagi. I have also visited Kyoto; I was just there last year for a month and a half. I also happen to be interested in Japanese culture from my university days, and I think I also share other interests with Professor Fukui. We get along very well.

How did the Nobel Prize change your life?

The Nobel Prize came when I was fairly young, as these things go, 44. Some consequences were pretty obvious, like many more invitations. In some of those invitations people were interested in my name rather than in me. This was fairly easy to sort out, but I had to learn how to say no to some things. In general, my life became busier. I had to work a little harder to carve out some time for myself, but I have managed to do that. In many ways it did not change things very much. That's a fortunate thing specific to being in the United States, where there is a substantial number of Nobel Prize winners and where society ignores scientists and their achievements anyway.

For Professor Fukui life was, I suspect, much more difficult, he being the first Japanese Nobel Prize winner in chemistry. He is under much more pressure. No one cares about me (I'm smiling). Even within the scientific community, there is not that much respect or value attached to the Nobel Prize. Just as many negative as positive feelings are engendered. I see occasionally in reviewers' reports, in the darkness, negative feelings come out. People say "I expected a Nobel Prize winner to do better". I have recently (four years ago) lost funding for about half of my research, the work on surfaces, which I think is actually going quite well. My being a Nobel Prize winner has no effect (at least in the United States) on my research support. Linus Pauling has had difficulty getting funding to do chemistry for 30 years. It's a very unforgiving world out there, a very competitive one, in our country.

Being a Nobel Prize winner has helped in no way to get my poems published. My last poetry collection has been making the rounds with publishers for four years, and individual poems go through many rejections before finding a place. What I resent then are assumptions by fellow scientists that things are easier for me. I'd love to show them all those rejection slips.

There is some opportunity to make a fool of yourself. The press is out there, looking for stupid things. So when William Shockley had some weird ideas on race, he found a willing public to listen to that.

There is some barrier that comes between a Nobel Prize winner and students, because students set you up on a pedestal, whereas your colleagues do not, at least not in the United States. This barrier is a negative thing. It stands in a way of informal communication. Fortunately, American students are not too respectful. When I teach first-year chemistry, I know how to overcome that barrier without any trouble. I cannot go to the library to look up a journal article two days before an examination, if I am teaching a first-year course. Immediately I would be surrounded by students asking me about the questions of the examination. I don't think that would happen in Europe.

I think I have had occasionally the opportunity to do things that I could not do before. I was asked to do the *World of Chemistry* films, in part because I was a Nobel Prize winner. What's working here is an image on the part of producers, one that maybe is reflected out there in the world. After making the *World of Chemistry* films, 26 one-half-hour films, we tried to raise money for three prime-time specials about chemistry on public television (PBS), and we failed. So you see, the name does not mean too much.

Following the Nobel Prize, maybe there were some raised expectations from the scientific community, and maybe some expectations from myself.

There has been added pressure, but often from interesting opportunities. But I think the pressure has been generated as much from myself going into different fields as it has by the fact of the Nobel Prize and external demands.

You have used an interesting expression "knowledge is permitted" in one of your poems.

The phrase "knowledge is permitted" comes from a poem I wrote about Sor Juana, a Mexican nun. She had to enter a convent in order to do what she wanted, which was to write poetry. I have always been interested in instances in human history where the spirit and knowledge has come through through periods of oppression or suppression, where it can't be expressed. This is what caught me about Sor Juana's story. Aside from her poetry, she wrote an incredible letter (the *Riguesta*) about the history of women speaking in a church. Her story moved me. In an unpublished lecture about Sor Juana, I remembered something from my vaguely leftist teenage years. It was a German hymn from the middle ages with the line "Die Gedanken sind frei" — "the thoughts are free." I think it was initially a Reformation song, a song of freedom. It was then taken up by the German freethinkers, and I learned it in a leftist context, as a protest song. It was also a leading song of the German opposition to the East German regime in the days of 1987–89, as they struggled toward freedom.

I've always been interested in similar situations. For instance, I've written about the Disputations of Barcelona (again, not yet published). These were staged debates between Christians and Jews. There was one in 1240 (I think) in Paris, in 1274 in Barcelona, and then one around 1290 in Tortola. The Church constructed these forced debates about the merits of the religions, between clerics (who were often converted Jews) and Jewish thinkers. The debate in Barcelona was the most free because the king of Aragon, James I, was powerful with respect to the Church at that point and could guarantee the freedom for the Jews to speak their mind. This debate was another monument to human freedom, at a time when free thought was not possible. There are other examples: K.F. Bonhoeffer's and Otto Hahn's statements at the memorial for Fritz Haber in 1935. I've always been interested in those things.

I think "knowing is permitted" in the context of the poem has several meanings. There is curiosity, a desire to know that cannot be stopped. There must be social responsibility. Should one censor knowledge that leads to evil and destruction, and who should censor it? Should one do research on things that are inimical to the fabric of society? Let me give an example.

Should one do research on new weapons or the difference of intelligence between the races? Scientists are doomed to create, and must bear responsibility for their creation.

I have been very interested in communist societies, or the thin overlayer which communist societies place on top of the nations. So I have been interested in Russia, China and Cuba. I spent a year in Moscow in 1960–61. I don't know why I am interested in those things, but I suspect it is mixture of the social justice in socialism and the fight that I have with the suppression of the freedom of expression in these system.

Did you experience any fall-out from the controversy about the theory of resonance while you were in Moscow?

That also has interested me a great deal. The proceedings of the 1951 conference constitute a beautiful case study of survival in the Russian scientific climate of that time. This is the Lysenko period, the last days of Stalin. Given Lysenko's success in questioning Russian scientific authority, in every field, opportunists, politicians, and publicists outside of science created Lysenko-type problems. In chemistry we had the controversy about resonance theory, condemned for being idealist. Really, it was opportunism and politics at work. In this particular case, what was under the table, and not obviously discussed, was that the opportunist attack was not only on the theory of resonance, but it was also an attack on the leadership of Soviet chemistry, on Nesmeyanov, who was then the head of the Academy of Sciences. The attack was led by a scientifically illiterate chemistry professor at the Moscow Military Academy, Chelintsev.

How this could have happened, and how Soviet chemistry was saved, and how Nesmeyanov kept his job, is a fascinating story. Some sacrificial lambs were offered in the persons of Syrkin and Dyatkina, and Vol'kenstein in Leningrad. Syrkin and Dyatkina lost their jobs, which was terribly sad. They still remained influential through their students.

Another reason that the proceedings of the resonance theory conference are interesting is that there are occasional hints, in them, in the printed text, of opposition. There are brave anonymous questions from the floor. It's also very interesting to see who gained ascendance in that period through the stance they took in that controversy. Some people, in fact, made their name through this, by supporting the wrong, irrational side. People such as Tatevskii in Moscow University, and the physical organic chemist Reutov. I have not forgotten them and Russian chemists have not forgotten them either.

Many of the people involved were forced to write abjurations of what they did, in various books. There is a sad episode around the time I was there. Pauling came to Russia. Pauling presented real problems to the Russians, because for political reasons he was a friend, an advocate for peace and disarmament, but here he was suddenly the subject of this controversy and on the "wrong" side. Finally, they invited him for propaganda's sake, and Pauling came. He quite insensitively spent most of his time in Russia criticizing those books by Vol'kenstein and others. Everyone knew why they had written these books, to save their positions. But Pauling just was not sensitive enough to this, and whereas everyone else wanted to forget about those books, eight years later, after the controversy, Pauling criticized those books. But that's the way he is.

I remember that during my stay in Moscow the classic book *Molecular Vibrations*, by Wilson, Decius and Cross came out in Russian translation. The Preface to it, I think, was written by Tatevskii, who somehow managed to drag in the theory of resonance. It was startling to me to see an opportunist keeping things alive nine years after the controversy.

There is no question in my mind that the resonance theory controversy kept young people from theoretical chemistry in Russia for a good number of years, at least for ten years. A country that was in a good shape in theoretical chemistry, was set back. The scientists suffered, which in itself was a great loss; I have already mentioned Syrkin and Dyatkina who both were very, very talented. Young Russians, for a long time after that, did not feel comfortable with theoretical chemistry. People negotiate their progress in those societies within the context of what is allowed and what is not. Sometimes this is done in subtle ways. A young Russian, talented, chemist or physicist, thinking about what he's going to do in his career, trying to decide between theoretical chemistry and something else, would in various unwritten, unspoken ways get the signal that theoretical chemistry was somehow a little dangerous. Even if that young person were doing molecular orbital theory, which was not the subject of the criticism directly, it was probably better to go away and do something safer, like solid state physics. I think that through that mechanism many talented people were lost to theoretical chemistry in Russia.

How about writing poetry. Do you need suffering for writing good poetry?

I think this is a piece of a popular myth, that suffering helps creativity. There is a poem about this subject in my first book, entitled "Admission Price." It is a romantic fallacy that to be creative in art you have to be at the edge of madness. I don't think you have to suffer to write good poetry. Nevertheless, human life is full of suffering, at every level.

Two Fathers

I suppose my stepfather was a good man. It's not
that I didn't like him, he just wasn't my father,
who was a hero. I don't really remember my father.
In photographs there is a man pushing a baby carriage,

a man holding up a laughing child dressed up in
a Carpathian costume. I heard stories from my mother
of how he was hazed as a Jew at Lwów Polytechnic,
I've seen him in Zionist youth group photos

with my mother. I read the notes he made in the camp
on a book on relativity theory, and I've heard
(again from my mother) how they went to Brody,
his first job as a civil engineer being to build

a cobblestoned street there, and how they stayed
in the house of the local priest. My mother sometimes
told these stories with my stepfather there. The war
came, we were in a ghetto, a labor camp, then toward

the end my mother and I were hidden by a Ukrainian
schoolteacher. My father was killed in an attempt
to organize a mass breakout from the camp. I was five
when the news came to us in the Ukrainian's attic,

and I cried, because my mother cried. That's when
my father became a hero, which he was. The war
ended, 80 of 12,000 Jews in our town survived.
In Kraków, where we went in 1945, my mother met

my stepfather, who had lost his wife in the war,
and they married. I was eight, and though my stepfather
tried and took me on carousels, I didn't want him.
Later I built up a theory that my mother remarried

to provide me with a father, not because she liked him.
But friends who knew them say they were in love.
In the U.S. my stepfather didn't try — he was busy
working, first in a luncheonette on Delancey Street,

and when that failed, as a bookkeeper. When he
was angry he raved in his room, then sulked along.
We never made up in our family. Any punishment
(I was too good a child for that) was left to my mother.

My father was talked about all the time, and that
is how my sister, born in Queens, found out she
and I had different fathers. When my stepfather
and I had a fight about my getting married

to a girl who wasn't Jewish (I think he was hurt
by this more than my mother) I told him
he wasn't my father. He died in 1981, and
when I get angry I see that I sulk like him.

Roald Hoffmann, *Gaps and Verges.* University of Central Florida Press, Orlando, 1990, pp. 22–23. Reproduced with permission.

Poetry has been important to me. At the beginning I was just reading, I did not start writing until I was about 40 years old. Right now the poetry is a little slow, and the essays and books are going much better. I just finished a book called *The Same and Not the Same*, which will be out next year. It is intended for the general public and certainly is about chemistry. I sort of like things falling into place; I don't start out writing a book with a theme but I let the theme come together from smaller pieces of writing. Perhaps this comes out of my poetry.

How does your language interact with your writing and how did you change from one language to another.

My first language was Polish. Ukrainian and Yiddish were also around and I knew them early on. Then came German, followed by Hebrew. Children learn quickly and forget quickly. By the time I came at age 11 to the United States, German was my dominant language. Then English took over. I was not very nice to my parents; I made them speak English to me.

Four languages were spoken interchangeably at home: Polish, Yiddish, English and German. Subsequently, I learned two other languages well, Russian and Swedish, for one reason or another, and another language, French, not so well because I studied it only at school. English is, however, the only language in which I can write; it *is* my native language. Native speakers may detect a slight accent in it, and I have just a few small problems in the writing occasionally, such as confusing "like" and "as" and "that" and "which". Occasionally my constructions are a little funny, but English is my language and I love it.

Roald Hofffmann with István and Magdolna Hargittai in 1982, attending a conference on inorganic structural chemistry in Reading, U.K. (photographer unknown).

I think more important to writing may be being an outsider. Knowing several languages puts you a little outside the one language you have been thinking in. You ponder of things more than the native speakers.

Being an outsider comes from being an immigrant in a society. And Jews have been outsiders in other ways. I have also switched subfields of chemistry; I felt like an outsider when I started in organic and in inorganic chemistry. I sort of like that feeling, for you get a different perspective. Coming from the outside, at first it's a little dangerous and difficult but I like the feeling of penetrating the walls built of jargon and custom around a field.

Please tell us something about the TV series you participated in.

Its title is *World of Chemistry*, and it is a series of 26 half-hour TV programs. It's now a few years since we made it. The programs have had a reasonable reception in the United States. Their main use has been at the high school level, although originally the series was designed at the junior college level. The telecourse is best used in conjunction with a good teacher talking before and after the films. On television in the U.S. it's on at various hours, whenever the cable channels put it on. The films have been very successful in Israel and Sweden; they are available for international distribution elsewhere, at some cost. I don't get anything out of the sales. I was paid a fee, but I used the fee essentially to replace my salary from Cornell for the year when this was being done, so I could take a leave. I am happy that we made the series. I was one cog in the wheel, we were six academic people involved, and the others in fact worked harder than I did on this. Television is one set of compromises with the material after another. But what a way to reach people!

We tried subsequently to make a three-hour PBS series, in which I was to play a more leading role as a scientific director, and I wrote the scripts for it. But in the end we were not able to raise enough money for the project (we need $1.35 million) in spite of tremendous help from the American Chemical Society. We were given good support by foundations but not by the chemical industry. I think industry is rather short-sighted in these things, and not only American industry for we tried Japanese and German companies as well. We had to give that up. This is one of my failures.

How about Chemistry Imagined*?*

This book and art exhibit came out of a meeting with a talented artist, Vivian Torrence. I thought that her impressions of chemistry were intriguing and would help form an image of chemistry for the intelligent lay public.

Vivian's work has a kind of intellectual feeling to it, and a deep understanding of art and science. I have always been interested in interaction with artists. I also thought of the book in terms of an audience of people who are around chemists. Something that a chemist would give to a non-chemist friend. Not to teach about chemistry directly, but to tell people of the spirit of what chemists do. I also thought of *Chemistry Imagined* as a work of art and literature. It has a novel construction, a collage of essays, collages and poems. I also thought about its similarity to emblem books of the Renaissance, these marvelous products of the 16th century where there was a motto, a picture, often enigmatic, and some text connected to the picture. As our book evolved, I saw in it also something about shaping interactions between art and science, about moving across borders.

I like to think that everything I do about chemistry shows chemistry as an integrated part of our existence, of the economy, social structures, art, of the intellect, creativity as well as utility.

How much does outside support determine what you do?

This depends. The poetry is a small personal enterprise. I could do it without any support, and I do do it so. Making the TV series would have been just impossible, as it cost $2.3 million. *Chemistry Imagined* is an in-between thing. Probably we would have done it without the $90 000 grant from the National Science Foundation that helped Vivian to live while she was working on the collages. The new book I am doing for Columbia University Press I would have done anyway.

I don't actually expect society to support my interdisciplinary activities very much. Of course, I get a bit unhappy when I reach out for some support and it is denied. There is a book I'm writing with Shira Leibowitz, on science and religion, for which we have failed to get support (despite trying) from the National Endowment for the Humanities and from the National Science Foundation. We have received an advance from the publisher but that's not enough to support Shira and me while we are doing it. So we'll do that project on our own.

Ideally it would be nice to have a Maecenas who would give me $100 000 a year to do what I want to do. I really don't expect that. My greatest disappointment is not that *society* does not support these things, but that *fellow scientists* don't. I am unhappy when scientists don't seem to be interested in a project as unique and original as *Chemistry Imagined.* We were unable to have it reviewed in *Nature* or in *New Scientist* or in *Science*, or in *Scientific American.* This is a failure of the individual chemists (editors) who could have gotten us those reviews. But then I think of the lukewarm reception in the chemistry community to absolute classics, such as Primo Levi's books, when they first appeared.

My poetry books had good sales, the first one (1987) 1,300 and the second one (1990) about 900. That's actually very good for poetry books. But I know that there are 150 000 members of the American Chemical Society out there, that $14.95 is not much money for a book, and I was hoping that a few more would buy them. There are not that many poetry books written by scientists.

How about education as we enter the twenty-first century?

Education is a conservative enterprise, and it does not change very quickly. I think the shift in chemistry education has to come from the recognition of the fact that 99.9% of the population are not going to be chemists. Recognizing this is especially difficult for the Europeans, as chemists there usually teach future chemists, whereas in American universities we teach the masses. The perception of the necessity is clear in the United States and will become clearer as government decreases funding or imposes utility as a criterion for support. We have to educate people outside of chemistry, who become our leaders.

Courses will have to be constructed for the general public. To be responsible citizens people have to make decisions which are, in part, based on technological and scientific issues. Those decisions should not be left to experts. We know very well that experts can be marshaled on any side of an issue that you want. You want to site a dam or an incineration plant and you can get experts who tell you that there is no danger at all and you can get experts who tell you that there is going to be damage from it. There is a role for experts, but the public has to decide by themselves. For this, they need to know a little chemistry.

Gay-Lussac said something, over 150 years ago, about the day when we shall be able to calculate everything in chemistry what we can only measure today. Then I heard people saying that experiments should be replaced by calculations. How do you view this?

Of course, I don't agree. I think chemistry has an infinity of ways of renewing itself. Our wonderful science is the making of molecules and their transformations. I take much heart from physics, which is a more theoretical science than chemistry. Has it been reduced to calculations? While accepting a theoretical ideology, physics remains an experimental science.

I think there are many things that can be calculated, and it is interesting to see how molecular modeling has indeed entered the every-day life of the organic chemist. But I think that chemistry will not be replaced by calculation. Maybe that's in some sense a hope that comes out of being older, but I don't think so.

In fact, I take heart in a perverse way from the fact that the more people calculate, the less they understand. To understand means to me to sense trends and to sense relationships, to know the physical mechanism that make an observable. To understand means to make an order-of-magnitude estimate, to talk qualitatively about the next member of the series. Then to use the computer to get the quantitative numbers.

Let's say that the dipole moment of water is calculated. Then you ask about the dipole moment of methanol or ethanol, and the computational chemist says, wait a minute, I'll have to go back to my computer and calculate it. That's far away from understanding to me.

How to assess excellence in research?

There are no good ways, although there are many criteria that are partially valid. Citation frequency and impact factors have something to do with it, but sometimes they have very little to do with it. Short term citation frequency may not be relevant. Once I was asked to comment by Eugene Garfield on the top ten chemistry articles in 1991 or 1992. They produced for me a list, based on recent citation frequency and *every one* of the top ten was on fullerene chemistry. What they wanted was to spot *the* hot new field. What they spotted was *a* hot new field plus fashion. Twenty years later, you may get a different picture. You probably will be better off in establishing excellence by polling a group of people, including experts and ordinary chemists, workers in the gardens of chemistry. In such an evaluation, of course, you must eliminate self-citation and citations by the same school. In any case, human beings and subjective judgments must be involved in the evaluation.

Finally, what question would you have asked of Roald Hoffmann?

I'm glad you didn't ask the question of what development I foresee in chemistry in 20 years or something like that. That would have been a terrible question. I just don't like to prognosticate. I think chemistry evolves nicely in its own random, chaotic fashion. The new chemistry, certain to be exciting, is also certain to escape my limited foresight.

Kenichi Fukui, 1994 (photograph by I. Hargittai).

16

Kenichi Fukui

Kenichi Fukui (1918–1998) was born in Nara, Japan. He was Director of the Institute for Fundamental Chemistry since its foundation in 1988 till his death. In 1981 Kenichi Fukui and Roald Hoffmann shared the Nobel Prize in Chemistry "for their theories, developed independently, concerning the course of chemical reactions."

Dr. Fukui studied at the Department of Industrial Chemistry of the Kyoto Imperial University from which he graduated in 1941. His first employment was at the fuel factory of the Japanese Imperial Army. He started his academic career in 1943 at the Fuel Chemistry Department of Kyoto Imperial University, rising to the rank of professor in 1951. He published his seminal paper in 1952. Kenichi Fukui became Professor Emeritus of Kyoto University in 1982; served as President of the Kyoto Institute of Technology between 1982–88, was President of the Chemical Society of Japan in 1983–84. Professor Fukui received the Japan Academy Medal in 1962, Person of Cultural Merits in 1981, and the Imperial Honour of the Grand Cordon of the Order of the Rising Sun in 1988, among many other awards. He was Foreign Associate of the National Academy of Sciences (elected in 1981), member of the Japan Academy (1983), member of the Pontifical Academy of Sciences (1985), Foreign Member of the Royal Society (London, 1989), and other learned societies. Fukui was very concerned about the quality of science education in Japan and, as the only Japanese Nobel laureate in chemistry, felt a great responsibility in such matters. When he died, Leo Esaki (Nobel laureate in physics, 1973, and currently President of Tsukuba University) said: "Japan should nurture better scientific researchers so they receive more

Nobel prizes and thus allow Fukui's soul to rest in peace" [*Ashah Evening News*, January 10, 1998].

In July, 1994, I was visiting Japan as guest of the Japan Academy and a visit with Professor Kenichi Fukui had been scheduled. We had agreed on an interview and I had sent him my questions from Budapest. I received the prepared answers during our meeting at the Institute for Fundamental Chemistry. A few additional questions came up during the actual conversation in the Institute and during the following social program, which were duly noted and the answers were delivered while I was still in Kyoto. What follows here contains the entire material and it was also published in *The Chemical Intelligencer.**

Could you please tell us about your schooling and teachers from early ages. How did schools and teachers influence you? How did your interest turn to chemistry?

It was in the middle school that I was impressed by the marvellousness of Nature. I was enthusiastic about outdoor activities as a member of biology-oriented group of pupils, excited by "Souvenirs entomologiques" by Jean Henri Fabre. At the time I was deeply attracted to literary works, particularly of the great Japanese writer, Soseki Natsume. These experiences no doubt affected various selections of mine in subsequent life. As a matter of fact, such as Nature-loving tendency was already rooted in the experiences I had in elementary school and in my rather early years spent in my native place. Later, in addition to these, in my high school years I was influenced by Jules-Henri Poincaré through his triology of science-philosophy.

The reason for the selection of chemistry is not easy to explain, since chemistry was never my favorite branch in my middle school and high school years. Actually, the fact that my respected Fabre had been a genius in chemistry had captured my heart latently. The difficult point of chemistry for me was that it seemed to require memorization to learn it. I preferred more logical character in chemistry. It was a certain Kyoto University professor who alluded to the hidden logical character implicit in the then prevailing chemistry, and I followed his suggestion after all.

---

*This interview was originally published in *The Chemical Intelligencer* **1995**, *1*(2), 14–18 © 1995, Springer-Verlag, New York, Inc.

How important is education in determining a future scientist's activities?

Education is a task in which one influences another through the function of both brains, the mechanism of which is still not yet clear. Therefore, what is an ideal education for producing most active scientists remains an extremely difficult question. Even with the use of auxiliary or subsidiary means in modern education, the direct contact with nature surrounding the pupil or student seems to me absolutely essential. Of course, in the surrounding nature the teacher is included as a part of nature. In this context the role of teacher is extraordinarily important. Also, appreciation of all sorts of beauties and goddesses will cultivate pupils' mind to turn to the pursuit of the truth of Nature.

Even in poor schools, such a good teacher will be available and such a teaching attitude will be applicable, to affect gifted children effectively for their future higher education stages.

What made you prefer theoretical work over experimental?

I originally started as an experimental chemist. I wrote more than 100 papers on experimental chemistry and was awarded a nationwide prize in 1944. Still now, I read papers on experimental studies besides theoretical ones. When I was in charge of supervision in Kyoto University, I assigned experimental work to some of my students as their thesis study. That was because I considered it important even for theoretically-oriented students to experience the complexity of chemical phenomena.

It was by chance that I became interested in the chemistry of hydrocarbons. The title of my thesis for graduation from Kyoto University was the chemical reaction of paraffinic hydrocarbons. The work I was engaged in during the first few years after my graduation was on the chemical conversion of olefinic hydrocarbons. And, finally, my first theoretical paper was on aromatic hydrocarbons. All kinds of hydrocarbons on parade!

It was a very important fact that I had learned quantum mechanics by myself up till then. For me it was natural to attempt to apply quantum mechanics to explain the mode of occurrence of some chemical reactions that had not been sufficiently interpreted theoretically. Aromatic hydrocarbons were just the material.

That was in 1952. After that, my own interest and involvement in theoretical work surpassed that in experimental work. The reason why I had studied quantum mechanics since my entrance into Kyoto University was partly due to my instinctive concern and partly due to the suggestion of a professor.

Who was that professor and how did you choose your subject of research?

The selection of my original research theme was almost determined at the instant of my selection of chemistry. Excited by the suggestion of the professor of applied chemistry, Professor Genitsu Kita, I entered his laboratory. His suggestion was that chemistry, then having strongly empirical nature, would become a logical science in the future. The theme given to me was the chemistry of hydrocarbons. Rather by instinct, I was strongly attracted by their chemical reactions. At that time their unique chemical behavior was not yet sufficiently interpreted theoretically. Therefore, in my case the original selection of research subject was a combination of chance and intuition. Further developments have been executed in due course.

Was the engineering environment the best one for your studies?

The Japanese university system is rather unusual, and in every large state university the faculty of engineering is larger than the faculty of science. The system of big engineering schools had long been a strange characteristic practice since the Meiji era. In Kyoto University, too, there were six chemistry departments in the engineering faculty, while there was only one chemistry department in the science faculty. In my student years, in 1938–41, however, we had the Department of Industrial Chemistry, which was the only department of chemistry in the Faculty of Engineering and larger than the Department of Chemistry in the science faculty. However, we attended the same lectures in fundamental chemistry together with students of pure chemistry. The pure chemistry students had, of course, more chances to listen to lectures of physics or mathematics than applied chemistry. I suppose that such a unique university system is a symbol of a later-developed nation. I had to overcome these disadvantages by self-taught study. The department I chose was industrial chemistry, not chemical engineering. The reason that I chose applied chemistry was very simple. I just followed Professor Kita's suggestion.

Professor Kita was the brother of my father's aunt, and had served as guarantor for my father when he was a college student in Tokyo. Professor Kita suggested to my father that his school would be most suitable for me to study, when my father told the professor that I liked mathematics and physics more than chemistry! Accordingly, the benefit which I might receive by studying practical science was absolutely not in my heart. As a matter of fact, Professor Kita often encouraged his students to study hard fundamental sciences rather than practical sciences.

How did you choose, eventually, your own students?

During the years many students and co-workers joined in our group. They were fully familiar with what was going on there. They were all those who wanted to take part in our work. I have regarded their wishes in selecting research themes for them as much as possible.

How do you view the future of computational chemistry?

It cannot be denied that the empirical character of chemistry changed because of recent unexpected progress in instruments for physical measurements and in theoretical chemistry, particularly in quantum-mechanical calculations, using high-speed computers. However, there exist, for instance, in mathematics a number of logical problems that cannot be solved practically on account of too long machine-time required. The complexity of chemistry expands without cease. The characteristic feature of chemistry lies in its complexity, which surpasses what can be achieved by logical deduction.

Please tell us about your relationship with Roald Hoffmann.

It was in 1964 at a scientific meeting in Florida that I met Roald for the first time. He was then a young researcher. It was four years after his marriage and two years after he got his Ph.D. However, he was well known as a theoretician by his work for his doctor thesis on a new method of theoretical calculation for molecules. Soon after, the historic papers in collaboration with R. B. Woodward were published. These papers, published in *JACS* in 1965, which became noted afterwards as the original papers of the so-called Woodward-Hoffmann rules, attracted my attention, since these papers pointed out that the direction of a chemical reaction was controlled by the phase ("symmetry") of *particular* MOs (molecular orbitals). In 1964, I had learned that the mode of occurrence of a type of chemical reactions was correlated with the phase of the particular MOs, the role of which in chemical reactions I had noticed since 1952. In this way, he and I shared our fate to develop our common interest in the same direction. He accepted two postdocs from my laboratory, and they are now active in Japan as excellent professors.

In every sense, our mutual relation has continuously been intimate. Even in the past 6 years, he visited our Institute twice and stayed for a while here. All of the members of the Institute enjoyed having Roald visit.

How did the Nobel Prize change your life?

Japan has only very few Nobel Prize winners. As the first awardee in chemistry, I would have become a "darling of the fortune" in journalism. The situation that Yukawa died just before my receiving the prize and Esaki did not live in Japan at that time, gave impetus to such circumstances. Needless to say, my academic life would have received disturbance.

However, fortunately or unfortunately, it was the time of my retirement from Kyoto University by the age-limit system applied to every professor without exception. I left the place where I had held an academic position for 39 years. It was in the year after receiving the prize.

Kenichi Fukui in a restaurant (photograph by I. Hargittai).

A privilege to me was that thoughtful people in Japanese industries built a splendid institute, this Institute for Fundamental Chemistry, for me, and I am able to continue my research in this wonderful site of science.

The big prize did not very much change my private life. Every day I walk from my house to the institute with the same suits, and have supper with my wife sharing a bottle of beer, after coming back home.

What visits to foreign countries and meetings with foreign scientists have influenced you?

I have visited so far more than 25 countries since 1963, attending many scientific meetings and lecturing in many places. For instance, in 1973 I visited 13 states and 18 universities and gave 20 seminars or lectures during my stay in the USA for 55 days. It is almost impossible to specify which was the most influential. All of them were impressive to me, since I had not experienced a long stay abroad during my studies.

How much did you try to communicate with the non-chemist public about your chemistry or about science in general?

I have rather not been eager to communicate with the non-chemist or non-scientist public about my professional work. This is because my field of chemistry is not suited to amateurs, and because it takes time. However, it is often very difficult to refuse a rush of requests to give talks to the public about chemistry or science in general. So, actually, I am trying to satisfy these requests as much as I can.

What is the perception of fundamental science in Japan? Is it supported adequately?

As is known worldwide, Japan has tried to catch up with the western countries since the beginning of this century by importing sciences from them. Now, however, people have learned the importance of promoting fundamental sciences. The materialization of our Institute for Fundamental Chemistry is one such instance. The financial support for fundamental sciences is still much poorer in comparison with that in the United States. However, it is gradually increasing, in contrast with the decreases often seen these last years in some western countries.

How is the image of chemistry in Japan?

View of the Institute for Fundamental Chemistry (courtesy of Kenichi Fukui).

Not much different from that in other countries. Chemistry and physics are both becoming not agreeable branches of study for the majority of pupils in high schools as well as in middle schools. Such a trend may come, in my opinion, from the human nature. Man instinctively hates a tremendous package of knowledge without limit that is required for study in the course of modern scientific education, which may exceed man's natural capacity. However, in order to cultivate the frontier of chemistry or basic sciences, we need only a fraction of selected students who love science from the bottom of their heart. The general public can utilize the results of science and technology, although they cannot necessarily create them.

How do you suggest to determine excellence in research?

The evaluation of achievements is always a difficult task in science policy. Truly original works are frequently disregarded by the majority of scientists. Obviously, however, hardly understandable works are not always good ones. Citation is no doubt an index that may be used to judge the value of research, but it is apt to be dominated by fashion and current, so that the number of citations is strongly dependent upon the field. The judgement of specialists may generally be of significance, while in some cases the intuition or farsightedness of talented non-specialists may be more reliable than the judgement of the mediocre majority in the field.

How much disadvantage is there for non-English-speaking scientists to get their papers published in the best journals and to get recognition for their work?

Nowadays, the circulation of a journal is one of decisive concerns for the contributors. Non-English journals, even those written in other European languages, are not favored by the international community of scientists. Therefore, the disadvantage for non-English-speaking scientists like myself is obvious. Additional efforts are usually required to get public recognition for their work. Evidently, this is a big handicap! It should be got over through international intercourses of all sorts.

How do you view the separation of the sciences and the humanities in our times? How can we decrease the gap?

The distances among human, social, and natural sciences are becoming shorter and shorter. This comes from both progress and necessity. The former reason is evident. The progress of physiology and medicine correlates with new methods for psychology and the science of the human mind. The mathematical and computational progress in statistics greatly influences economics, sociology, and politics. Physicochemical sciences promote the progress of archaeology. The examples are numerous.

The latter reason is more important. It comes from the self-accelerating character of science and technology recently becoming appreciable. Science produces new technology, while the results of new technology promote the progress of science. The human native desire intervenes between them. The mutual acceleration of science and technology in this way brought about conveniences and comfort to human life. On the other hand, the uncontrolled development produced serious problems. Nature on the earth and environment of the earth were seriously changed, and terrible inequality in the distribution of civilization on the earth was caused. To affect these circumstances by controlling the perpetual desire of human beings for growths and advance, the cooperation of natural, social, and human sciences is a necessary prerequisite.

How should we be getting prepared for the next century in education?

The generations living in the next century must carry the burden to ameliorate the conditions in which the earth is presently put. For this purpose, they have to make the level of science much higher and apply it to amend the situation, and simultaneously they have to contribute to harmonize the human mind between tolerance and self-restraint. All of these aims can only be expected to be achieved by appropriate education. The importance of education in the future is immense, and the timing is imminent.

In order for the future education not to be hated by subsequent generations, we need some techniques. The package-knowledge type teaching is evidently inadequate.

Please, tell us about your interest and activities outside chemistry.

I was originally an experimental chemist, much interested in organic chemistry, and engaged in some periods in chemical engineering. So, I am not indifferent to industry. To the contrary, I once indulged myself in physics, particularly in theoretical physics in my youth. In my middle school ages biology and literature were my favorite subjects. So my field of interest is rather wide, although my activities outside chemistry were limited.

I understand that you are very much involved with the celebration of the 1200-year anniversary of Kyoto.

In July 1984, the Heiankyo 1200th Anniversary Memorial Foundation was established and the Chairman was Professor Emeritus of Kyoto University Takeo Kuwabara, a noted man of French letters, who was Person of Cultural Merit and Honorary Citizen of Kyoto City. Unfortunately, he suddenly passed away in April 1988. The Governor of Kyoto Prefecture and the Lord Mayor of Kyoto City, who were both Vice-Chairmen of the foundation, asked me to be the Chairman, and it seemed to me extremely difficult to refuse it in the circumstances we then had.

What would be the question you would like to be asked?

I have already mentioned in the beginning of this conversation that various experiences in my early ages affected my life, and how they actually occurred to me. That I was a lover of Nature — forests and rivers, plants and insects — in my childhood, presumably had some connection with my selection of science as major in my schooling.

That I was fond of reading Soseki's books as teenager happened to possess a closer connection with my selection of a research theme much later. As I mentioned before, the chemistry of hydrocarbons became my speciality after my graduation from the university. The then prevailing theories for chemical reactions were not convenient for explaining those of hydrocarbons. Several papers tried, but they did not satisfy my mind's pursuit of a more *natural* theory, since in my memory there remained a story of dream by Soseki, which had captured my mind. An ancient famous sculptor's art was so free that it seemed as if he only had dug up statues that had originally been buried in timber. Reminded of this story by other scientists' papers which seemed to me not so natural, I decided to search for a more *natural* theory. That was the beginning of my first paper on the quantum mechanical interpretation of chemical reactions of hydrocarbons.

Milton Orchin, 1997 (photograph by I. Hargittai).

17

Milton Orchin

Milton Orchin (b. 1914 in Barnesboro, Pennsylvania) is Emeritus Distinguished Service Professor of Chemistry and Director of the H. S. Green Laboratory of Catalysis at the University of Cincinnati. He received his degrees from Ohio State University (Ph.D. 1939). He was the first graduate student of the late M. S. Newman. After having worked with the U.S. Food and Drug Administration and the U.S. Bureau of Mines, he has been with the University of Cincinnati since 1953. His primary research interest has been in homogeneous catalysis by transition metal complexes but he is best known, worldwide, for his symmetry in chemistry books co-authored with the late H. H. Jaffé. We recorded our conversation in Dr. Orchin's office on May 13, 1997 and the text was finalized in August 1998.

Your interest in symmetry started very early and your book was one of the first about symmetry in chemistry.

The first book that Jaffé and I published was on ultraviolet spectroscopy and that's where my interest started. I came to the University of Cincinnati from the United States Bureau of Mines in Pittsburgh, working on synthetic fuels. I went there before the American entry into the War, in 1941. The Government was concerned with the short supplies of transportation fuels. They were afraid of a German blockade and were looking for sure, indigenous supplies of transportation fuels.

There is coal in 36 of the 50 states and there was a lot of political backing for trying to convert coal to liquid fuels like the Germans had been doing. There was a Synthetic Fuels Act passed by Congress and that set up a program in the Bureau of Mines. The technologies were the Fischer-Tropsch process and the direct high-pressure hydrogenation, called the Bergius process.

I had a position at the Bureau from the start of that program, being in charge of the organic chemistry part. We wanted to find out the composition of the fuel, as well as the mechanism of the conversion from coal to liquid products. In the Fischer-Tropsch reaction, carbon monoxide is reduced with hydrogen producing CH_2-groups which polymerize to hydrocarbon fuels. Then here was another program of converting oil shale to shale oil in Wyoming which was a thermal process.

One of the identification techniques for polyaromatic compounds is by ultraviolet absorption spectroscopy and I got interested in it. Eventually, I got together a book with Gus Friedel, the head of analytical chemistry at the Bureau, on the ultraviolet spectra of aromatic compounds. This was the first compilation of spectra of aromatic compounds in a systematic way. There was also a substantial introduction to the book explaining the origin of ultraviolet spectra.

When I came to the University of Cincinnati I was still interested in ultraviolet spectroscopy but I recognized that I didn't know much of the theory. A plot of the ultraviolet spectrum of a compound involves a measure of the energy, usually as the abscissa, plotted as a function of the intensity of absorption at each particular wavelength, the ordinate. This is the energy required to promote an electron from one occupied molecular orbital to a higher energy empty orbital, while the intensity depends on selection rules. The energy requirements involve knowledge of molecular orbital theory while selection rules are dependent on symmetry considerations. I didn't know much about either molecular orbital theory or applications of symmetry such as group theory. Fortunately for me, Hans Jaffé was at this Department and he was an outstanding theoretical chemist and had a broad approach to chemistry. I decided I needed to be educated and we began a collaboration. Our first book was *Theory and Applications of Ultraviolet Spectroscopy* published in 1962.

As I started getting involved in symmetry, I started to appreciate, perhaps in the same way you did, the tremendous applicability of symmetry considerations not only to chemistry but to almost everything. Our next book was *Symmetry in Chemistry* and that came out about the same time that Cotton's book on group theory came out. Then the next was our little book on antibonding orbitals. Both these books became very popular.

The ultimate condensation of all the material was in a text book, M. Orchin/ H. H. Jaffé, *Symmetry, Orbitals, and Spectra* (*S.O.S.*), Wiley-Interscience, New York, 1971, intended for class room teaching. We wanted to make it simple for an organic chemist without doing injustice to theoretical chemistry. When we developed our concepts, Jaffé made it correct and I made it understandable. That was our division of responsibility.

The book sold about 10 thousand copies and was translated into Chinese, Russian, Spanish, German. It was designed for the first year American graduate students. We have a quarter system here and it was a 10 or 11 week course. All graduate students in all chemistry fields used to start with this course. It was intended to integrate all graduate students of the most diverse backgrounds into a common background of understanding. All fields benefited from this and the advanced courses in each subdiscipline did not have to spend time explaining the basic concepts.

This was implemented when I was Head of the Department and I had more clout. Now I have been Emeritus since 1981 and this common introduction of graduate students through symmetry has become victim of conflicting territorial interests.

I retired in 1981 because my wife was getting Parkinson disease, and I didn't want my caring for her competing with other responsibilities. So I kept working but not at the same level as before. She died five years later, and after she had died, I came back to concentrate more on my work.

You have Chaim Weizmann, David Ben-Gurion, Golda Meir on your wall in this office and a Coca Cola ad in Hebrew.

I get a kick out of the Hebrew ad of Coca Cola. I collected it when I went to Israel before it was Israel, in 1947. The United Nations partition decision was voted in on November 30, 1947. I was in Palestine then, working with Chaim Weizmann, not directly with him though because he was very busy but with his sister, Anna Weizmann who was also a chemist and with Ernst David Bergman who was a close advisor to David Ben Gurion. I had a Guggenheim Fellowship which gives complete freedom of what you want to do. My wife was a very strong Zionist and she felt very bad about returning during the disturbances but I was at the Bureau of Mines and I had a strong commitment.

You were 30 in 1944. How much did you know at that time about what today we refer to as the Holocaust?

I was largely overtly unaware of that. We had heard these stories and didn't know how much credence to put in them. It may even be that I didn't hear it because I didn't really very much want to hear it. I think I was very typical of the Jewish community. I have changed over the years and have become much more Jewish, mainly because of my wife who had a much stronger Jewish background than I did. It is partly because of the increasing recognition of that period of all the destruction and then the creation of Israel in my lifetime. To be alive and witness the resurrection. I am not a religious Jew but I am a very affirmative, committed Jew.

When there is a crisis in Israel today you are concerned about what you can do.

I am.

But in 1943, 1944, this was not the case.

Absolutely not. The great upsurge of Jewish identity came in America with the '67 War. That's when there was a fundamental change in the attitude of American Jews.

Do you ever look back to 1944 and think we should have done something?

Yes, I do, many people do. There is also a lot of finger-pointing. At that time the Jewish leadership did not even make a special plea.

Britain took in a lot of children when the tragedy was approaching. This was not characteristic of U.S. Immigration.

On the contrary. They turned the ships away, and did so without much protest from the Jewish community. I ascribe this to three factors. First, there was a lack of information about it. Secondly, they were sold on the idea that it was more important winning the war than saving the Jews. Thirdly, they didn't want to believe the terrible things that were emerging. The Jews were not terribly forward. They didn't want to risk anti-Semitism.

Did you know the famous organometal chemist Morris Kharasch at the University of Chicago? He was Jewish and had Jewish students and would tell them to change their names.

I got my Bachelor's degree at Ohio State University in 1936 and wanted to go to graduate school some place else. I applied to all the big ten state universities and also Northwestern University for an assistantship and got turned down by all of them. I was number one in my class of 170 students at Ohio State and couldn't get an assistantship. So I went back to Ohio State and applied there, too, but the Department Head told me, I can't encourage you, and it is for your own good. You won't be able to get a job because of your being Jewish. What's the point in studying? So he didn't give me an assistantship. But he said, I recognize your achievement and recommend you for something that doesn't require any contacts with students, a scholarship. I got an OSU scholarship which was a big honor but I would have preferred to teach. At that time Melvin Newman came to Ohio State University. He became my mentor. I was his first graduate student. He was a wonderful person and a very good chemist.

Melvin Newman around 1960 (from the *Proceedings of The Robert A. Welch Foundation Conferences on Chemical Research*, Houston, Texas, 1960).

When I finished the year and I got my Master's degree, I came back to the department head and said, now I would like to be an assistant, do some teaching. But he couldn't give it to me so I quit school, and returned home to Cleveland. It wasn't really my home, and I didn't really have a home, I was raised in an orphanage. In Cleveland I looked all summer for a job, I would go to any company that had the word chemical in its name but couldn't find a job.

Does Orchin sound Jewish?

No, but I look Jewish. It was also the deep Depression years and there weren't any jobs anyway. I finally went to the Cleveland Clinic and I told them I would work for nothing just to keep my hands in chemistry. They took me and paid my car fare as salary.

When the academic year came in September, I said, what's the point, I don't earn money anyway, I might as well go back to school, and I did. I conferred with Dr. Newman and said to him, Dr. Newman, I'm ready to eat crow if you can get me back my scholarship. He did and I finished another year and I came back again asking for a teaching job but he couldn't get me an assistantship either. Instead, I got an appointment as University Fellow which was even more prestigious than the University Scholar.

This story may give you some impression about those times. There was this feeling among American Jews that always there was a threat of anti-Semitism, and you wouldn't like to make a fuss although you may be shot anyway.

When did this change?

Towards the end of and after the War, mostly as a matter of absence of manpower. They needed people badly. That's also when Truman issued the Executive Order about integration of the blacks into the armed forces. Then it was interrupted by the McCarthy era.

Interrupted?

That surge towards integration was side-tracked because of the anti-Communist scare. Then with Kennedy and Johnson there was a big emphasis on civil rights.

But back in 1956 I remember I got a letter from the University of Florida to be distributed among our Seniors because they were looking for graduate students. On their paper they had this question about race. At that time I was Head of the Department and I thought that question was outrageous. So I wrote back saying that I am not going to circulate your request for students because you ask what the race is and I think it is inappropriate and can have nothing to do with qualifications. Well, I got a stinging rebuke back. That just gives you the setting we had at that time. At that time, they were asking this question to exclude people.

Today, there is a different approach but this has become nonsensical. I am not much involved with hiring but whenever I hire a postdoctoral fellow I have to deal with a lot of nonsense. When I got the latest questionnaire, I called up this office, and I asked, suppose Tiger Wood would apply, how would you classify him? They said, you ask him what he is, what does he want to be classified as? But this is only part of it. When you fill out the form, you have to decide what your perception of that person is. He may claim that he is black, or he may claim he has a surname that is Hispanic, but you don't view him as a Hispanic.

You don't have to make measurements.

Well, you got the idea.

I see some more photos on the wall.

Melvin Newman is there. Then there is Henry Storch who was my boss at the Bureau of Mines. He was the head of all the scientific effort for converting coal into fuel. He was my idol, an ideal research director. He believed in the people who worked with him. He was a liberal in the old sense. He believed that every person could work up to their ability if they have the proper environment. He was educated at the City College of New York, and he never forgot that he went to school free. Then there is also Hans Jaffé from the time we were working together on our first book. Then there is Martin Luther King who demonstrated the greatness of the American system by achieving change without advocating violence.

Hans H. Jaffé (1919–1989) (courtesy of Milton Orchin).

F. Albert Cotton, 1996 (photograph by I. Hargittai).

18

F. Albert Cotton

F. Albert Cotton (b. 1930) is W. T. Doherty-Welch Foundation Distinguished Professor of Chemistry and Director of the Laboratory for Molecular Structure and Bonding at Texas A&M University. He has produced 100 Ph.D.s and has had more than 130 postdoctoral and other senior coworkers. His research publications number over 1270. His most well known textbooks are *Advanced Inorganic Chemistry* with G. Wilkinson (over 500 000 copies sold) and *Chemical Applications of Group Theory*. There have been 30 foreign language editions of his books. His many distinctions include memberships of the National Academy of Sciences, the American Academy of Arts and Sciences, the Royal Society (London), the French Academy of Sciences, the Russian Academy of Sciences, and the Academia Europaea. He received the first ACS Award in Inorganic Chemistry (1962); the National Medal of Science (1982); the Award in Chemical Sciences of the National Academy of Sciences (1990); the Robert A. Welch Award in Chemistry (1994); the first F. A. Cotton Medal (1995); and many others, one of the latest being the Pristley Medal of the American Chemical Society in 1998. Our conversation was recorded in Professor Cotton's office on March 8, 1996 and it appeared in *The Chemical Intelligencer*.*

*This is a slightly modified version of the interview originally published in *The Chemical Intelligencer* **1997**, *3*(2), 14–21 © 1997, Springer-Verlag, New York, Inc.

Let's start with your early years.

I was born in Philadelphia. Two years after my birth my father died and my mother was left to raise me. It was tough during the great depression, but she was absolutely marvelous about it, and I had a very happy childhood. I grew up in a normal environment with kids on our street that I played with. I was always a problem in school for disciplinary reasons. School was too easy to get me engaged, but that changed when I got to College. I scared my mother to death because for a while I was determined to become a jazz guitarist, but I outgrew that at about the age of 17 when I entered college.

What attracted you to chemistry?

I must have been born to be a chemist; there was no epiphany. At about eight I began playing with chemistry sets, haunting book stores for chemistry books. I well remember, in about 5th grade, sitting in the Principal's office where I'd been sent for disciplinary reasons and passing my time reading a chemistry book.

Do you remember any teacher who was a great influence?

Absolutely the reverse. The chemistry teacher I had in high school was so embarrassed because she knew that I knew more than she did that she let me play around in the prep room so long as I took the tests.

What does the F. in your name stand for?

Frank, but I prefer to be called Albert or Al. If you really want to know, I operate under a false name. I was christened Frank Abbott Cotton. Frank Abbott was the name of the family doctor who delivered me and he was a friend of my parents. My father's name was Albert Cotton, and when he died, my mother suggested that I might want to be known as Albert. I agreed and it's now my legal name by usage. I had no siblings. My mother died at the age of 91, four years ago.

So she witnessed your success.

Oh, yes. When I received the National Medal of Science, I took her to Washington and she was in the White House. She was a great admirer of President Reagan, and when she saw President Reagan shaking hands with her son, it was a great day for her. For me as well.

Albert Cotton and his wife, Diane Cotton with the John Scott Medal in 1997 (courtesy of Albert Cotton).

Your present family?

I married at the tender age of 29 (to someone who was well worth waiting for). We have two daughters. One of them is rather independent and we don't see a great deal of her. The older daughter, Jennifer, is just a jewel, and she now manages our ranch for us.

How did your career begin?

I went directly from my Ph.D. at Harvard in 1955 to MIT, where I became a full professor in 1961. After two years as an instructor I became an Assistant Professor in 1957, Associate Professor in 1960 and then had an offer of Associate Professor with tenure at Harvard.

Ever since I left Harvard, it had been an unquestioned life goal to be a Professor at Harvard. So when I first received the offer, my immediate reaction was, of course, I'd go. We just had to work out the details. Well, the details didn't work out quite as well as I'd hoped and I was realistic enough to say, look, I'm not chasing dreams, I'm chasing realities and the reality is that I might be better off staying at MIT. I had better facilities there and an atmosphere in which I felt more comfortable. On top of this, MIT certainly wanted to keep me and they offered me a Full Professorship if I stayed. So I went back to Frank Westheimer who was Chairman at Harvard, and I said, Frank, MIT has offered to make me a Full Professor. He said, a little reluctantly I thought, that "we will do that, too." Then it was really weighing two offers which both carried the same rank. I am basically not a romantic, and I thought I would be better off at MIT. I have no way of knowing whether I made the right decision or not but it certainly was not possible to lose either way.

In mid-career you came to Texas A&M from MIT. What was your main motivation?

There were a lot of forces, some repulsive and some attractive, and some were personal, things about a lifestyle one could have here, which MIT could do nothing about; for example, being able to have a large amount of land and enjoy outdoor activities. I had been very active in the Norfolk Hunt Club, where I was Master of Fox-hounds, and enjoyed that thoroughly. During my last years at MIT, I spent about 20 hours a week on a horse. But the development of the suburbs and even the exurbs of Boston into bedroom communities for Boston led to the closing out of that sort of activity. There were many other reasons. One of the repulsive ones was that MIT was extremely dominated by political liberals in those days, and I've always been a political conservative. Although it didn't prevent me from doing my work, it irritated me.

You were spending 20 hours a week on horseback and you've also been one of the most prolific chemists. How do you manage your time?

In addition, never in my life have I been able to function properly with less than seven hours sleep at night. I don't waste time, that's for sure. My subconscious mind is always going. When I say I sleep seven hours a night, it rather means that I am in bed seven hours a night but very often wake up for an hour or two in the middle of the night and think about problems because I care about them. The next morning I come in and fire orders to people about what we need to do. Also, I am linguistically endowed; I can write things very rapidly that require little or no subsequent revision.

Do you drive your students hard?

No, but I make it very clear to them that they can make me happy only if they work very hard and produce. I do not go around and see them often in the lab, but I tell them that they're free to see me any time. There is nothing higher on my agenda than talking about their work when they feel they need to talk to me. I've always had postdoctorals in my laboratory and they not only do their own work but they are there to work with the group. They teach the students the basic techniques. My group over the years has fluctuated between 10 and 25 with a 1:2 postdoc/student ratio. I myself stopped doing lab work about the time I became an Associate Professor at MIT in 1960.

Your theoretical capabilities are unusual for a synthetic chemist.

My research director for the Ph.D. was Geoffrey Wilkinson who taught me a lot about preparative chemistry, including the fact that making new compounds is the most exciting part of inorganic chemistry. However, I was also more mathematically inclined and my best friends in graduate school were mostly physical and theoretical chemists, working for Bright Wilson or Bill Moffitt. Through them I got an informal introduction to theoretical chemistry and especially symmetry and group theory. This began in my second year of graduate work. I remained so imbued with the idea that symmetry analysis of electronic structures and spectroscopy problems

Professor Sir Geoffrey Wilkinson (1921–1996) of Imperial College, London, shared the Nobel Prize in Chemistry in 1973 with Ernst Otto Fischer of Munich "for their pioneering work, performed independently, on the chemistry of the organometallic so-called sandwich compounds." While being an Assistant Professor at Harvard University, he was Albert Cotton's research director. Later they co-authored the highly successful text, *Advanced Inorganic Chemistry*. [See also, Malcolm H. Chisolm's remembering Geoffrey Wilkinson, *The Chemical Intelligencer* **1997**, *3*(3), 50–51 and 54.]

was important that when I went to MIT I set up a course in group theory and symmetry, which ultimately led to the book *Chemical Applications of Group Theory*.

My enthusiasm for symmetry-based theory was strongly enhanced in 1954 when I spent the Spring in Copenhagen with Geoff Wilkinson, who had recently acquired a Danish wife. This was at the time when C. J. Ballhausen and C. K. Jørgenson were graduate students in J. Bjerrum's lab and they, as well as Leslie Orgel in England and Schläfer in Frankfurt were rediscovering, as chemists, the prewar work of Bethe and Van Vleck on crystal field theory. Because I was already conversant with group-theoretical arguments I became one of the earliest converts to what is now called ligand field theory. A great deal of my research in my early days at MIT, say from 1955 into the early 1960s, involved the study of spectroscopic and magnetic properties of tetrahedral complexes, which I interpreted by ligand field rather than valence bond methods.

I've heard that you tell your students it can take the same amount of effort to solve an umimportant problem as an important one. How do you decide about the importance of a problem?

Well, first let me say that what you've heard is true. It not only takes as much effort but also as much money, and grant money is scarce these days. If at the end of several years you expect to get your grant reviewed, the reviewers will have to be convinced you've done something significant.

More important, though, is that if you have the ability to do research and you want the respect of other researchers, you need to discover things that don't so much close a door as open a new one — or many. With such criteria in mind, before you plunge into something you should consciously ask yourself whether it will, in fact, be worth the effort. You can't tell for sure, obviously, but at least look before you leap.

What do you consider your most important achievement in chemistry?

It is the discovery of how important the formation of direct metal-metal bonds is for transition metal chemistry. I began to think about metal-metal bonds, or the scarcity of them, in my very first years as an Instructor at MIT. As early as 1958, I had begun thinking of why there were no compounds with direct metal-metal bonds. In fact, there were a very few, but they looked like anomalies. By the early 1960s, I was really thinking about this quite a lot. By a strange coincidence, a sort of serendipitous intersection of different interests, I got into actual experimental work on metal-metal bonding because of a remark in Leslie Orgel's little book on ligand field theory. He had noticed that there were compounds with the general formula (alkali metal)$ReCl_4$ that were diamagnetic. Of course, you have four electrons on the rhenium(III) ion and he said, maybe, just maybe, this could be an example of a low-spin tetrahedral complex. There was none known up until that time and there has never been one recognized as such. But I thought, let's check that out, and make this compound and see if it has tetrahedral $ReCl_4$ ions in it. It didn't; it had triangular metal clusters with Re=Re double bonds. These were the first metal-metal double bonds ever recognized. Then, in our efforts to pursue that chemistry further, we came across the fact that under certain conditions, when we reduced perrhenate to rhenium(III) in the presence of HCl, instead of getting a deep red solution we got a deep blue solution. So we got the blue material out of solution and found that it contained the $Re_2Cl_8^{2-}$ ion with an incredibly short distance between the two big rhenium atoms, only 2.24 Å apart. The eclipsed configuration was also a stunner. I asked myself, why doesn't this silly molecule have the brains of, say, the hexachloroethane molecule and rotate half way around? But it doesn't. That's when I tried to think of an explanation. Because I was one of the very few inorganic chemists at that time who was used to thinking in terms of the symmetry properties of the orbitals, I recognized that you could have a delta bond which would restrict the rotation. I also realized that here I was with actual experimental work to do on metal-metal bonds of totally new types, so my old dream of working on metal-metal bonds had suddenly become reality.

This was a field to be pursued. I also realized that if rhenium were to be the only element that does this, people would properly say, well, that's a nice little piece of work to put on the shelf and admire, but what does it have to do with the broad ongoing development of chemistry? So I wanted to see if it branches out to other transition elements. Indeed, it has over the years branched out to almost every one of them. The only remaining exception now is manganese, but as of this date I think we may know how to make one.

I like to refer to this field as non-Wernerian transition metal chemistry. I have read very carefully Werner's books. The idea of direct metal-metal bonding never occurred to him. Wernerian coordination chemistry or transition metal chemistry is based on the idea that all the bonds are between metal atoms and ligands, or ligands and ligands, but never is the concept of metal to metal bonding addressed in his books. And it wasn't for many,

many years. Even when I got to that point there were no more than a handful of compounds in the literature that were recognized to have direct transition metal to transition metal bonds. People tended to be either totally ignorant of their existence, or, if they looked at them or thought about them, they would say, yes they're very curious compounds, but let's get back to the main stream. To me also they were very curious, but I felt that Nature normally does not work in such a way that it does something only once and never again.

How do you put your work and the Russians' pioneering work in perspective?

The perspective, I should say, is quite clearly this: the Russians had made a few such compounds before I did. They had done little crystallography. They did know the structure of the $Re_2Cl_8^{2-}$ ion, although they did not have it formulated correctly. They had the wrong oxidation state, and they did not understand the bonding.

Were you, in any way, influenced by the Russians' work?

No, because we only discovered the early Russian literature in this area (apart from their structure of "$Re_2Cl_8^{4-}$") after we had independently made the $Re_2Cl_8^{2-}$ ion, done the structure of the potassium salt, and I had proposed the quadruple bond. Then, of course, I went back into the literature to make sure that it was not already there somewhere. We then discovered at least half a dozen papers from the Russian literature of the 1950s, extending back as much as ten years before our work, in which they had been making strange low-valent rhenium compounds. It became clear to us that these probably were compounds with at least triple and maybe quadruple bonds, and that's indeed what they are. Some of them were bogus, but many of them were quite real. The Russians just didn't have any conception of what they had.

A direct metal-metal bond looks like the beginning of a cluster.

That's true. There are of course metal-metal bonds and metal atom clusters in the realm of the carbonyl and organometallic chemistry of the transition metals, but my choice of clusters has been particularly those with no carbonyls or cyclopentadienyls, because those were already known.

I wanted to know if you'd have metal-metal bonded species in chemistry that could even be aqueous. One of the big events in my career has been my friendship with Zvi Dori in Israel because he and other people over there had just began looking at lower-valent molybdenum chemistry and had the idea that trinuclear clusters were important. Then Zvi and I got together along with Avi Bino who today is Professor of Inorganic Chemistry at The Hebrew University of Jerusalem. Zvi and Avi have both spent years here at A&M where we all worked together. Of course we've worked separately as well. We have shown, for example, that much of the ordinary chemistry of molybdenum in its lower oxidation state is cluster chemistry. It's amazing to me that over all the years before, nobody seemed to be aware of this other world of transition metal chemistry and limited themselves to the world of Werner transition metal chemistry. The realm of non-Wernerian transition metal chemistry is very extensive, full of new molecules and new bonds. It's been exciting, like being the first one on a new continent.

Albert Cotton with Zvi Dori and Avi Bino in 1980 (courtesy of Albert Cotton).

In addition to metal-metal chemistry, what else would you like to mention?

I've had a lot of fun with other things. As a graduate student I became interested in metal carbonyls. One of the things that spurred my interest in applications of group theory occurred in Wilkinson's lab during my second year of graduate work. He and other graduate students were making compounds such as cyclopentadienylmetal di-, tri-, and tetracarbonyls. I can very clearly remember the day when Geoff said to me, we have this molecule, cyclopentadienylvanadium tetracarbonyl and we're sure that the four carbonyls are all equivalent in their relationship to the metal, but still we have more than one CO stretching frequency. If the COs are all the same, why don't they vibrate at the same frequency? Of course, that's a question that someone with a knowledge of group theory and quantum mechanics can answer easily. There is a coupling between the carbonyl groups and you get symmetric and antisymmetric vibrations that don't have the same frequency. I was able to recognize that immediately. Together with a graduate student of Bill Moffitt's, Andrew Liehr, we wrote several papers in which we explained how to interpret these spectra. Then I began thinking about metal carbonyls more generally, and later developed a simple but effective method of analyzing metal carbonyl spectra which is now known as the Cotton-Kraihanzel method. Kraihanzel was then a postdoc who now teaches at Lehigh University.

I presented this method before it appeared in print at a Gordon Conference and the pleasure that I got out of this was that I felt briefly like Linus Pauling. Pauling had that genius for finding an intuitive theory which would seem to be so horribly oversimplified that it could never work, but, it did work! If you look at the way I analyzed the infrared spectrum, from the point of view of a real purist, you might say, my God, that's a ridiculously oversimplified theory; it will never work. In fact, it works like a charm, as everybody knows. Today people use it in metal carbonyl chemistry without thinking twice about what they are doing.

Another thing that I became interested in at about the same time was the use of NMR spectroscopy for studying metal carbonyl molecules. I became interested in the possibility that structures of the metal carbonyl species could be nonrigid, that is, fluxional. Not long after that, it became clear that for organometallic compounds as well fluxionality could exist.

That proved to be a very exciting period. Alan Davison and Jack Faller and I published the first paper showing that one could use NMR to determine not only the rate and activation energy of a fluxional rearrangement, but its mechanism or pathway. Our work dealt with "ring whizzing" in η^5-$CpFe(CO)_2(\eta^1$-$C_5H_5)$.

It was also around 1962 that it became clear to me that there were two main directions opening in crystallography. One would be to work on very small systems with extraordinarily accurate data and try to map electron density. The other one was in a sense the opposite, namely to look at huge molecules, like hemoglobin, whose structure just had been solved. I chose the big molecule direction. I believe I made the right choice because I don't think that electron density mapping has ever proved to be of much importance. What they do get a map of is the electron density with very high errors and then note that it agrees with the theory. On the other hand, the other direction which I chose, the study of large biological molecules has absolutely taken off like a rocket. We did one of the very first truly high resolution enzyme structures, staphylococcal nuclease. Today that structure is really a workhorse for people using site-directed mutagenesis to explore the enzyme mechanism. However, it was also clear by the time we got to high resolution that in this field you would have to become a real biochemist as well as a crystallographer if you wanted to be a leader. If you didn't master biochemistry you'd be nothing but an expert technician for the real biochemists. So I decided to drop it. My heart was in inorganic chemistry.

Are you popular among you colleagues?

How would I know? I've never had a difficult relationship with anybody that I thought was my intellectual equal. I've always had difficulty with people that I think are stupid.

You were running for ACS President some years ago and some people say that you were running more like in a political campaign than in a scholarly organization.

That is an opinion that some might hold, and I couldn't comment on that. But let me tell you what I think the issue was — why I didn't get elected. First of all, there are many different kinds of people in the American Chemical Society, ranging all the way from scholars like myself to people without Ph.D.s. In fact, much more than half of the membership of the ACS consists of people without Ph.D.s. So you're not talking about a homogeneous, scholarly electorate.

It was (and, I suppose still is) an unwritten tradition in the elections for President of the ACS that the Presidency should alternate between an industrial person and an academic person. The year before I ran there were industrial candidates and it was their so-called "turn to be in." However, there were many people, even in industry, who were unhappy about both of the candidates and they persuaded Fred Basolo to run, and Fred won. This enormously exasperated the people in the American Chemical Society who were not too concerned about scholarship and who think of it more as a professional organization than as a scholarly organization. There are lots of people who tend to make almost a career of being on the Council and being chairmen of committees, and so on. They were very annoyed by the fact that this academic had come in when he shouldn't have, in their opinion. Secondly, Fred was not shy about telling everybody what he thought was wrong with the ACS, and that he was going to try to fix it. He was horribly naive in thinking that in one year you can fix any organization with this kind of inertia, and he didn't succeed in fixing anything. But he sure did succeed in convincing a large part of the membership that they didn't want to have another character like that as President for a long time to come.

The next year should've been an academic's year, and I became one of the academic nominees. George Pimentel was the other one. George and I talked it over very frankly. We said that if there should be another candidate by petition (which is very easy to do, and we anticipated it) one of us would drop out of the race. We wouldn't want to divide the vote for the academic candidates. George was at that time in the midst of what would later become known as the *Pimentel Report*, and he decided that he would withdraw from the race. This left me in against the industrial guy who had become a candidate by petition. I knew that the people who supported him were out to get me because they didn't want another academic again, and they felt that if Basolo was bad (in their view) Cotton would undoubtedly be worse. They put on a campaign like you have never seen. They wrote about twenty thousand letters. I sent out a few hundred letters to academic people, basically stressing that they shouldn't fail to vote. One thing I said in my letter was that my opponent was "a mediocre industrial chemist." This was a precise description of him. Unfortunately, I shouldn't have coupled those two adjectives. It was used then to suggest that I was implying that, by virtue of being an industrial chemist, you were automatically mediocre. That is something I never have believed and wasn't saying. But, it was used very effectively against me.

There was an article in *Science* after the election was over, pointing out that I'd lost and so on, and pointing out that my opponent had the highest number of votes of anybody who had ever been elected President of the Society. I wrote a letter back pointing out that I had a higher number of votes than anybody who had ever before been elected President. My number of votes would've beaten everybody in history, because there was such a tremendous turnout for this election.

There is a new F. A. Cotton Award. How did it come about?

Six sections of the American Chemical Society have had gold medals for achievement in any and all fields of chemistry but no medal of that kind existed down in the southwestern part of the country. So my friends and colleagues in the Department chose the double occasion of my 65th birthday and the graduation of my 100th Ph.D. student to raise the necessary money and establish an F. A. Cotton Medal and Lectureship. I was chosen for the first award in 1995. The award is sponsored jointly by the Texas A&M Section of the American Chemical Society and the Department of Chemistry of Texas A&M University. I did participate in setting up the rules, because I'd made it clear that if there was going to be a medal in my name I wanted it to be run properly. Since I'd received all of the other six medals, I knew how they all worked, and I formed some definite opinions about what were good and bad procedures for handling the medal, that is, to select the winner, and so on.

I suggested that for it to have stature that it would be very important in the early years to establish major precedents. We should also give it to someone who is foreign to prove that it is not limited to Americans, give it to people in different areas of chemistry, to show that it's open to all, and, finally, make sure that at an early date it goes to a woman. The choice for 1996 was George Olah. By pure coincidence, I am chairman of the local Section in 1996 and thus, ex officio, the Chairman of the Selection Committee for 1997. We have already chosen somebody very distinguished but it cannot be announced at this time.

I used the Cotton-Wilkinson text, Advanced Inorganic Chemistry, *in my graduate studies. How does it work writing the new editions jointly?*

Geoff and I became very close at Harvard. He was an Assistant Professor and he was my Research Director. He was not married for the first couple of years and therefore he and I were closer than usual for such a relationship.

He was teaching the one-year course in advanced inorganic chemistry, which I took the year he first taught it. He complained constantly about the fact that there was no textbook which he found satisfactory. By the time I had finished my Ph.D., he and I had firmly agreed that we would write the appropriate book. By 1962, it came into print, and we've kept at it ever since. The work goes smoothly because we know and trust each other.

We are very different. He is a relatively nonquantitative person, much more of an intuitive chemist, and I am more theoretically inclined. This difference has been a strength. Between the two of us we manage to maintain a good balance and we're working now on the sixth edition.

Your number of publications has recently surpassed the number of papers by Linus Pauling.

Averaged over the years I've been in business, it's been 30 publications a year. I'm talking only about research papers. Right now the total is about 1275.

Sometimes I worry a little bit that people will condemn me for publishing so many papers. Many of my papers could be described as superficial, that is, there is no new message or there is no deep thought in them, just the preparations and structures of new compounds. The ultimate reason I have so many papers is because I and many coworkers do a heck of a lot of work, and I think when something has been done, and it's publishable, it should be published. The granting agency has basically funded you to produce new knowledge, but it won't do anyone any good if it isn't available.

Can you name one person or a few who were the strongest influence on you, your life, your career?

In forming me into the human being I am, my mother was the most dominant influence. As for my chemistry, let me say this. There was a socio-psychological tract written 30 or so years ago where the author divided all human beings into two broad types, outer-directed and inner-directed, that is, people who were motivated by stimuli that came from outside and people who had some sort of built-in stimuli which just drove them, regardless. I've always been a classic inner-directed person. I've never done anything because I thought that somebody else would think the better of me for it or because it was currently "the done thing." I have just always had a drive to be a scientist, and in particular, to be a chemist. Chemistry has always fascinated me. Still does.

Any message?

Yes. In recent years I've become a little involved in science policy; I've been on the Committee on Science, Engineering and Public Policy of the National Academy, and I've been on the National Science Board now for ten years, and will be for another two years, so I've had a participatory view of what's going on in American science policy. Like most people of my type, namely, pure scientists, I'm terribly concerned about the fact that this country is abandoning support of basic science. The long run consequences will be very bad for the country.

There are powerful political forces in Washington which are willy-nilly, throwing good things overboard to save a little money. It's revolting what they're doing in trying to balance the budget in this country. If you want to do that you have to cut about two hundred billion dollars. To make cuts of that magnitude, you have to go after the biggest components of the budget, because, as Slick Willy Sutton said, "That's where the money is." These are the entitlements such as medicare-medicaid and the defense component, primarily. The idea that you should cut a little three-billion dollar a year program like the NSF when you're trying to make a two hundred billion dollar cut in the budget is so ludicrous, you'd think a five-year-old child could understand that it won't get you anywhere. It's a spit in the ocean. They want to cut something that is vital to our future, and even if it weren't, it's almost at the noise level of the trillion and a half dollar Federal Budget. You've got to have the courage to cut the big things, but, of course, none of the politicians have the courage to come out and say, I'm going to cut medicare or defense and I'm going to cut deeply. The alternative is to raise taxes, and, of course, none of the politicians have the courage to propose that either, because they'd lose the next election.

They're hurting the future of this country, because the intellectual vitality of our economy comes from technical innovation. The roots if this tree are the training of future scientists and prosecution of pure scientific research, and they're cutting both. I think it's politically motivated irresponsibility. Of course, the average voter doesn't know that it hurts the scientists and it hurts the country, nor are there many scientists with votes, and politicians count votes. The politician's attitude to this is very much like the response of Fredrick the Great when somebody told him that the Pope objected strongly to something he was doing. Fredrick the Great said "Oh, really, and how many divisions does the Pope have?"

19

The Beginnings of Multiple Metal-Metal Bonds

The first conspicuously short metal-metal bond was reported in 1935 in the $W_2Cl_9^{2-}$ dimeric unit[1] but was not described by a triple bond until much later. According to Al Cotton, in the conversation (see previous chapter), the Russians had been making low valent rhenium compounds in the 1950s, as early as 10 years before his own work. They were probably compounds with triple and quadruple bonds but the Russian scientists did not know that.

From the Russian perspective, there was extensive work going on in this area as was described by Petr A. Koz'min in his account, titled "Quadruple Metal-Metal Bond: History and Outlook."[2] A. S. Kotelnikova and V. G. Tronev had obtained samples of several low valent rhenium compounds and reported their findings in 1958.[3] On the basis of various properties they assumed direct rhenium-rhenium bonding but did not foresee the existence of multiple bonding. The first report on the structure of the dimeric $Re_2Cl_8^{2-}$ unit was to a meeting in 1961 by Koz'min and Kuznetsov.[4] They found the ReRe distance to be much shorter than in metallic rhenium. By the mid-1960s, the time was ripe for extended X-ray crystallography of these systems and Cotton *et al.* started by reinvestigating the structure of the $Re_2Cl_8^{2-}$ unit.[5] They stated, "The very short Re–Re distance reported

Petr A. Koz'min
(photograph courtesy of Petr Koz'min).

Ada S. Kotel'nikova
(photograph courtesy of Petr Koz'min).

by Kuznetsov and Koz'min initially led us to suspect that their structure determination was in error, especially when the pronounced tendency of this compound to give twin crystals (as reported by Kuznetsov and Koz'min and confirmed by us) and the trial and error method of refinement were taken into account ... We have concluded, however, that the work of Kuznetsov and Koz'min is correct in all essentials. ..." Cotton *et al.* suggested, for the first time, that the Re–Re bond is a quadrupole bond in the dimeric unit. Al Cotton then continued his pioneering research on the nature of multiple metal-metal bonds.[6]

I met Dr. Ada Kotel'nikova (1927–1990) once but was unaware of her important contribution to the science of metal-metal bonding. She certainly did not volunteer any such information. She was a pleasant, unassuming person. At the time of her work mentioned above, she was in her late twenties, probably in the position of junior scientist, the equivalent of a postdoctoral fellow or research associate. Since her co-author was the head of the laboratory, Professor Tronev, it is a safe assumption that their 1958 paper reported mostly Kotel'nikova's work. Tronev's laboratory was in the N. S. Kurnakov Institute of General and Inorganic Chemistry of the Academy of Sciences of the USSR (today, Russian Academy of Sciences). That the importance of the research on $Re_2Cl_8^{2-}$ was recognized early was witnessed by the Soviet stamp issued in 1968 to commemorate the 50th anniversary of the Kurnakov Institute, displaying the square prismatic structure of this unit.

Soviet stamp, 1968, with the dirhenium octachloride ion and the Kurnakov Institute of Inorganic Chemistry of the Russian (then Soviet) Academy of Sciences. The stamp was issued in celebration of the 50th anniversary of the Institute.

References

1. Brosset, C. *Nature* **1935**, *135*, 874.
2. Koz'min, P. A., *The Chemical Intelligencer* **1996**, *2*(2), 32.
3. Kotel'nikova, A. S.; Tronev, V. G., *Zh. Neorg. Khim.* **1958**, *3*, 1008.
4. Koz'min, P. A.; Kuznetsov, V. G., *Abstracts of the All-Union Conference on Crystal Chemistry*. Shtiints: Kishinev, 1961, pp. 74–75.
5. Cotton, F. A.; Curtis, N. F.; Harris, C. B.; Johnson, B. F. G.; Lippard, S. J.; Mague, J. T.; Robinson, W. R.; Wood, J. S., *Science* **1964**, *145*, 1305.
6. Cotton, F. A.; Walton, R. A., *Multiple Bonds between Metal Atoms*, 2nd Edition; Oxford University Press, 1993.

Herbert C. Brown, 1995 (photograph by I. Hargittai).

20

Herbert C. Brown

Herbert C. Brown (b. 1912 in London, England) is Richard B. Wetherill Research Professor Emeritus of the Department of Chemistry, Purdue University, in West Lafayette, Indiana. He shared the Nobel Prize in Chemistry in 1979 with Georg Wittig (1897–1987) "for their development of the use of boron- and phosphorus-containing compounds, respectively, into important reagents in organic synthesis."

Dr. Brown's family moved to Chicago when he was two. He was a student at the University of Chicago and started his career there. In 1943 he moved to Wayne University, Detroit, and in 1947 he became Professor of Inorganic Chemistry of Purdue University. He retired in 1978. Dr. Brown is member of the National Academy of Sciences, Foreign Fellow of the Indian National Science Academy and many other learned societies. His numerous awards include the National Medal of Science (1969), the Priestley Medal of the American Chemical Society (1981), the Perkin Medal of the Society of American Industrial Chemists (1982), and the Gold Medal of the American Institute of Chemists (1985).

Our conversation was recorded in Dr. Brown's office in the H.C. Brown Building of the Department of Chemistry at Purdue University on November 8, 1995. It appeared in *The Chemical Intelligencer.**

*This is a slightly modified version of the interview originally published in *The Chemical Intelligencer* **1997**, *3*(2), 4–13 © 1997, Springer-Verlag, New York, Inc.

Can you single out a compound that has been the most successful chemical you ever made?

Sodium borohydride is the most successful compound that I made while working as research assistant to Professor H. I. Schlesinger at the University of Chicago. Hydroboration is the most important reaction that I discovered, later at Purdue University. My Ph.D. thesis, under the direction of Professor H. I. Schlesinger and Dr. Anton B. Burg of the University of Chicago, involved the reaction of diborane with carbonyl compounds, and we discovered that aldehydes and ketones react with diborane in a matter of seconds, to give alkoxy boranes. When these are hydrolyzed with water, the alcohols are obtained. This provided a new method of reducing aldehydes and ketones to alcohols and was the beginning of hydride reduction in organic chemistry. That was the start.

In the course of studying ways of increasing the reducing properties of sodium borohydride, Dr. B. C. Subba Rao and I ran experiments on the reducing properties of sodium borohydride enhanced by aluminium chloride. We used 56 representative organic compounds, and we tested them against each new reducing agent. We would mix them together, four hydrides to each molecule. We'd let them stand for one hour at room temperature, hydrolize the product with water, measure the hydrogen evolved, and we could tell whether or not hydride was used for reduction, one hydride per molecule, two hydrides per molecule, and so on. Doing this, we found that aldehydes and ketones took up one hydride, esters took up two, carboxylic acids utilized three (one for hydrogen evolution), in the course of reduction to the alcohol.

One of the compounds we tried was ethyl oleate and that took up 2.37. The work was done by Dr. B. C. Subba Rao who had taken his Ph.D. with me and stayed on as a postdoc. When I asked him, why this was 2.37, he said, remember, ethyl oleate is not one of our standard compounds. (All our standard compounds had been carefully purified and kept from other people touching them.) When Dr. Subba Rao told me about his success in achieving the reduction of esters to alcohols, I suggested to him to try an unsaturated fatty acid ester, and reduce it. He obtained ethyl oleate from the stockroom, but didn't bother to purify it. He thought the unpurified ester must contain peroxide impurities giving the high results. He said, why not forget about this one odd result and publish all of our other 56 results.

However, fortunately, it is the research director who is in a position to insist upon high standards; he doesn't have to do the work himself.

So I thought for a couple of seconds, and said, no, go back, redistill the ethyl oleate, and repeat the experiment. He repeated the experiment and it was 2.37 again. Then he let it stand for three hours at room temperature and it went up to 2.9; for six hours it went up to 3.0; 12 hours, 3.0; and 24 hours, 3.0. Clearly, there was something else in the molecule that was taking up precisely one hydride per molecule of ethyl oleate.

We decided that the C=C double bond was adding a B–H bond, and that is how we discovered the hydroboration reaction. This was in 1956.

You were the humor column editor of your high school newspaper.

I still have all my clippings. I entered Englewood High School in Chicago in 1924 when I was 12 years old. Then my father died in 1926. There was no Social Security in those days. We had a small hardware store, and I left school to take care of it. I was 14. However, I am afraid I neglected the business and spent a lot of time reading books. I enjoyed studying. Finally, my mother said that she could take care of the hardware store during the day, so I could go back to school, and I would go to the store in the afternoon and take over. So I went back to school and I was there three more semesters before I graduated.

I began working for the school newspaper and I was given the humor column there, modeled after the famous humor column of the *Chicago Tribune*, "A Line of Type." Ours was a weekly and I used to write much of the column myself, and it won a prize. I tend to be humorous, and I consider most things jokes, and I rarely get excited about interactions that excite others.

I graduated in 1930, and for two years I tried to find a job, without success. This was the time of the Great Depression. Then I decided to go back to school. There was Crane Junior College, operated by the City of Chicago. It did not involve payment of any tuition, but they were very selective. Fortunately, I was accepted and I went there to become an electrical engineer, because somebody had told me that electrical engineers made good money. At that time that was the most important thing in my life.

The first year of the electrical engineering curriculum required the study of chemistry, and the subject fascinated me. I had a very good memory and could learn these things better than anybody else in the class. After the first year, I decided that I could write a book on chemistry. I paid my younger sister five cents an hour to type it. The book was about general chemistry, my own interpretation. It's still around somewhere in my archives.

Julius Stieglitz at the University of Chicago (photograph by I. Hargittai).

Also at Crane I met a young girl, 16 years old, who was registered as a chemical engineer. At that time this was unheard of. There were student engineers at this Junior College by the hundreds, but only one girl, Sarah Baylen. At first she hated me because she had been the brightest student in the chemistry class until I joined, and she didn't like to be second.

But after one year, Crane Junior College was closed because of lack of funding. There was a Dr. Nicholas D. Cheronis, originally from Greece, at this college, and when our college closed down, he did a very nice thing. He operated a small commercial laboratory at his home, in an expanded garage, called "Synthetical Chemicals." He used to make costly chemical indicators and other high-priced chemicals as a side business. He invited about 10 of us students to come out and do what we wanted to do in the laboratory, to keep us off the streets. Sarah was there, and I went there.

I registered for a correspondence course on qualitative analysis at the University of Chicago. They sent me the unknowns, I did the analysis in the laboratory, and I sent the reports in. The course was given by someone called Julius Stieglitz. He was the professor and one of the graduate students actually ran it. One time I got an unknown that just didn't make sense, it didn't behave the way it should. I called up Dr. Stieglitz and described the problem. He then said that the unknown must contain phosphate, although it was not supposed to have phosphate in it. He said that he would ask his assistant to send me a new sample. That began my acquaintance with Julius Stieglitz, one that changed my life.

After a year, three new Junior Colleges were opened in the City of Chicago. Dr. Nicholas Cheronis was appointed Head of the Physical Sciences Division at Wright Junior College. I went and so did Sarah. In the meantime we had become good friends. At Wright Junior College Nicholas Cheronis taught many of the chemistry courses; I took his course on Organic Chemistry. He was writing a book on doing Organic Chemistry on a semimicro scale. He gave us all the experiments to test to make sure they were all right. That was the first book that carried out laboratory experiments on a smaller scale than the usual procedures.

They gave me a free hand there, and I was allowed to experiment with equipment not yet used in classes. I was permitted to publish a new paper, the *Physical Science Monthly*. It described various experiments. I was the editor and often the reporter who did all the work. They were very good to me at this college, and they told me, when I was approaching graduation, that I should take the examinations for Competitive Scholarships at the University of Chicago. I signed up for this, but when I saw the examination, I thought it was hopeless. At that time the University of Chicago had a president, Dr. Robert Maynard Hutchins, who thought the way to produce ideal college graduates was by educating them in the great books, "the hundred Great Books", with the education based primarily on philosophy, psychology, literature, etc., with not much attention paid to chemistry, physics, mathematics, and so on. But all of my education had been in these specialized areas. However, I did the best I could and to my great amazement, I won a scholarship. In 1935, I went to the University of Chicago. Another idea of President Hutchins was that the students should not spend longer time than they needed to get their college degree. It cost no more money to take 10 courses a quarter than to take the usual 3. So I took ten courses per quarter and finished in three quarters, in 1936.

I didn't apply for a teaching assistantship because I wanted to marry Sarah, and in those days one didn't get married without a job.

Professor Julius Stieglitz was now Emeritus. He had been Head of the Department for many years. He had a very famous twin brother, Alfred Stieglitz who married an artist, Georgia O'Keefe. Julius Stieglitz was teaching the Advanced Organic Course. I and Sarah were in that class. He used to have a custom of speaking for about five minutes, and then he would

Herbert and Sarah in 1995 (photograph by I. Hargittai).

ask the class a question, a mechanism to keep them alert, I guess. I was always volunteering with the answer. Apparently I made an impression upon him. He called me into his office, and said, Mr. Brown, I know that you haven't applied for a teaching assistantship which would allow you to go on for a Ph.D. degree. Why haven't you? I am confident that you would do very well as a Ph.D. I told him that I'd met this girl and I wanted to marry her, and I didn't see how I could possibly manage both a Ph.D. program, and the marriage. He said, I think you are making a bad mistake. Why don't you discuss it with your girlfriend and see if you shouldn't let the marriage wait until you get your Ph.D. So I did discuss it with Sarah, and to my great astonishment, she said, he is probably right.

It didn't make my life very easy because they paid me $400 a year for a half-time teaching appointment. I had to pay tuition, $100 a quarter for three quarters, but not the summer quarter; that was free. So I didn't have much to live on. I did things like tutoring, and managed to survive. Then one week-end, when Dr. Stieglitz wasn't looking, Sarah and I went down to City Hall to get married secretly. We were such ignoramuses; we didn't know that they published these things in the daily newspapers. There are people who read these long columns of names. While no one recognized H. Brown, they recognized Sarah Baylen. There were only two Baylen families in the City of Chicago. The secret was out and Sarah's mother told me that now that I'd married her, she was my responsibility, and she should move out of the house. So she moved into the room I rented near the University, and my rent went up from $3.50 to 4.50 a week.

Was she also studying at that time?

Yes, at the University of Chicago, but she was still an undergraduate. She hadn't moved as fast as I had.

We were on a point of going bankrupt when she got a job at Billings Hospital of the University of Chicago, paying $90 a month, and that made life for us much easier. But she hadn't yet finished her courses for a bachelor's degree. So I would go to her courses and take notes, and she would study those notes and do the laboratory work at night, and that's the way she finished her bachelor's degree.

She stopped at the bachelor's level. In those days it was next to impossible for women with higher degrees to get jobs in chemistry. In fact, it was considered a shame if a wife worked. Sarah worked for about six years, and then stayed home, since by that time my salary was sufficient to support her. It may be of interest to mention that during this period she became a research assistant to Dr. Charles B. Huggins, who received the Nobel Prize in Medicine in 1966.

How did you choose your research area?

This question came up in the graduate school. Sarah had bought a book *Hydrides of Boron and Silicon* by Alfred Stock and gave me this book as a graduation present, in June 1936. I read this book and became interested in the hydrides of boron. At that time, these were very rare substances. They could be made in only two laboratories in the entire world; in Alfred Stock's laboratory in Karlsruhe, Germany and in the laboratory of H. I. Schlesinger at the University of Chicago. That's how I happened to select the hydrides of boron for study, and that's how I happened to work for H. I. Schlesinger.

Why did Sarah pick out this particular book of hundreds of chemistry books available in the University of Chicago bookstore? This was the cheapest book in the bookstore. We were very poor in those days. This book cost $2.00 with six cents sales tax. I still have the book.

Sarah had great faith in me. When we graduated from Wright Junior College in 1935, I took my Year Book to my friends and they signed it. In that book she wrote "to a future Nobel laureate." She may have been prophetic, but I have another explanation. She had been the brightest student in class until I joined. She became No. 2 and she didn't like that, but she figured that anybody who could beat her in chemistry was sure to win the Nobel prize.

What was your relationship with Georg Wittig?

He invited me in 1963 to spend a summer at Heidelberg and he had been here as a visiting lecturer, so we had been together a number of times. We never did any work together. He had trouble with English. His wife did the translation for him.

How did you happen to be at Purdue?

In the 1930s, universities usually waited until the tenth year to make a decision on tenure. I'd seen at the University of Chicago a number of cases where people were told in the tenth year that there was no future for them at the University of Chicago. This happened to Anton B. Burg, who was Schlesinger's research assistant, and he moved to the University

Herbert Brown receiving the Nobel Prize in 1979 (courtesy of Herbert Brown).

Georg Wittig and Herbert Brown at the University of Heidelberg, Germany, celebrating the Nobel Prize in December 1979 (courtesy of Herbert Brown).

of Southern California where he did very well. In fact, he has always been mad at me because he thought that if I had not been available at the University of Chicago he might have gotten tenure. He used to review my papers for *JACS* and say the most nasty things about them. Finally I requested Noyes, the Editor of *JACS*, not to send him my papers, and he agreed.

Consequently, I had seen this happen to a number of people at the University of Chicago. I was there as an instructor, I was teaching any class that nobody else could do. During the four years I was there I had taught general chemistry, qualitative analysis, quantitative analysis, physical chemistry, and organic chemistry. This was in addition to the large World War II projects in which we had become involved. Then I decided that I would not wait the usual ten years, and in my fourth year I asked Schlesinger to find out from the Department whether I had any future there. But they said, no, they didn't want to make any exception to the usual 10-year process.

This was in 1943. At this point Professor Morris Kharasch helped me. He had a very good friend at Wayne University, at Detroit, Michigan,

Professor Neil Gordon, who had just become Head of the Chemistry Department there. He was also the Gordon who had started the Gordon Research Conferences. He offered me a job at Wayne University as Assistant Professor. This was a school without a Ph.D. program. Neil Gordon wanted me to help them develop a Ph.D. program. I went there. The teaching load was 18 hours, but he promised me 12 hours to give me a chance to do research. Wayne operated an evening school. I arranged to teach 6 to 10 o'clock in the evening, three days a week (Mon., Wed., Fri.), a course on undergraduate organic chemistry. This gave me the whole day free to do research. Unfortunately, the other people in the Department went to see Gordon and said, why do you give this new man 12 hours? We want to do research too. Neil Gordon asked me for a solution, and I suggested one. I said, tell the staff that you'll give any person, on a rotating basis, a year sabbatical in residence when they'll have only 12 hours of teaching. Then, for every paper they publish in a recognized journal, the teaching load will be reduced by two hours. The first year I published six papers, so my teaching load was six hours the following year. Unfortunately, most of the others had been away from research for several years, and it was very difficult to get it going again.

In 1946, I was invited to visit Ohio State University. It was Mel Newman there who wanted me to go there. I went there and said that I was interested in organic, inorganic, and physical chemistry, so that I took students in each of these areas to do research. The Head of the Department, Ed Mack, vetoed my appointment. He said, nobody can be that good.

My wife was very unhappy in Detroit. She is a very intelligent person, and she liked to participate in university life and activities. In Detroit, the university was in the central part of the town, a place where you couldn't live. Thus the staff spread out all around the suburbs of Detroit, so there was no evening university life. In 1947, after four years at Wayne, I was invited to Purdue University. The Purdue offer was very attractive; it had an established Ph.D. program and a lovely campus with evening activities, ideal for Sarah.

How did you get into war-oriented research?

I was Schlesinger's research assistant, and one day he got a request from Washington, from the National Defense Research Committee, to investigate the existence of new volatile compounds of uranium compounds which would be stable at a temperature where they had a pressure of at least 0.1 mm. They knew about UF_6 but it was highly corrosive. They were

Herbert Brown in front of the plaque of the Herbert C. Brown Laboratory of Chemistry at Purdue University (photograph by I. Hargittai).

encountering serious problems in using it. We went to work on this with four coworkers. First, we made acetyl acetonates of uranium, and we ended up with the hexafluorinated derivative. The complex was stable enough and volatile enough to satisfy the requirements. When we reported this to Washington, they said, sorry, they had failed to tell us one thing. It was important for the molecule to have a low molecular weight. Our compound had a molecular weight 1,066. We had to find a new approach. By then we had made aluminium borohydride, with a boiling point of 56°. We had also made beryllium borohydride, a solid, with a sublimation point of 93°. These are the most volatile compounds of beryllium and aluminium. We thought that maybe we could make uranium borohydride, and we did. When we reported this to Washington, there was great excitement; they asked us to bring a large group together, and gave us money for 25 people, and they wanted us to make a hundred kilograms for testing. We went to work and, ultimately, we developed a practical process. By that time though, we were told that they'd learned how to handle UF_6 and $U(BH_4)_4$ was no longer needed.

At this time we got a phone call from the Army Signal Corps. They had a problem of generating hydrogen in the field. They were using a World War I method. They took a big strong steel cylinder, and in it they would place ferrosilicon, caustic soda, and a controlled amount of water, and close the top. Caustic soda would dissolve in the water and the temperature would go up by about a hundred degrees, and the hot caustic soda would start reacting with ferrosilicon, forming sodium silicate and hydrogen, raising the temperature to about 200 degrees. There would be about a thousand pounds pressure of water vapor and two thousand pounds pressure of hydrogen. After they cooled, hydrogen could be withdrawn as needed. The cylinders had to be very strong to withstand the pressure. Consequently, they were very heavy. It would be difficult to ship them to the many battle fields around the world. There was another problem. After the hydrogen would all be used up, they'd have to clean out the sodium silicate, which coated the walls like glass. It took the average soldier a full day to scrape all the walls clean. They had orders to bury the scrapings, but the soldiers were so tired then that they'd just dump the material behind the nearest bush. The cattle seemed to love the stuff, and they would go there and eat it and the Army was getting all kinds of lawsuits for the damaged cattle. The Army badly wanted some other way of producing hydrogen in the field. We recommended our sodium borohydride; it tested out and worked well. The Army issued a contract to a large corporation in the amount of $ 25 million to build a plant to produce sodium borohydride in quantity. This was just before the Battle of the Bulge, in December, 1944, and there was confidence that the war would soon be over. Orders came from Washington not to build any new war plants. So this plant was not built either.

You have made many patents that must have generated a tremendous amount of revenues.

The army encouraged us to patent these things, giving the Army a royalty-free license. There was a company, Metal Hydrides, which manufactured calcium hydride and used it to make pure uranium for the Manhattan Project at the University of Chicago. But with the war ending, their business started drying up. They were facing bankruptcy, when they decided that they could manufacture sodium borohydride and lithium aluminium hydride and explore their markets. They turned to us, and we gave them licenses to make sodium borohydride, aluminium borohydride, and these then became important reagents for reduction. This and my Ph.D. thesis showing diborane to be a good reducing agent, started the hydride era for reductions in organic chemistry.

It was our good fortune that the Army persuaded us to patent these things and give them a royalty-free license, so that no one else would patent these procedures.

Your war-time research on sodium borohydride was published much later.

All the work done during the war was under secrecy, and had to be declassified before it could be published. However, with sodium borohydride I ran into an additional problem with Schlesinger. He had been the professor, I was an instructor. I thought that he should write the papers, and he thought that I should write them. Finally, in 1951, I was invited to spend a summer at UCLA teaching, and wrote up the papers. I had 18 papers in the first issue of *JACS* in 1953 (11 on the research involved in the synthesis of sodium borohydride and uranium borohydride, and 7 on strained homomorphs).

Please, tell us about your studies on steric effects, strain, and crowded molecules.

When I was at the University of Chicago, I tried to see how I could use the results of my classified boron work to produce some worthwhile new chemistry that could be published. We had a reversible system. One can take trimethylamine and trimethylboron, put them together, forming the addition compound, $Me_3B:NMe_3$. At higher temperatures, the addition

Cartoon by D. Sattler in the *Journal and Courier*, Lafayette, Indiana, on Saturday, October 20, 1979. Reproduced by permission. On October 25, the newspaper carried Herbert Brown's Letter to the Editor, saying "I read your cartoon with a sinking feeling. Sarah has always brought the garbage out and cartoons such as you published can only create difficulties in an idyllic arrangement. You should understand that in our long, very happy marriage I have assumed total responsibility for the chemistry, and Sarah has assumed responsibility for everything else. Please, don't sow doubts in a wonderful cooperative arrangement."

Cartoon by Hsiupu Daniel Lee dated December 26, 1979. Reproduced by permission. Hsiupu Lee was then Herbert Brown's postdoctoral fellow, now he is head of the Microanalytical Laboratory at Purdue University.

compound volatilizes and partially dissociates. Careful measurement of the pressures at several temperatures permit the calculation of K_{eq} and the determination of ΔH, ΔS, ΔF. This represented a new approach to steric effects. I studied these and decided that Ingold was wrong. He said that all effects were polar, electronic, with no room for steric effects. The fact that the nitration of isopropylbenzene gave much less of the ortho isomer than did toluene, was not attributed to the larger steric requirements of isopropyl compared to methyl; instead the change was attributed to different electronic contributions. I disagreed with this interpretation and attributed the phenomenon to the steric requirements of alkyl groups. Recently, there was a celebration of the 100th anniversary of Ingold's birth and I gave a talk there "C. K. Ingold Meets Steric Effects as a Factor in Chemical Behavior."

When I went to Wayne, they had no shop, no glassblower and so on, but I wasn't discouraged. I decided that I could have my undergraduate students make highly branched alcohols by the Grignard reaction. We could isolate the alcohols, treat them with hydrochloric acid, make the chlorides, hydrolyze them, and correlate the rates with the steric effects of the branched alkyl groups. In this way, we embarked on studying chemical effects on steric strains.

When I came to Purdue, I continued the study of steric effects but returned also to boron chemistry.

Did understanding the diborane structure help you much in your work?

At the beginning, everybody assumed that diborane had an ethane-like structure. Simon Bauer and Linus Pauling published an electron diffraction study of diborane, and they came out with an ethane-like structure. At that time, Schlesinger asked me to supervise Norman Davidson's work on methylaluminium. Davidson, a Rhodes scholar, had just got back from Oxford when England shipped back all the Rhodes scholars at the outbreak of World War II. He wanted to continue in the same area. Schlesinger was not interested in this area, and asked me to supervise Norman Davidson's work. So he was my first student, although he had been my classmate three years earlier.

Trimethylaluminium, $Al_2(CH_3)_6$, is a dimer, as is aluminium chloride, Al_2Cl_6. The X-rays showed chlorine bridges in Al_2Cl_6. We made $Al_2(CH_3)_6$, $Al_2(CH_3)_4Cl_2$, $Al_2(CH_3)_2Cl_4$, and Al_2Cl_6, and measured their heat of dissociation. All were dimers. The data revealed a smooth curve with no irregularity. From this I came to the conclusion that diborane must also have a bridge structure. I went to Schlesinger with this idea. He said, this is not our field, I'll contact Linus Pauling. This was about 1942. Pauling said the idea was nonsense, diborane clearly had an ethane-like structure, so Schlesinger discouraged me from pursuing the idea. Then Davidson took $Al_2(CH_3)_6$ to Lawrence Brockway at Michigan to do electron diffraction for comparison with hexamethyldisilane, $Si_2(CH_3)_6$, which we were confident was ethane-like. We asked Brockway to consider the possibility that $Al_2(CH_3)_6$ and $Si_2(CH_3)_6$ might have different structures. Brockway concluded that all had ethane-like structure, so I lost confidence in physical chemistry as a way of solving these problems. Of course, the problem was eventually solved and the hydrogen-bridge structure of diborane established.

How about the classical/nonclassical carbonium ion controversy. Was it very important?

It was important at one time. Saul Winstein observed that if you took exo-norbornyl brosylate and solvolyzed it in acetic acid, you got the exo-acetate. If you took the corresponding endo derivative and solvolyzed it, you also got the exo-acetate. He attributed this to bridging between C6 and C2 (the cationic center in the classical structure for the cation).

We had subjected norbornene to hydroboration-oxidation and realized 99.4% exo-norborneol. Yet the reaction did not involve any cationic species. The reagents simply attacked the more exposed exo site. You could describe the intermediate in the solvolysis as a pair of rapidly equilibrating enantiomeric ions, which substituted solvent predominantly at the more exposed exo face.

This is an old problem in organic chemistry. When I took organic chemistry I was taught that benzene was an equilibrating pair of two Kekule structures. Now we consider it to be a resonance hybrid. I was also taught that cyclohexane was a planar structure. That accounted for the known stereochemistry. Now we consider it to be an equilibrating pair of chair structures. We did a number of experiments that convinced me that the solvolysis of both secondary and tertiary 2-norbornyl derivatives must proceed through an equilibrating pair of 2-norbornyl cations.

Has computational work been helpful in resolving this problem?

At the beginning, the nonclassical ion was computed to be some 69 kcal/mol more stable than the classical one. Over the years this difference kept diminishing, and, finally, Michael Dewar, who had proposed the nonclassical ion in the first place, came out with a 2 kcal/mol difference in favor of the classical norbornyl cation.

How did the Nobel Prize change your life?

It didn't. I know that some people become citizens of the world, who pontificate on all of the problems facing the human race, but I was never tempted to abandon either of my two great loves.

What has been your main research interest since the Nobel Prize?

We discovered the hydroboration reaction in 1956. Systematic exploration of the characteristics of this reaction led us to apply it to alpha-pinene, an example of a highly labile olefin. This compound underwent hydroboration without rearrangement, but only two olefinic units add to borane, forming a new, optically active hydroborating agent, Ipc_2BH (Ipc = isopinocampheyl). Indeed hydroboration of *cis*-2-butene with this reagent provided after oxidation (+)-2-butanol or (–)-2-butanol in up to 98.4% optical purity, the first non-enzymatic asymmetric synthesis in high ee (1961). Hydroboration made organoboranes readily available for the first time. Systematic exploration of the chemistry of organoboranes revealed that these compounds are highly versatile reagents. It was discovered that substitution reactions of organoboranes proceed with retention of configuration, in contrast to the course of substitution in carbon compounds. It finally dawned upon us that we had the basis for a general asymmetric synthesis.

Following the Nobel, we explored asymmetric hydroboration and demonstrated that we could make many asymmetric organoborane intermediates. We developed a number of new reactions (24 major ones) to convert these intermediates into asymmetric organic molecules. Indeed, we estimate that we can now synthesize 130 000 pure enantiomers, using simple demonstrated methods. We have extended these procedures to asymmetric reduction of ketones, asymmetric allyl- and crotylboration of aldehydes, asymmetric opening of meso-epoxides, asymmetric enolboration, and others — a truly general asymmetric synthesis in high ee (1961), this field has become a major area of research for organic chemists. Many new synthetic procedures have been developed. However, none possesses the generality of our approach, "Asymmetric Synthesis Via Chiral Organoboranes."

Who supports your research?

For many years it was NSF and NIH, the Army, and the Navy. Since I reached 80, both the NSF and NIH had stopped supporting it. But others have stepped in to support the Borane Research Fund. That provides half of my research support. The other half comes from the Office of Naval Research and the Army Research Office. The University supports my secretary, my research assistant and gives me all the space we need.

How about your son?

Charlie was born in 1944. He also graduated from high school at 16. Sarah objected to sending a child of 16 away, so he stayed to get his B.S. at Purdue. Then he got his Ph.D. at Berkeley, working independently. (After all, he came to Berkeley with 10 *JACS* publications.) He then did postdoctorate work with Carl Djerassi at Stanford. In 1970 he went to Cornell, where he emulated his father and failed to get tenure. He wanted to return to the Bay area, and he now is a Scientist with IBM in California. We have two granddaughters, Tamar, a student at Smith College, and Ronni, still in high school.

Please, tell us about your own origins.

Both my parents were born in a town, Brovari, between Kiev and Zhitomir in the Ukraine. My father's family name was Brovarnik after that town. His family decided to emigrate. The way it was done, usually the father and eldest son would go first. So my grandfather went with his eldest son to the United States. They would work and make enough money to bring the next batch over. My father was the second son and he went with his mother and another son and the youngest daughter in the next batch. When they got to the United States, the young girl, who was five years old, had a virus infection of the eyes, and was refused admission, so the four members of the family had to return. They didn't go back to Russia; they went to London, England. My father was a cabinet maker and he got a job and got acquainted with my mother whose family was migrating to England. They were married in 1908, had a daughter in 1909, and I was born in 1912.

What language did you speak?

They spoke Yiddish, and my father picked up English. Then, in 1914, my father decided that they should join his family in the United States. My mother didn't want to go there but women didn't have much voice those days. We all came to Chicago. My father was one of seven children, two girls and five boys. We started very poor but kept moving up and, finally, my father opened up a hardware store in the south side of Chicago and we lived above the store.

Was it a religious family?

No. When I was studying for my Bar Mitzvah, I had a rabbi who wanted me to consider becoming a rabbi, but we weren't a religious family, so I didn't do that.

Concluding our conversation, what would you like to add?

Eight years ago I got a call from my granddaughter Ronni. What is it, Ronni? And Ronni says: "My class has asked me to invite you to speak to them, to give them a lecture." I say, Ronni, I don't understand. The class is made up of eight-year olds, why do they want to hear me? She says, "We were studying Martin Luther King, and the teacher explained that he was a Nobel laureate. So I stood up and said, my Saba is also a Nobel laureate." The teacher says, Ronni, you shouldn't imagine things.

The Herbert C. Brown Medal of the American Chemical Society for creative research in synthetic methods. Its first recipient was Herbert Brown in 1998 (courtesy of Herbert Brown).

Ronni says, oh, but he is. Why don't you ask my mother? So my daughter-in-law confirmed it. The class got together and decided that they should invite me to come there and give them a lecture. Will you do it? I did not answer her as I thought how I could get out of this invitation without hurting her feelings. I've lectured to a lot of classes, but never to eight-year-olds. But my wife was on the other phone, and I heard her saying, oh, Ronni, he'd be delighted to. So I had to go there and gave them a lecture. I spent a lot of time worrying about it. But it went well. I brought them the Nobel Medal and passed it around for them to look at. I explained what chemistry was about. Then I gave them a book of souvenirs and the *Aldrichimica Acta* where I was featured. At the end, one of the boys stood up and asked me, are you rich? I said to him, we live in a wonderful society where anybody who does original things gets richly rewarded.

When I got home I received a letter from the teacher saying that 60% of the class had declared they wanted to become scientists.

George A. Olah in Budapest, 1995, in the author's office (photograph by I. Hargittai).

21

George A. Olah

George Olah (b. 1927, in Budapest, Hungary) is Professor of Chemistry and Director of the Loker Hydrocarbon Research Institute at the University of Southern California in Los Angeles. He was awarded the Nobel Prize in Chemistry in 1994 "for his contributions to carbocation chemistry."

George Olah graduated and received his doctorate from the Budapest Technical University. He started his research career as a research assistant to Professor Géza Zemplén, who himself had been a student of Emil Fischer in Berlin. In the early 1950s, Dr. Olah built up a research group at the Institute of Organic Chemistry at the Budapest Technical University and, simultaneously, was also among the founders of the Central Research Institute of Chemistry of the Hungarian Academy of Sciences in 1954. In 1957, he moved to Canada and worked for the Dow Chemical Company in Canada and the United States until 1965 when he returned to academia. The chemistry departments of Western Reserve University and Case Institute of Technology in Cleveland, Ohio, were joined under his leadership. In 1977, he moved to the University of Southern California, where he established what eventually became the Loker Hydrocarbon Research Institute. He has received many distinctions. He is Member of the National Academy of Sciences (1976); Honorary Member of the Hungarian Academy of Sciences (1990); received the Alexander von Humboldt Award for Senior U. S. Scientists (1979); the F. A. Cotton medal (1996); and many others. The University of Southern California (USC) inaugurated the George A. Olah Lecture Series in 1996. It was my being the First George A. Olah Lecturer at USC, on February 20, 1996, that gave me the opportunity to record this conversation.

The Chemistry Building of the Budapest Technical University. It was built in 1902 and houses about one half of the chemistry faculty.

What is the most important scientific achievement in your career?

In all probability it was the realization, in contrast to the generally accepted concept expressed by Kekulé and going back well over a hundred years, that carbon cannot bind more than four atoms simultaneously, that under certain conditions in electron deficient systems carbon can indeed bond five or six and in some cases may be even up to eight groups or atoms. This opened up very exiting new perspectives of the chemistry of carbon which is so central to our terrestrial life.

When did you come to this idea?

I wish I could tell you that this was an idea *deus ex machina*, or that this came as a revelation in my sleep. In reality, it was a long process. In science we need to have both concepts and facts. Science still is very much based on findings. It was a long thought process putting together many observations going back to the classical/nonclassical ion controversy which was one of the major chemical controversies of our time, which allowed me to realize this rather fundamental new aspect of organic chemistry.

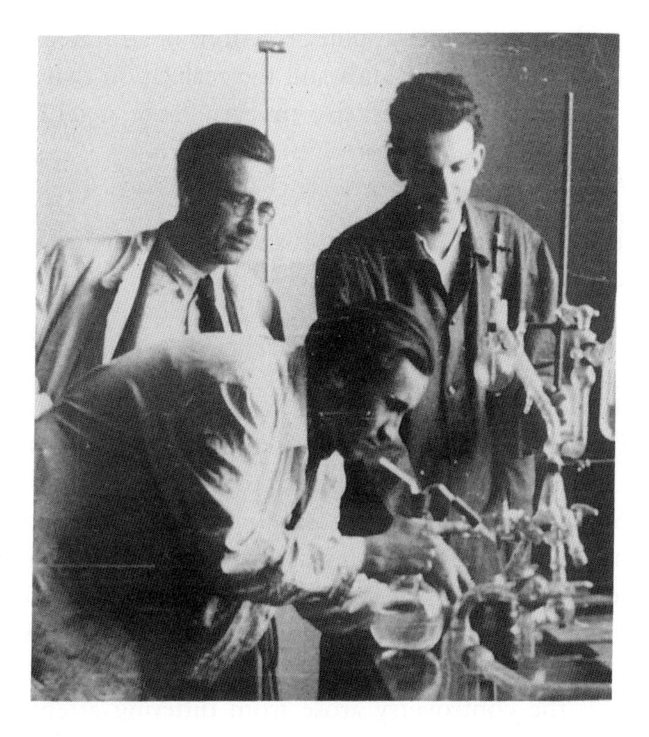

George Olah, standing in dark coat, in the laboratory, Budapest Technical University, 1948 (courtesy of George Olah).

Was this controversy truly so important?

No, *per se* it was not. Frankly, whether the norbornyl cation has a bridged structure or an open equilibrating structure, this would have, *per se*, not affected in any way the future of chemistry. It would have, however, certainly affected the egos of the involved scientists. I came into it because around 1960 I discovered methods to generate positive organic ions, called now carbocations, as long lived species, and we were able to take all kinds of spectra and establish their structure, including that of the norbornyl cation. In the course of this work I realized, however, that the problem has much wider implications. In the norbornyl ion, the C–C single bond acted as an electron donor nucleophile. In this particular case this happens within the molecule, that is, intramolecularly. This delocalization, which had been originally suggested by Winstein, was indeed there and we were able to see it directly for the first time. Later came, what I thought was a logical idea. The question I asked myself one day was, if this can happen within the molecule, why can't it happen between molecules? This led to the discovery of a wide range of electrophilic reactions of saturated hydrocarbons, that is, of C–H or C–C single bonds and the realization that carbon, under some conditions, can indeed bind five or even more neighboring groups.

When was this?

This was in the mid- and late sixties, and it was further generalized in the early seventies.

Were you a latecomer into the controversy?

I wasn't a latecomer; it was just that the controversy had started some years before I got interested in it. It started in the mid-fifties. I discovered what is now called stable ions, superacidic chemistry, in the late fifties, early sixties. So I was in the right position to enter and decide the issue. People who had been "fighting" this controversy before had only indirect methods at their disposal, such as kinetics and stereochemistry. We in contrast could prepare stable, long lived carbocations and study them directly by spectroscopy and other means.

In a nutshell, the controversy arose from differing attempts to explain the same experimental facts. Winstein and others have found that if they solvolyzed esters of norbornane (bicyclo-2.2.1-heptane), the exo- and the endo isomers reacted with different rates. The exo isomer consistently reacted hundreds of times faster than the exo isomer. Winstein suggested that the explanation for this is that the exo reaction is accelerated and with great foresight indicated that the reason for this was neighboring C–C bond participation. In other words, solvolysis is accelerated because a carbon-carbon bond, the electrons stored between the C_1 and C_7 atoms, can interact with the developing carbocationic center. On the other hand, H. C. Brown, while looking at the same data (which incidentally were never in doubt, and this is significant) as Winstein, however, said that there is no acceleration or bridging in the intermediate ion. He suggested that the high exo/endo rate ratio is not due to the fact that the reaction of the exo is fast but in fact endo is slow. Because he was always much interested in steric interactions, he suggested that the reason for this is that the approach at the endo side of the molecule is hindered. The intermediate carbocation was considered by Brown as a pair of rapidly interconverting classical norbornyl cation and not as Winstein's bridged nonclassical cation.

Here is an example where people were looking at the same set of data, made very sure that the data were correct, but one explained the rate

George Olah and István Hargittai at the University of Southern California, in George Olah's office, in 1996 (photograph by M. Hargittai).

ratio difference of $k_{exo}/k_{endo} = 300$, saying that the exo rate is faster by so much and the other said, no, the same result is due to the fact that the endo is slower due to steric reasons. It was the interpretation of the data which was argued, not the experimental facts. In this regard, I have a favorite quotation from a fellow Hungarian and Nobel laureate who won his Nobel prize in medicine for his studies of the inner ear, George Bekesy. He wrote that what all scientists need is to have a few good enemies. When you do your work and write it up and you send it to your friends, asking for their comments, they are generally busy people and can afford only a limited amount of time and effort to do this. But if you have a dedicated enemy, he will spend unlimited time, effort, and resources to try to prove that you are wrong. Bekesy ended up saying that his problem in life was that he lost many of his good enemies who became his friends. I don't want to comment on whether the situation is the same in chemistry or not, but I can assure you that many people have worked hard in this field to try to find experimental errors and enormous efforts were spent to establish and check the facts. I believe in contrast to many that this was no waste. One starts out to do research and initially you may have no idea of the importance of your interest. I came into this problem because I had a method which enabled me to look directly at intermediate carbocations. People challenged me and told me, if you can look at these ions, tell us, in the norbornyl controversy Winstein or Brown is right? I did it which, *per se*, was perhaps rather simple. In the process, however, I found the general basis of the electrophilic reactivity of σ-bonds which can not only be intramolecular (as σ-participation in the norbornyl cation), but can involve σ-bonds in intermolecular (between molecules) reactions. This represented the key for electrophilic reaction for alkanes (including their C–H or C–C bonds). We eventually even challenged methane, the simplest hydrocarbon molecule and we found that you can protonate (or otherwise attack by electrophiles) methane. The intermediate CH_5^+ is not only real but there is a broad chemistry of CH_5^+ and its homologues. Coming back to your original question, I was fortunate to do something quite significant in this field.

Was it an important setback in your career that after getting out of Hungary in 1957, you had to start all over in North America?

I was still rather young when all this happened. Even when I first presented my new carbocation chemistry with reference to its significance in 1962 at the Brookhaven Organic Mechanism Conference, I was only 35 years old. It's true that I was forced by events outside of my control to work in an industrial environment which some may think could not have been the most conducive for scientific research. But I believe that sometimes it is good to be challenged. I started after all in a poor country after the War under very difficult conditions. In retrospect it did me no harm because it forced me to do things that otherwise I wouldn't have done, and from which I later benefited a great deal. When I got out into, as we say, the big world, I worked for seven years in the industry. But, again, I always felt this was a challenge and a wonderful experience. I did a lot of practical chemistry too but in turn, I had also the freedom to pursue my own interests.

I rejoined academia in 1965 when I moved to Cleveland as Professor and Chairman of the Chemistry Department of Western Reserve University. I was able to maintain my research and teaching while doing also university administration. Across from our building, on the other side of a parking lot, not more than about 50 yards away, there was another sizable chemistry building belonging to the Case Institute of Technology. The two universities were really back to back, as were their two chemistry departments. In two years, we managed to merge the two departments into a single one. It worked out so well that it catalyzed the subsequent merger of the two

George Olah with biochemist professor Lars Ernster of Stockholm University (photograph courtesy of Lars Ernster).

universities to what became Case Western Reserve University. Having achieved my goal, with great pleasure I rotated out from being chairman. I told you this just to show that I always liked challenges.

As I was walking around the Caltech campus yesterday, I was thinking of your career in contrast with the ideal conditions at Caltech.

I am really not envious of anybody's conditions which may be better than others. I have a wonderful institute here at USC. At the moment, I'm also involved in what's probably the most enjoyable project in my life. At the urging of my wife and my friends I'm in the process of writing a book which is not going to be only just an autobiography. Let's face it, chemists are not, in general, the most interesting people, and telling about their lives and achievements in chemistry in great detail is not the most readable thing for the wider public. However, I've always had substantial interest in broader aspects of science. I believe in the universality of science.

Chemistry, although a central science, is considered "below" physics and mathematics but perhaps "above" biology. It's fascinating to me to consider how chemistry fits into the broader context. It's also inevitable to try to think about the deeper problems of existence, life, etc. Obviously, I'm doing also a lot of reading these days in writing my book.

Talking about Caltech, I've been reading recently some books about Linus Pauling, written by himself, and some by others. These books also give a good background of Caltech. It started as a small local school. A few remarkable people really built it up. Millikan was one, Noyes was another. When we talk about Caltech, of course, we realize it is now a world renowed institution. It's one of the ideal places for research, a large research institute with a small number of excellent students. What better can you have? All this developed from practically nothing, over a period of just a few decades. America is still a young and dynamic country. Here you still can do things. It's perhaps not so easy as it used to be, but if you have imagination and desire you can still achieve much. Just take my own Hydrocarbon Institute. We started from absolutely nothing in 1977 when I moved here. The University of Southern California started as a local school 115 years ago and became better known, besides its football team, for its professional schools. In the sciences though it hasn't been on a par, say, with Caltech or Berkeley or UCLA. Thus it gives me great satisfaction that during the past 20 years, I've been able to build up our Institute. It is a very unusual part of American science, where not many university basic research institutes exist. It's even different from Caltech's Beckman Institute. The Beckman Institute is a framework where scholars of different disciplines come together. We at USC built a more closely focused Institute as a framework for doing research in hydrocarbon chemistry. Hydrocarbons represent a significant part of organic chemistry and there are also great practical interests. The Institute was created through the help of friends and it wasn't financed by any public money. We just added a new wing and renovated the original Institute building, which will greatly help our future work.

Let's get back a little to your family background.

I was born in 1927. My father was a lawyer and to my best knowledge nobody in my family ever had any interest in science. I had no interest either as far as I can remember in my early school years. I was lucky to get a very good education. I was a student of the *Gimnázium* [liberal arts school at the junior high and high school level] of the Piarist Fathers in Budapest. I started it in 1937. I did well and I always read a lot. I was interested in literature, history, languages. In retrospect I am very glad that I got a very good general liberal arts upbringing. One of our science teachers was József Öveges who was teaching us physics. He was a very stimulating man. I feel embarrassed but I don't remember my chemistry teacher. If I had been asked at that time, I would have selected literature or history for my career. However, it was the end of the War and we lived in a poor and ruined country, so it was clear to me that I had to choose something practical and signed up for chemical engineering at the Budapest Technical University. There seems to be a tradition for this, for example, I believe that Eugene Wigner had done the same in the early twenties and there are other examples. Let me stress that chemical engineering is probably not the proper term for what we really studied. The chemistry education at the Budapest Technical University was built on the example of the German Technical Universities. In the U.S., chemical engineering means something closely related to mechanical engineering with very little "real" chemistry being taught. We studied chemistry, with minimal engineering.

As to the beginning of my chemical "career", I remember from my high school years that I had a friend and we set up some chemistry experiments in the basement of their house which ended up in a small explosion and disastrous stink. My friend's parents closed down our "lab" and this ended my early experimentation.

It was when I started University that I fell in love with chemistry. Probably it impressed me because it had such a wide breadth. Chemistry on one end of its scope represents the foundation of what now is modern biology, but it is also essential to the health sciences. On the other end, there are the practical aspects of making materials, plastics, pharmaceuticals — all the man-made compounds and synthetic materials essential to modern life.

Parents and children?

My mother died very early, in her late forties. My father lived until 1979 and could witness some of my progress in the field. I had an older brother who died at the end of the War in Russian captivity.

We have two wonderful sons. Fortunately, they are not in chemistry. I really think that children should find their own way in life and it is generally a mistake if they try to follow in their parents' field, particularly if their parents did reasonably well. Our younger son was a chemistry major at Stanford and then became a medical doctor. Our older son is an M.B.A.

The Olah family, 1963 (courtesy of George Olah).

and works for a financial company. They both live in this area. We have two adorable grandchildren, a boy and a girl who just arrived before Xmas.

You have written and edited quite a few books. What would you care to say about this experience.?

Over ten years, I had produced more than a dozen books. Even in my edited books, I wrote a significant part of them. Writing or editing books has forced me to organize my topic well and also do a lot of initial thinking. It's a most useful thing. My books are all related to my own work. The latest, *Hydrocarbon Chemistry,* deals with a very significant broad part of carbon chemistry. Up to this point, however, very little exposure was given to hydrocarbon chemistry. If you open the textbooks of organic chemistry, many aspects of new hydrocarbon chemistry are not even mentioned. Maybe my Nobel prize will help a little to change this. You find in Cotton and Wilkinson's *Inorganic Chemistry* much more about contemporary carbon chemistry than you find in the organic texts. This is in a way wonderful because I don't believe in dividing chemistry (organic, inorganic, physical, etc.) or even in "hyphenated" chemistries (physical-organic, bio-inorganic, etc.). Who can honestly define the difference between organic and inorganic chemistry? I wonder whether you can?

I can't even classify myself.

Neither can I. The best definition people generally give is that organic chemistry is the chemistry of carbon compounds. But I wrote a whole book summarizing work on hypercarbon chemistry discussing numerous carbon compounds that are not part of traditional organic chemistry. Norman Greenwood gives his definition of inorganic chemistry as the chemistry of the elements with some de-emphasis of the element carbon. This is correct but surely isn't a very scientific definition. I consider myself as a chemist who is interested in the chemistry of the elements with preference for some. I'm not, however, only a physical organic chemist. We are all using, of course, physical methods. If the tools are available, you would be a fool not to use them.

Going back to Hungary, in my time there was practically no modern instrumentation. In the whole country at the time I left in 1956, not a single NMR or even IR instrument was available. So obviously for me, using all the new techniques which came my way was a wonderful thing. We are generally using now quantum mechanical calculations. I am not a theoretician, but I have some very good students and co-workers who use theory. For us, this is a tool to supplement experimental work. As an example, we are continuously learning new things about CH_5^+ and its homologues including CH_6^{2+} and recently even CH_7^{3+}. There has been a lot of discussion of the structure of CH_5^+. What is really the meaning of structure? All what we know is generally where are the nuclei and we try to figure out the electrons. It's now realized that some structures, such as that of CH_5^+, are very flexible. The barriers for a proton shifting from one C–H bond to another are very low, making it to look symmetrical and it's a flexible system. There is, however, two electron three center bonding involved in it. Of course, in a strict sense there is no such thing as the chemical bond, only chemical bonding. Pauling's famous book *The Nature of the Chemical Bond* I really think is about chemical bonding. In CH_5^+, what we know is that five protons are simultaneously interacting, i.e., bound with one carbon atom.

We are in 1996 and pride ourself by our advanced knowledge. However, we can't define science without putting a historical marker on it. A few hundred years ago, which in the history of the Universe is not even a fleeting moment, the best scientists of the time thought that the Earth is flat and you should be very careful not to fall off the edges. They also believed that we are the center of the Universe. Now we look back and

George Olah in his Cleveland office, 1970. The photographs on the wall are (from the top, clockwise) Meerwein, Ingold, Brown, Whitmore, Winstein (courtesy of George Olah).

say how ignorant and naive these people were. We won't be around but in another few hundred years, future generations will look back at us and rightly may say the same about us. I only hope they will also say, at least, we tried to our best ability.

Could we return to the book you're currently writing?

As I said, it'll be more than just an autobiography including my thoughts about chemistry's role in a broader context. Some of my friends are publishing strict autobiographies in an ACS series. This is not what my book will be. I'm not criticizing anybody but it's somewhat disappointing that very few chemists really try to think in a broader sense to look also outside their own field and see how chemistry fits into the broader scope of our existence. Physicists and mathematicians are much more inclined to do so. Even biologists. There are, of course, exceptions among chemists, such as Ilya Prigogine, Manfred Eigen and others. Very few chemists, however, care about all this or even open their minds wider. There is, however, no reason why chemists shouldn't try to think more and I'm trying to collect my thoughts on a broader scope.

Back at the time of the Greek thinkers, the natural sciences were part of overall philosophy. What's happened is perhaps understandable but unfortunate. Science became so complicated that most thinkers (philosophers) had just given up on it. That's why people with scientific backgrounds have a role to express themselves. This keeps me interested and gives me a challenge. I hope what I am writing will be of some interest and even readable.

John D. Roberts, 1996 (photograph by I. Hargittai).

22

John D. Roberts

John D. Roberts (b. 1918) is Institute Professor of Chemistry, Emeritus, and lecturer at the California Institute of Technology in Pasadena, California. He graduated from the University of California at Los Angeles with an A.B. degree in 1941 and a Ph.D. degree in 1944. After initial appointments at Harvard and MIT, he came to Caltech in 1953. His research has been mainly concerned with the mechanisms of organic reactions and the application of nuclear magnetic resonance spectroscopy to organic and bioorganic chemistry and biochemistry. He has numerous memberships and awards, including the memberships in the American Academy of Arts and Sciences and the National Academy of Sciences; the Priestly Medal of the American Chemical Society (1987); and the National Medal of Science (1990).

Our conversation was recorded in Dr. Roberts's Caltech office on February 19, 1996, and was first published in *The Chemical Intelligencer.**

You made a moving statement in your autobiography that you spent years on the center stage and then you started experiencing that fame fading.

I think that happens to everybody, it even did to Pauling. There are a number of changes that you experience. The first one is that you can't

*This interview was originally published in *The Chemical Intelligencer* **1998**, *4*(2), 29–33 © 1998, Springer-Verlag, New York, Inc.

go on taking graduate students, and it diminishes your influence among young people. I've been able to get around some of that by two things. One is by still teaching. I give a course about NMR which I try to keep up-to-date. The other is by doing research with undergraduates. Postdoctoral fellows are expensive. So it's quite natural that although your colleagues respect your opinions, they have their own agendas, and the future of the Division is in their hands. Once you reach retirement here, you are not invited to participate in the selection of new faculty although you are invited to participate in the normal business of the department. Then the same thing happens to you nationally. An exception is the National Academy where I've stayed very active, and the Organic Synthesis, Inc. where I've been member of the Board.

Then, of course, what happens with everybody is that if you do some important work, it is identified with you for a certain length of time and then, unless it's something like the Friedel-Crafts reaction, you gradually lose that identification. I recently saw a textbook of organic chemistry discussing our work on benzyne for two pages without mentioning our names.

Which of your results would you like to be most identified with?

For elementary textbooks, I think people would pick out work on carbonium ions, on benzyne, on cycloadditions, all the stuff we'd done on NMR. I wouldn't put any of those above the other.

One thing is to be remembered for your work that goes into the textbooks, the other is the general influence that you have. We had, for example, the first digital NMR spectrometer that I know of, and that's now sort of a standard, and there is no reason for me to be identified with it particularly. But what is most important is what your students do.

About the carbonium-ion controversy, does it seem as important in retrospect as you felt it was during the actual discussions?

If you look at the great controversies in organic chemistry, they were interesting because in most of them you had to decide what it is that you have to write down on paper to represent chemical structures. It's hard sometimes to reach satisfaction about that. If you go back to the really old days, they had a lot of trouble in formulating organic compounds and being satisfied with what they had got. They had problems with benzene and resonance hybrids of various kinds. The carbonium-ion discussions represented a step further beyond that, in which you're getting a kind

J. D. Roberts with friends of thirty or more years at the First Annual William S. Johnson Symposium at Stanford University in 1988: (from left to right) Albert Eschenmoser, William Johnson, Gilbert Stork, Derek Barton, Carl Djerassi, and Konrad Bloch (photograph by Eric Leopold, courtesy of J. D. Roberts).

of behavior that was really very difficult to rationalize in terms of the usual simple structures. George Olah's work has certainly extended that enormously. But even before that, we had a lot of structures that just couldn't be made to be consistent with the old ideas. Some people will fight very hard against that; others will go along with it. In this particular case, there was some justification for saying that maybe those structures aren't right, and there ought to be experimental verification. A result of all that was a lot of interesting chemistry. Herb Brown prolonged this discussion because he was unwilling to accept the new ideas, and I never quite figured out what motivated him, except that I have the feeling that he felt that he wasn't part of the establishment.

Was he?

No, he wasn't part of what you might call the fashionable establishment anyway.

Once I was set up with him for a debate at Columbia University on a fairly short notice. The Columbia people arranged his lecture and my lecture at the same time. There was an interesting and lively discussion, but it didn't make much progress. In any case, he did provide impetus for doing a lot of work. Some of it was very interesting work. I find it fascinating that over the years he has never allowed that there had been any change in the situation in spite of all the new results, and has still felt that he was right. At times, I got so discouraged with this debate that I decided that I'd work on something else. But then I'd come back when something new came along, like George Olah's work on superacids.

Was Olah part of the establishment?

Well, early on, when Paul Bartlett and Saul Winstein and myself and some others who were part of the establishment met together, we'd discuss each other's work. George didn't seem to be too much interested in entering discussions of the details of other peoples' work, so he didn't often participate in those discussions, and I don't know why. But later, he had a lot to contribute, and that continues to be true. When George says that you've got a situation where you can put sodium bicarbonate in acid and you don't get carbon dioxide coming off, that requires some real restatement of what you teach in elementary organic chemistry or even in freshman chemistry. I think, on the whole, people have not moved in his direction as fast as they should.

At one point of your career, you became Chairman of the Chemistry Division of Caltech. This is the position Linus Pauling had had before.

Yes, but more important to me was that I was the professional successor of Howard Lucas, a very fine organic chemist. Early on, Howard was not thought of so highly by the people here. He was very unusual. He did not have a Ph.D., and I don't know of anybody else elected to the National Academy of Sciences in chemistry who didn't have a Ph.D. He wrote the first modern textbook of organic chemistry in 1936. Winstein was his graduate student. He did some very interesting work which Winstein followed up on later on.

Last April, in 1995, I attended a session honoring the memory of Linus Pauling, at Anaheim, as part of the ACS meeting. The session was organized by Caltech, and some people were surprised to hear that Pauling felt more or less forced out of Caltech.

I don't believe that at all. It may be a question whether he was forced out as Division Chairman. I've been researching this question, because he got me in the parking lot once, about 1956 or 1957, and he asked me if he ought to resign. I said, no. The trustees were pretty upset with him, but they were not going to kick him out of his professorship, that had been established. I remember one of them telling me that Pauling had cost them millions of dollars in gifts. My counter to him was that schools with principles got more millions of dollars in the long run. When Pauling talked to DuBridge, according to one story, DuBridge told him that the trustees could not fire him as a Professor, but they could fire him as Division Chairman. To which Pauling said, if they can do that, I might as well resign. Then when Linus resigned, he wrote a very nice letter of resignation. He did not say, I've been forced out.

As to his being forced out later on, that's not true either. He resigned. That was under my chairmanship. He came in and he said, I've decided to resign. I think he had reached a stage of his life in which he wanted to be more involved in politics than he wanted to be involved in chemistry, and he probably felt that Caltech wasn't a good place to do it. It's certainly true that there were a lot of physicists around here, including the President and the Provost, who felt that nuclear testing was important to the security of the country. They didn't want to see it stopped. So they were very antagonistic to Pauling and his getting the Peace Prize. But so were a lot of other people and *The New York Times* was editorially highly antagonistic to his getting the Peace Prize. Pauling may have wanted everybody to jump on board. We had a meeting to honor him, and I went to the meeting, and I believed that he was doing the right thing, and I feel that even more strongly now. You see the results of what not stopping testing altogether has done. I think this gave him a good excuse to leave and it was probably an appropriate choice. On the other hand, when we offered him the chance of being connected with the Institute as a Research Associate, a quite honorable title, he accepted it graciously. This continued up to when he became 70 at which point we made him Professor Emeritus and he accepted that. Later on, we organized birthday parties for him at 85 and 90 and they were big successes.

Do you have a museum or some memorial to Pauling here at Caltech?

There is a Linus Pauling lecture hall, and there was a picture of him somewhere up on the wall but the Chairman put it in his office.

What turned you to chemistry?

In the sixth grade I read the biographies of a number of scientists, and then I read Paul de Kruif's book, *Microbe Hunters.* It was a very popular book. It described some of Pasteur's experiments on spontaneous generation of life and so on. Then, a little later, another book came along to me, called *Creative Chemistry*, by Slosson. It was partly organic and partly inorganic, and it first came out before the War, that is, World War I, and I don't know whether it has ever been revised or not. It was very inspiring, but I was already experimenting with a chemistry set before I got this book. I was 12 or so when my parents gave me a chemistry set. I found some of the experiments rather dull so I started reading some chemistry books because I wanted to liven up my experiments, and I bought some equipment, and my mother gave me a cabinet in the kitchen to store my stuff. My father was a handy person. He didn't know much about chemistry but he was interested in mechanical things and he liked to make things. We did some forays into glass-blowing but not with much success, although later on I did become a very good glassblower. I also had a cousin who worked in a scientific supply house in Los Angeles, and he would give me things that were discards and so on. Later I moved more into the garage, so I did quite a bit and did a lot of reading of textbooks in high school.

You have written about the history of physical organic chemistry. How is this field evolving?

It has changed some because of new instrumentation and better theories. Much of it has been absorbed into synthetic and bioorganic chemistry. The inorganic guys have also understood finally that those methods are helpful for them. So in many ways it has spread through chemistry. What's left in the middle of the field is not pursued so much these days. I feel that is what's going to revitalize, and I am already positioning myself for it, that it will go back in the middle where we don't really know as much as we thought we did. There again, more and more people are getting interested. It's a nice area because this is where calculations using quantum theory ought to be able to work, and predict things.

Conformational analysis used to be not very actively pursued in physical organic chemistry because there wasn't very much you could do. It all changed when NMR came along, and I decided at the early stages that I wanted to work on NMR and conformational analysis. We did some of the very early experiments to show that you could stop rotation around a carbon–carbon single bond. Now I have returned to conformational problems. There's still a lot that we don't understand about conformations, particularly in water solutions. There are ions involved, and hydrogen bonding is involved, so the whole situation is very complicated. We don't understand the details of solvation very well.

When we started off in physical organic chemistry, we couldn't do any of this. Now you can spend your whole life doing nothing else, and that's a big change. It's almost scary that we're in a situation in which people can be so specialized. This makes it more difficult to move people to do completely new things because everything is getting harder and harder to learn as things become more complicated.

When I was starting, we were using all the same techniques that people used for at least the last 50 years. The only instrument we had in the laboratory was a refractometer. When the refractometers were inadequate, there wasn't much you could do. The work that I did on the Grignard reagents we couldn't solve except by looking at the reaction products. We came up with nearly the right solution from that work but there were still a lot of surprises when eventually we could determine the Grignard structures by NMR. In some ways, it was much easier in the old days, but on the other hand, you didn't find out nearly as much, and it was so much more work. It took a long time merely to find out what your products were, and we had to use much larger quantities than today.

Do you consider yourself a rather unusual synthetic organic chemist, writing books about NMR and molecular orbital theory?

I don't really think so. Those books I wrote aren't really high level. On the other hand, Michael Dewar did some experimental work in organic chemistry in the early days, and Ken Wiberg has done both theory and experiment for many years. I don't think all this is so unusual. It was the product of the times. The reason I got into it was because I wanted to be more quantitative about these things. One thing I've enjoyed about my recent book-writing is that I've learned enough about the mathematics of NMR to actually be able to calculate what different kinds of conditions will do to the spectra and make graphs and so on. The calculations are very, very accurate and reproduce the machine behavior very well.

How many books have you published?

Nine. I'm working on a book right now. It's about NMR, intermediate NMR, between what you learn about it in organic chemistry and what you need if you want to write pulse sequences for multidimensional NMR. My molecular orbital book went to 16 printings. People still ask me for permission to xerox portions of it for classroom use. It should be updated but it hasn't been high on my priority scale.

What's your experience with publishers been like?

On one occasion I helped start a publishing company. William Benjamin published my first NMR book with McGraw-Hill. He was an editor there, and that worked quite well. Then, I was sort of an adviser to the company for a while on a chemical series. The next step came when Benjamin and I met in a hotel in Cleveland during a meeting, and he said, "I'm thinking of starting a company of my own, would you like to get involved?" So he and I and Konrad Bloch, David Pires and a lawyer sat down and had the first board meeting. Eventually, the company got big enough to move into quarters on Broadway in New York. The company went pretty well for quite a while, and got to be too big for Bill to handle by himself. Because he was so bright and so impetuous, it was hard for people to work with him in a large-scale organization.

When I became Division Chairman, it became harder and harder for me to find time for the company. The company was doing very well, and finally it was bought by Addison-Wesley, and at that point I got out of it altogether, except they did publish the second edition of my organic book. I still get royalty checks. Small ones.

One of my books, a shorter version of *Basic Organic Chemistry*, was pretty popular, and I gave the copyright to my children for their college expenses, and that worked out extremely well. They still get a few dollars twice a year, 20 years later.

In addition to your involvement with the publishing company, did you have any other outside activity?

I consult for Du Pont. I've been doing it for 46 years now. I keep myself pretty busy.

Richard R. Ernst, 1995 (photograph by I. Hargittai).

23

Richard R. Ernst

Richard R. Ernst (b. 1933 in Winterthur, Switzerland) is Professor of Chemistry at the Laboratorium für Physikalische Chemie der Eidgenossische Technischen Hochschule, ETH-Zentrum, Zurich, Switzerland. He received the Nobel Prize in chemistry, 1991, "for his contributions to the development of the methodology of high resolution nuclear magnetic resonance (NMR) spectroscopy."

He studied at the Swiss Federal Institute of Technology (Eidgenossische Technische Hochschule, ETH) in Zurich. After having received his diploma as a "Diplomierte Ingenieur-Chemiker" he did his Ph.D. work at ETH between 1957–62. After that he worked for Varian Associates in Palo Alto, California. He returned to ETH Zurich in 1968 and took over the NMR research group at the Laboratorium für Physikalische Chemie. At some time he also served as president of the research council of ETH Zurich. His interest in Tibetan arts is getting more emphasis as retirement age is approaching in 1998. The interview was recorded on August 28, 1995, in Budapest and it appeared in *The Chemical Intelligencer.**

Would you give us a glimpse into your contributions to the development of NMR methodology?

Let's start with what NMR is all about. In science, we have to make measurements to learn something about nature, and for this we need sensors.

*This interview was originally published in *The Chemical Intelligencer* **1996**, *2*(3), 12–17

Fortunately, nature has built spies into all molecules. These spies are the atomic nuclei, which sense what is going on around them. We have the means to communicate with these nuclei, to ask them questions and to record their responses. The nuclear spies are small gyroscopes which are precessing and rotating and they are magnetic. So when they are in a magnetic field, they assume a particular precession frequency. By measuring the frequency, we can determine the strength of the local magnetic field, and this is our major source of information. The local magnetic fields inside matter are influenced by their surroundings. There are shielding effects of the environment which affect the local magnetic field. We measure the residual magnetic field at the nucleus and can make inferences about the shielding coming from the environment. This information tells us something about the internal structure of molecules, materials, and biological objects. In addition, there are magnetic interactions between the nuclei by which we can determine the distance between the two nuclei participating in this interaction. There is also another kind of interaction, called *J* coupling, which is a quantum mechanical interaction that depends on the bond angles. Thus the entire molecular structure emerges from these measurements. All this was already known when I started working in this field, although it was not much utilized because the sensitivity was not good enough to observe the NMR signals. They involve really very weak interactions, which are very difficult to measure.

I started my Ph.D. work in 1957 and finished in 1962 at the ETH under Professors Hans Primas and Hans Günthard. Then in 1963 I went to Varian Associates in Palo Alto, California. I have always been interested in sensitivity and it always bothered me how difficult it was to distinguish the signals from the noise in the experiments. I was sure that this problem had to be solved if the technique was to be of any practical use.

In Palo Alto I was involved in instrument building and was working with Weston A. Anderson. Together we came to the idea that we could use a multiple-channel concept involving a Fourier transformation for enhancing sensitivity. The idea is simple, we can compare it with tuning an old piano. Let's assume that it has the full keyboard but only a few strings left in it. You can go from the left-hand side to the right-hand side, pressing one key after the other, just to find out which strings are in place. This is a tedious experiment; you press a lot of keys and at the end you may have found, say, four strings altogether. There are better ways to do this, for example, you press a whole range of the keyboard with your arm at once and you get a superposition of different responses.

If you can't disentangle the various sounds, then you do a Fourier transformation in order to separate them. This is exactly what we are doing in Fourier transform spectroscopy. We don't go through the spectrum looking for individual responses, one nucleus at a time, but we get the responses of all the nuclei at once, and then we unscramble them mathematically. So we introduced the Fourier transform technique to NMR spectroscopy.

First we tried to publish our work in 1965, and our manuscript was twice rejected by the *Journal of Chemical Physics*. The objections were that it had nothing to do with molecules, it was not of fundamental interest, and that it was not really original. Finally we published it in the *Reviews of Scientific Instruments* [**1966**, *37*, 93]. We also took a patent out on this concept.

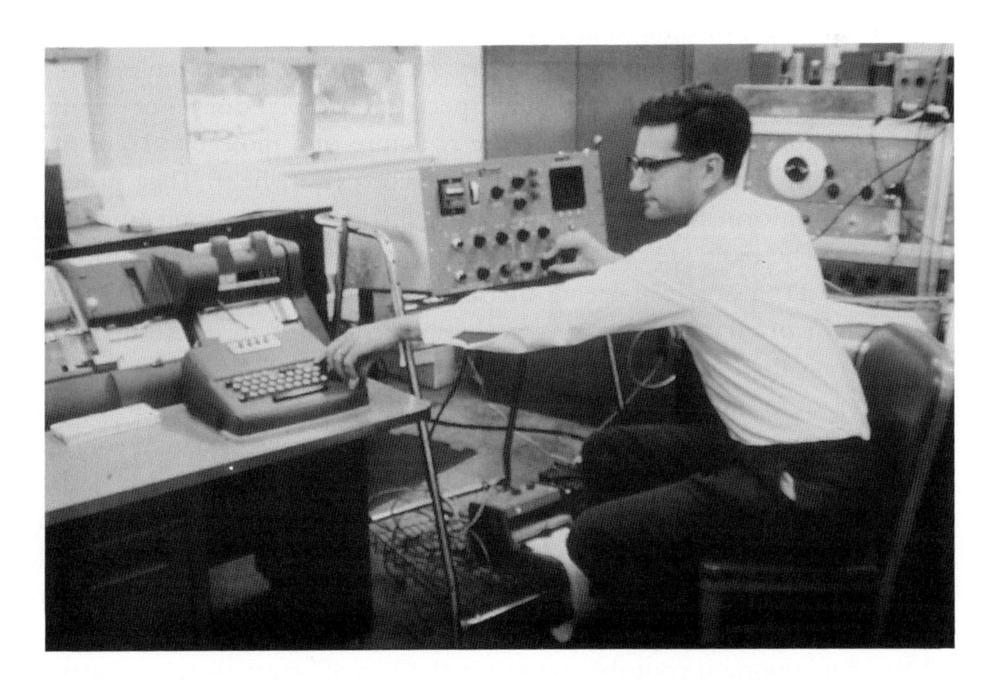

Richard Ernst in 1965 (courtesy of Richard Ernst).

Was the concept applied right away?

I was working at the leading instrument company at that time, insofar as NMR spectroscopy is concerned, and the company was just developing a new series of instruments, but the designers didn't want to incorporate the Fourier transform concept in these new instruments. They used field modulation for suppressing noise at that time, and whenever you use field modulation, you can't use pulse excitation at the same time because of the synchronization problems. They didn't recognize the importance of Fourier transformation and they didn't want to use it. Finally, it was their competitor, Bruker Analytical Instruments, that put Fourier transform NMR spectroscopy into practice and produced the first such instrument in 1969. The two companies had a patent agreement and everything went very smoothly between them.

Would you like to single out anything else from your work?

In essence most of our contributions were connected with Fourier transformation. First we introduced the one-dimensional Fourier transformation. Then came the two-dimensional Fourier transformation, and it was enormously important for determining the structure of biomolecules.

A one-dimensional conventional spectrum simply does not contain sufficient information to determine the structure of a molecule. Here two-dimensional and three-dimensional spectroscopies can help. They produce correlation information which measures distances and bond angles in molecules. The basic idea of two-dimensional spectroscopy was contributed in 1971 by Professor Jean Jeener of Brussels, and Professor Kurt Wüthrich was much involved in its adaptation to biomolecular structure determination. We did our first two-dimensional experiments in 1974.

At the same time we introduced Fourier imaging for potential medical applications, which was, again, based on the Fourier transformation concept. The imaging concept itself was introduced into NMR in the early 1970's by Paul Lauterbur. We just made the technique more efficient by using Fourier transformation of the response to time-dependent magnetic field gradients within one experiment. This is how one can get the image of a human head or a human body.

Another important contribution was in the area of heteronuclear resonance. Thus, for example, carbon-13 resonance today is as important as proton resonance. It is essential because the carbon atoms are where the action is in the molecules whereas the protons are often peripheral. The carbon-13 nuclei are also very sensitive to changes in their environment. However, carbon-13 is very rare, only one in a hundred carbon atoms, and their magnetic moment is very small. This leads to a low signal-to-noise ratio, and remedies are required to enhance the signal strength. One possibility

Four men who had a profound influence on Richard Ernst's career (photos courtesy of Richard Ernst): Hans Günthard and Hans Primas under whose supervision he completed his Ph.D. work at ETH Zurich (1957–62), Weston A. Anderson with whom he worked at Varian Associates in Palo Alto, California in the mid-1960s, and Jean Jeener of Brussels whose idea of a two-pulse sequence that produces, after two-dimensional Fourier transformation, a two-dimensional spectrum, in 1971, gave impetus to further work in Professor Ernst's laboratory.

is to decouple the carbon-13 spins from the proton spins to concentrate the numerous weak signals in a few stronger ones. For this we can use spin decoupling by applying radio frequency fields, which is an old well-known technique. What we introduced was a broad-band decoupling technique by which all the protons can be decoupled at once from carbon-13 and vice versa. This actually went back to my thesis research in Zurich, where I was working on stochastic resonance using random noise for the excitation, but we never did experiments at that time.

Both X-ray crystallography and NMR spectroscopy provide information on the three-dimensional structure of proteins. How do they compete or complement each other?

There are practical aspects. In order to do X-ray crystallography, you have to grow a single crystal. When this can't be done, there is NMR spectroscopy. An important aspect is that NMR spectroscopy works in solution, which is the natural medium of these biologically active molecules. Crystallization may change the conformational properties of the molecule. On the other hand, NMR spectroscopy has its limitations. X-ray crystallography can be applied to any size of molecule; there is no limit to molecular mass, it may even be one million or more. NMR is restricted to relatively small molecules. The limit is somewhere between 20 and 30 kilodaltons at the moment. For large proteins, it has also been found that they don't change their structure much upon crystallization. There is a lot of water in crystalline proteins, 50% or even more, so, in a way, the molecule is also in its natural medium. This is not so, of course, for crystals of small molecules; in their case, intermolecular effects may be much more important.

An advantage of the NMR work in solution is that it is possible to follow structural changes as a consequence of changes in the pH, changes in the salt concentration, adding a substrate to see whether it binds or not, and so on. You see these changes immediately in the NMR spectrum. With X-rays, this is not possible. A general advantage of NMR is in the study of the dynamics of molecules, and this is especially interesting for floppy molecules. The time scales that can be studied by NMR range from very slow processes down to 20 or 30 picoseconds, but the important biological happenings are relatively slow, more in the range of microseconds or even milliseconds. Under these conditions, NMR spectroscopy is unique. We can also look at individual sites in the molecule by addressing selected spies and determine whether they participate in the changes or not. So it is a very selective technique and this is one reason why it is so useful.

How did it all begin for NMR?

It was the work by Felix Bloch, another Swiss, who went to the United States and worked at Stanford, and by Edward Purcell at Harvard, and they did the first NMR experiments in the condensed phase simultaneously in 1945. The very first magnetic resonance experiments were actually done by Isidor Rabi in the gas phase in molecular-beam experiments in 1939. From that, one could infer that a similar experiment should also be possible in condensed matter.

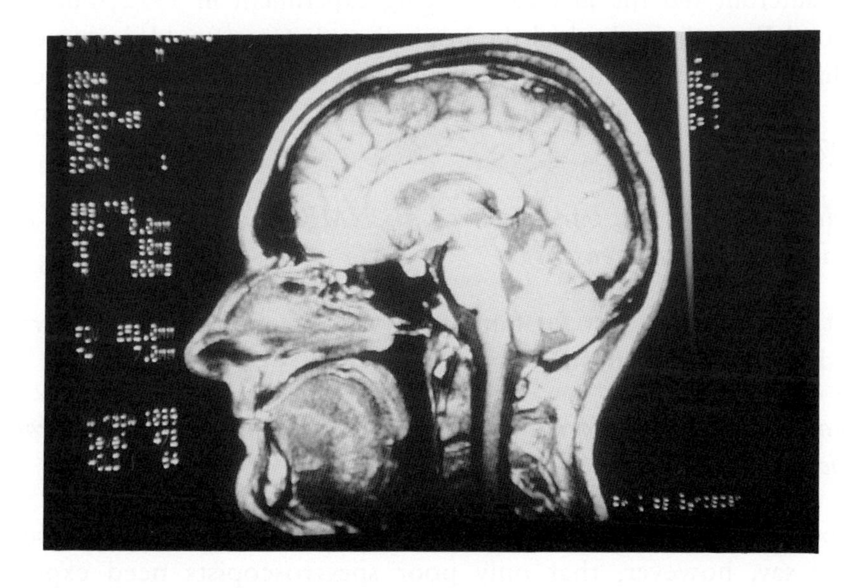

Nuclear magnetic resonance image of Richard Ernst (courtesy of Richard Ernst).

In the case of X-rays, the first applications were medical and then came the scientific applications. With NMR it was the other way around.

Edward Purcell was asked once in 1946 whether NMR could ever be used in chemistry, and Purcell said, never — it was completely useless for chemistry. At that time, it was a basic physical technique used for measuring nuclear magnetic moments very accurately to learn more about nuclear structures. The transition to chemistry and biology was very slow. Then, finally, slowly, NMR came to medicine. This was the natural way, if so to speak. X-rays, on the other hand, jumped directly into clinical applications, and within two years all the hospitals had these crazy Röntgen machines for imaging people. It happened extremely rapidly. With NMR it took several decades. Again, there was the sensitivity problem because it is really difficult to get the signal out of the noise. In the early 1970s, it was unthinkable that the human body could be imaged by NMR. Even as we were introducing Fourier imaging in 1974, I couldn't have believed it.

Who was the first who put a human into this huge magnet?

Paul Lauterbur did the first real imaging experiment in 1972, but it was Raymond Damadian who had the inspiration that one could do something like that, although he didn't know how to do it. There is some controversy about his claims. He is a medical doctor and he did a rough human scan in 1977.

I understand that you had prepared an NMR image of your own head. Was it for medical reasons?

No, it wasn't, it was just for fun, and I was disappointed that everything looked completely "normal".

In many NMR laboratories, the higher the field strength is, the prouder the people are.

It shouldn't be like that but it's easy to think in terms of numbers. I usually say, however, that only poor spectroscopists need expensive spectrometers. It depends, of course, on what you need the spectrometer for. If you want to invent new experimental schemes and new approaches, then you can also work with weaker fields. However, if you explore very complicated molecules, then you need strong fields.

What's the highest NMR frequency available?

At the moment 800 MHz. The first instrument is just being installed in Frankfurt, in the lab of one of my former coworkers. This is the maximum, while 750 MHz is today more or less routine. There are about 20 such instruments in operation around the world.

Would you like to go even higher if you could?

Of course, it would be nice to have higher magnetic field strengths but it's a question of balance between effort and profit. With brute force it might be possible even to go to 1,000 MHz, but that would be very expensive. That's about the limit with presently available materials. For going higher, first there'll have to be advances in superconductor materials.

Where is your main interest at this moment?

If we stick to NMR, what interests me for the moment is developing hands-on methods to investigate dynamic features of molecules. The structure itself is determined by a relatively small set of parameters. It is much more complicated to describe the dynamics of a molecule or an object. One has to follow the time course of a molecule, and the measurement techniques are difficult and a huge set of data must be collected. We always use simplified models to describe the dynamics of a molecule. I am presently interested in how to best approach this problem.

There is then the solid-state NMR, which is still somewhat less developed with regard to routine applications than NMR of liquids. Complicated anisotropic properties come into the picture. That's also something I'm interested in at this moment.

We are presently working on ordering phenomena in disordered materials, such as local order in glassy materials. There is a lot of development to be done in these areas, and we are very active in them.

Coming back to your personal history, how did you get back to the University?

The superficial reason was that I got an offer to come back to Switzerland. My thesis adviser, Professor Primas was an ingenious NMR spectroscopist, but he was not satisfied by the experimental work and became increasingly theoretical and moved even in the direction of philosophy. As his research group needed somebody to be in charge, I was called back to the ETH.

I had a lot of hesitation to accept the job, which was not a secure job and was not a tenured position. I had to work to overcome many problems, and after one year I had a nervous breakdown. I really liked it much more in the United States.

So what was the real reason for your coming back?

I really wanted to return to my home country. I felt attached to it. It makes more sense to work for the country in which you are born. I didn't want to be somebody just escaping, taking the easy way out. I wanted to return, though not necessarily to a university but to Switzerland.

At Varian it was more or less an academic atmosphere in the research department. We could do what we wanted to, so the step was not so great from industrial research to university research. But in Palo Alto we had some motivation. We were working for the company and for the profit of the company, in spite of the fact that my most important results were not used by the company.

Eventually, however, my return to university life worked out fine, but it was very hard, especially the first few years.

What was your main problem?

It was mostly human relations. Switzerland is small, and we didn't have the space and the means that you could take for granted in the U. S. I felt a lot of pressure, and I felt that I couldn't put the gifts I had to good use.

Moving away from NMR spectroscopy, what are your other interests?

There are two other interests. One is music and the other is Asian art. I used to play the cello in small orchestra, but it never became very serious. I may have played with the idea of becoming a composer as other children play with the idea of becoming a steam engine driver. Also, my interest in chemistry started early when I found a box of chemicals in the attic of our house. It was left by an uncle who had died before I was born. I also read all the chemistry books I found in the town library. I was then about 13. I was fascinated by my chemical reactions. Fortunately, I survived them — and became a chemist.

What's your family background?

My father is an architect and mother worked around the household. My family has lived in Winterthur since the fourteenth century, and I feel also attached to the place, and it's so strange now to be a local celebrity there.

How did you become interested in Asian art?

I have always been interested in religious art, not because I am particularly religious. It is always very emotional art, expressing the deepest feelings of people. The artists don't do it for becoming famous; they are just expressing their profound belief. Astonishingly, you find the same motifs in all the different religions. Being a scientist and having to worry about abstract concepts, so remote from our own feelings, I miss something and I can find this in the religious aspects of art, even without having to practice it.

Tibetan art is very explicit in its visual expression, so I find it easier to approach than other religious art. Our first exposure to Tibetan art was when we were returning home from the U. S. in 1968. We sent the children directly home and traveled through Asia. This is when we bought the first pieces of what has since become quite a collection of Tibetan art. Now we have a full house of it, mostly thangka paintings, so our house has more of a Tibetan atmosphere than a Swiss one.

Richard Ernst on board a plane surrounded by flight attendants, just minutes after the pilot had communicated the news about Professor Ernst's Nobel Prize, October, 1991 (courtesy of Richard Ernst).

Does Mrs. Ernst share this interest?

She likes it although she is not interested in it intellectually. She's an elementary school teacher by training and she is very much interested in music. She sings and plays the violin. She stopped having a paid job when we left for the U. S. We have three children, two daughters, 30 and 27, and a son, 22. Our oldest daughter is a kindergarten teacher and interested in art, the younger daughter is an elementary school teacher and interested in music, and our son is studying psychology at the University of Zurich. They don't want to do science, and they don't like too much the publicity around me.

Do you have any involvement in public issues?

I have got involved in the issue whether or not Switzerland should join the European Community research projects. We organized a letter to the government signed by five Swiss Nobel laureates. We told the government that they should be careful in diverting research money, normally used to support Swiss science, to Brussels in the hope that more than 100% would be coming back; this is what they were aiming for. We were skeptical. It would be a different situation if the Swiss scientists were isolated from the rest of Europe. But this is not at all the case. There is nothing more international than science, and we are very much involved with international cooperation, but we don't like the idea of programming it from the outside. I am also involved in the support of science through the Swiss National Science Foundation.

Is there anything you'd like to add to this conversation?

There is one problem that concerns me at the moment. It is the relationship between the general public and science. I think it determines the future of mankind. Although science has created problems, it's wrong to blame science for all these problems. Science stems from our very natural curiosity, and without science and technology the human society would never have developed into what it is today. Without science we would have remained in a very primitive state of development. In the future, we have to face many serious problems of survival, problems, that can only be solved by even more science. Science and technology is part of the human nature which we cannot deny. I think that it is very important to improve the public's understanding of the indispensability of science. We scientists have to become more active in informing and educating the general public and the politicians. We can't afford to remain safely in our ivory tower. We have to engage ourselves in the dispute with the public for the future of mankind.

How long have you been concerned with this question?

Since I received the Prize. But I had always a somewhat ambivalent relation to science. I was fascinated and saw its indispensability, but I saw also the inherent dangers. We have to learn to live with them as much as we have to learn to live with our own negative sides.

Eiji Osawa, 1994 (photograph by I. Hargittai).

24

Eiji Osawa

Eiji Osawa (b. 1935 in Toyama, Japan) has been Professor of Computational Chemistry since 1990 at the Department of Knowledge-Based Information Engineering of Toyohashi University of Technology, Toyohashi, since 1990. He studied chemical engineering at Kyoto University where he got all his degrees, including D. Eng. in 1966. He then spent three years as a postdoctoral fellow in the United States, at Wisconsin and Princeton. He was Associate Professor from 1970 till 1990 at the Department of Chemistry of Hokkaido University in Sapporo. He is most famous for having predicted the stable C_{60} molecule of truncated icosahedral shape in 1970, long before it was discovered in 1985. He did this entirely on the basis of symmetry considerations, without calculations. Professor Osawa became a computational chemist later in his career. Our conversation was recorded in July, 1994, during my visit of a few days in Toyohashi and it appeared in *The Chemical Intelligencer*.*

Your department is called Knowledge-Based Information Engineering. What does this name mean and what does it imply?

Actually, nobody knows what knowledge-based information engineering means. The reason for using this strange name is to attract and educate young people as software engineers. Such a department is usually called

*This interview was originally published in *The Chemical Intelligencer* **1995**, *1*(3), 6–11

Plaque of Eiji Osawa's department (photograph by I. Hargittai).

information science. However, there are many departments with that name. Originally, we felt that it would be easier to persuade the government to invest in some new and urgent field so the name information engineering was invented.

Do you get good chemistry students?

This has been a problem. Few chemistry students have applied because of our name saying nothing about chemistry. On the other hand, having chemistry in the name would perhaps not enhance our ability to attract students. Students don't like chemistry because of the bad publicity it has received. We have a chemistry department here which is called materials science, but it doesn't get good-quality students. If you look at a lower level of education, in this case at the technical high school, you see the chemistry departments declining. All good students go into either bio-related fields or information science. So chemistry is going to die [Professor Osawa laughs heartily]. There is then another problem in our name not containing the word chemistry. Companies in Japan usually select a department from which to recruit young professionals, and we find it difficult to attract chemical companies to come here and recruit. Since we educate information scientists rather than chemists, I also have to look for graduate students in other places. About the other part of the name, that is, "knowledge-based", please, don't be bothered about it, I'll change it.

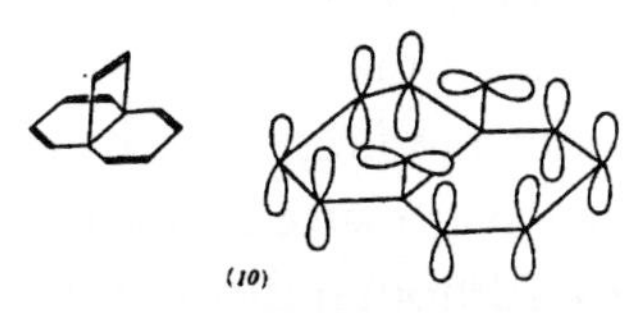

偶数系のビシクロ π 電子化合物は barrelene のほかには知られていないが，[4.2.2] propella-2,4,7,9,11-pentaene (11)[10] が非常に密接な関連をもつことは構造式から明らかである．(10)においてはトリシクロ系骨格によって分子の inflexibility がビシクロ系に比べてさらに増しているため図に示したような π 電子の重なりがさらに大きいと考えられるが，これもやはり熱力学的安定化は特別に起こっていないようである[10]．

(11)

(10)のビニローグ(11)[11]はまだ合成されていない．

2-3 corannulene

多数のベンゼン環の縮合した型のいわゆる"縮合多環式芳香族炭化水素"は典型的な平面分子である．これらの代表的なベンゼノイド芳香族が球状分子の型をとったなら超芳香族性を示さないだろうか？ たとえばサッカーの公式ボールの表面に描かれている幾何模様を思い浮かべてみよう．それは正多面体として cube のつぎに小さな正二十面体 (eicosahedron) (12)の頂点を全部切り落として正五角形を出したもので，truncated eicosahedron とでも称されるべき美しい多面体である(13)．図ではわかりにくいところもあるので，もし手もとにサッカーボールがあれば手にとってながめていただくとはっきりするが，五角形(黒く塗ってある)の間には規則正しく六角形がうずまっている．一見これらの成分多角形はたいして曲がってもいないし，各辺はすべて同じ長さにすることができる．もしこれらの頂点を全部 sp^2 炭素

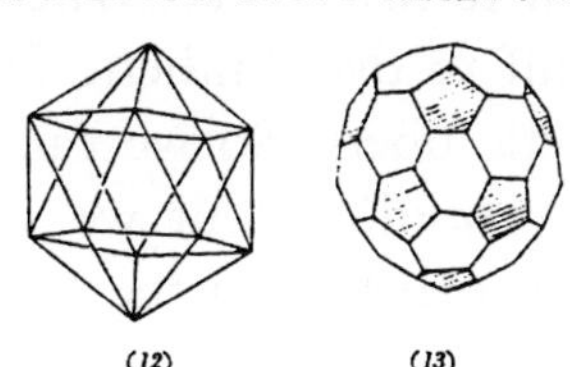

表

	π 結合エネルギー[a]
平　面 corannulene	28.924
非平面 corannulene	26.306

a) $\beta_0 = -1.7515$ eV

で置き換えることができれば球面状共役が実現できないだろうか？

実はこのサッカーボールの表面模様の最小反復単位を切り出した形の炭化水素が合成された[12]．それが corannulene (14)である．この分子はいろいろの点で非常に興味深い．まず (15) のような dipolar 構造の共鳴混成体である可能性がある．というのは(15)においては周

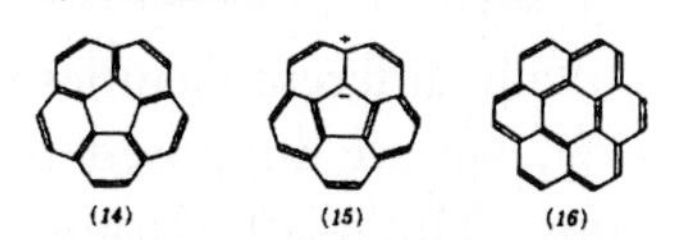

辺が 14π 系の環状共役，中央部が 6π 系の環状共役を形づくり，いずれも 4n+2 則を満たしているので強い芳香族性の発現が期待される．つぎに coronene (16)(平面分子)と比べてみると明らかなように(15)においてはベンゼン環が一つ足りない．したがって中央の 5 員環部に非常に強いひずみがかかり，その結果として分子が平面性を失っている可能性が高い．そのアニオンラジカルの ESR スペクトルには11本の等間隔線が現われ，10個の等価なプロトンの存在することが確かめられている[13]．したがってアニオンラジカル(そしてたぶん中性分子も) において非平面構造をとっているとすると，それは平面型における 5 回対称軸をそのまま保持した bowl 型[13] あるいは basket 型[14] とでも称すべき浅い朝顔型洗面器のような形であろう．分子模型を組んでみると明らかに非平面型となり，底の 5 角形面とその周囲に突き出ている 5 本の結合とのなす角度は38.75°にも達する．もしこのような非平面構造をとっているとすると，一体このような分子で共鳴安定化が可能であろうか？ この問いは(13)のような超共役分子の存否のかぎを握っているという意味で非常に興味深い．

この問いに答えることは実ははなはだむずかしい．たとえ水素化熱(まだ測定されたという報告はない)が出てもそのうちひずみによる寄与を差し引かねばならないが，その評価は一般に困難である[14]．非経験的 MO 計算を行なうにしても分子構造パラメーターの実測値が不可欠

Two adjacent half pages from Osawa, E. *Kagaku* **1970**, *25*, 854.

Do we really need more chemists?

Of course we need chemists, but in my country there is an overproduction of synthetic chemists. Synthetic chemists do not know anything about physical chemistry and theoretical chemistry. They just mix reagents, and they know how to purify and how to read NMR spectra, and nothing else. We have too many of them because chemical companies and pharmaceutical companies need thousands and thousands of derivatives in order to find drugs or agricultural chemicals. I think that the future of chemistry depends on how we can transfer this activity into more intellectually attractive science.

I think this is related directly to computerization. When synthetic chemists mix reagents and they fail to produce the desired product, they move onto something else, another mixture, without understanding first why they failed, or why they succeeded in other cases, for that matter. What I want to do is to use theory to understand these things. Although theory is incomplete, we have nothing else that can help.

I noted that you studied chemical engineering rather than chemistry. What was the difference in your preparation?

We have totally different faculties for science and for engineering. In chemical engineering, we are supposed to prepare industrial chemists interested in applications. The distinctions are disappearing, however. The science environment is better for the student to acquire a very broad perspective, and I like it this way. I find the engineering environment less useful. Technologies change so quickly. I myself was educated as a chemical engineer in an engineering environment. After graduation, I started working as a chemical engineer. I was employed by a chemical company producing fibers. I found out soon enough that all that I had learned was completely outdated. I had learned, for example, about the conditions of coke production in great detail, and this was useless. My professors were old and knew nothing about theory, and all they taught us was how to make coke. When I entered industry, they told me to find some dyestuff to dye synthetic fiber. It was a difficult problem to give color to synthetic fibers. So we came up with some ideas and wanted to patent a new procedure, but we soon found out that similar procedures had already been patented in the United States and Germany.

What is the main task in preparing chemical engineers, to be creative or to be able to follow instructions?

The Ministry of Education thinks that it must produce a certain number of chemical engineers, professionals who can read patents and understand what other people are doing in the advanced countries and follow their procedures. In the past, say, 20 years ago, our main task was to keep up with the United States. Now we have done this and the atmosphere has completely changed in Japan. On average, one fourth of the work force in our chemical companies is made up of researchers; some have Ph.D.s but most have Master's degrees. Their main task is the development of new products. They must be creative; otherwise, Japan cannot advance. The whole chemical industry, like all our industries, depends on the invention of new products. This is why the education of chemical engineers must change.

How about the language problem?

We must be able to write in English. Otherwise, nobody will pay attention to Japanese chemists. Many people of my age understand German, French, and English. A top level chemist in Japan has no problem with languages. The situation is different for the majority of the Japanese people, including industrial chemists. They don't speak English. At most, they can read scientific papers. However, they don't use English at work. The level of Japanese industrial development is so high that you can do a good work even if you don't understand English. It's a different matter though if you want international recognition. Then you must publish in English.

Yet you published your crucial paper on C_{60} in 1970 in Kagaku, which appears in Japanese only. Then, you republished your predictions about what is called today the buckminsterfullerene molecule, again in Japanese, in your 1971 book on aromaticity.

This is a contradiction indeed. I have published all my important results in English, except this one. At that time I did not realize its importance. We also write a lot of articles in Japanese, particularly articles for introducing the Japanese reader to the latest developments from outside of Japan, primarily the United States, because nobody reads the papers in English in this country. I have about 170 papers in English and about 140 in Japanese.

Please, tell us about the background of this 1970 paper?

At that time the nonbenzenoid aromatics and the general concept of aromaticity were very popular. I was also very excited about this new field. There were a lot of papers about it and I was just back from the United States and without a job. I was in the laboratory of Professor Yoshida from whom I had gotten my Ph.D. So this was my old laboratory in Kyoto. But I was unemployed. Professor Yoshida was a very unusual man. He gave some of his salary to me because I had no income. He gave me ¥30 000 from his pocket. His salary was about ¥200 000. He gave me more than 10% of his salary, and I never returned it to him. This was a very strange situation. He was just keeping me until I'd find a job. He was also interested in the problem of aromaticity, and he was organizing a special issue in *Kagaku*. He proposed that I should write something for this journal. I thought about three-dimensional aromaticity, and came to this interesting problem, which seemed to me unimportant but maybe worth putting into a Japanese article. Then Professor Yoshida wanted me to write a book and this idea got into this book as well. This is an introductory-level book, except the last chapter, which is original work. But I did not think it was extremely good because I was unable to formulate the three-dimensional aromaticity. I was not good at mathematics. I thought about the soccer ball, but this is not really three-dimensional. It is the extension of two-dimensional aromaticity. The only difference is that it is endless. I moved to Hokkaido in 1970 and there I thought that I should start something new, and I changed my research interests.

In retrospect do you regret that you published this paper on the C_{60} molecule in Japanese only?

Yes, very much. I think that this was the second biggest mistake in my life. The first was that I gave up my job when I was an assistant at Kyoto University. My professor was very successful and influential, and the Industrial Chemistry Department in Kyoto was very powerful. I was on the right track in the Japanese sense, and my professor liked me very much. I was very young, however, and I thought this was a very conservative system. The relationship between me and my mentor, Professor Yoshida, was a very special one. I have never respected his academic attitude. This is not to blame Professor Yoshida. This is my defect, a hereditary one. My mother and my grandfather did the same thing, that is to say, criticized their superiors. This is a very atypical and not a very respected attitude for a Japanese. I got very special treatment from Professor Yoshida because I was a very diligent student. When I am interested in something, then I really work very hard. I was very good in getting results and writing them up. Professor Yoshida was very impressed, and that's why he called me back. In this respect, I still regret my attitude.

Have you ever discussed with him your relationship?

No, but he understands it, I hope. The Japanese don't discuss this kind of thing very openly. When the fullerene fever came back, then our relation recovered very nicely. I praised him for giving me the chance to think of this molecule, and the first plenary lecture at the first fullerene symposium in Japan was given by him.

When did you first learn about the work of Kroto et al.?

From *Nature*. Immediately when it appeared.

How did you feel?

Worst day in my whole life. I was very much shocked when I looked at the picture. I thought this was my baby, and now somebody has stolen it, without telling me. I looked at the article and I was very much impressed. That was a very dark day.

This was in the fall of 1985. How soon did it then happen that Harry Kroto and others recognized that you had done something pioneering before?

They contacted me. I never wrote anything about this because it would have been ridiculous, that somebody in the Far East thought about this a long time ago but never worked seriously on it. Just wrote a small Japanese article. So this would have been out of the question. But they wrote to me. Very soon after the *Nature* paper, they wrote to Professor Yoshida, and he forwarded the letter to me. This letter was from Richard Smalley. He knew that we had a book in Japanese with something in it related to the discovery. He asked me to translate this part into English. I did this and sent it to him. Kroto was jumping around all over the world all the time, and he came to Japan, and we got to know each other quite well.

Do you like the name buckminsterfullerene?

It's too long, but otherwise I have no objections.

Did you use any name for this structure in your paper and in your book?

We called it soccer ball, at that time.

Some cynics say that computational chemists calculate everything and then, when and if something is produced, they can say that they had predicted it.

First of all, I did not calculate this molecule. The Russians did this. At that time Russia was prosperous and the Russian scientists had access to big computers. I did not. There has been a lot of misconception that I had predicted the stability of the molecule by theoretical calculations, but the truth is that I just had thought about it. This is a point for which Professor Fukui is criticizing me. He told me in Japanese that I was so stupid, because I did not calculate it. He is very much concerned about the Japanese doing more original work in science.

There were several stages of the C_{60} work. First the prediction, then the calculation, then the discovery, then the production of C_{60} in measurable quantities, and then a whole new fullerene chemistry. Which stage was the most important, and who should receive the Nobel Prize?

There are lots of rumors about Nobel Prize. One rumor is that only three people can receive it. Everybody likes to nominate three important people. I also like to nominate. It could come any time. Maybe it will be this year. I think that if there is some important development, I mean in terms of industrial applications, that would be very good. Everybody still has a chance. If you find high-temperature superconductivity or something similar, that would help.

Looking back, do you consider your role in this story as a triumph or as a big missed opportunity?

It was not really a triumph but the fact that I envisaged this molecule rather clearly in my mind and wrote down something about it, although in Japanese, and then this molecule was discovered, this was, if not a triumph, quite a remarkable achievement in my life. I really like it, and I am almost satisfied with it. Many people, including Fukui, told me that I was so stupid that I missed the Nobel Prize but actually this would be too much for me. I am quite satisfied at this stage. Once in your life you thought of something that you thought would never be realized, but it was realized, even though by somebody else, that's something very nice for a plain scientist like me, so I am quite satisfied by this.

You made your prediction in Japanese. Are you angry with your Japanese colleagues for having never made this molecule?

To some extent, but you cannot blame others. They were primarily not interested in somebody else's work. Another good example is Fukui's work. I still remember that nobody took it seriously. Not even in Kyoto, in his own university. It's also interesting that he was elected to the Japan Academy only after he had received the Nobel Prize. There is also a Decoration by the Emperor to outstanding scholars, which brings with it a lifelong income. Fukui received this award also after he had received the Nobel Prize.

In my recent meeting with the President of the Japan Academy I was pleasantly surprised that he and his colleagues, including astronomers and others, knew very well your work.

That's a rather interesting phenomenon. My first article appeared in the chemistry journal *Kagaku*, published by a small company in Kyoto. They were very much pleased that it appeared in their journal. So they came to me after this fullerene fever started, and they had a whole issue devoted to my "memory." I got interviewed, I appeared in newspapers and on a TV program, and I got pretty well publicized in Japan. Then I thought that I should do something for my countrymen. I had bitter memories that while I was an associate professor, and this was for a long, long time, I had to depend on other peoples' funding. Our funding mechanism is such that the government gives a very big fund to a senior professor. Then the senior professor supports the research of younger people from this funding. This is one possibility. There is then also the possibility for direct support. In the framework of the first possibility, I had received support from many professors, and I thought that now this was my time, and I applied for a big grant for fullerene research. This worked out nicely, and now I am supporting about 100 professors. This is one of the largest projects funded by the Ministry of Education. My funding application was greatly helped by the publicity that I have mentioned and also by another factor. This may be a little embarrassing to discuss. We in Japan have a certain "originality complex." The Japanese are very sensitive to criticism and we have been repeatedly told by Westerners that the Japanese lack creativity, that we are not innovative, and we only make modifications. Thus, we have developed this kind of an inferiority complex. So my story probably made an impression on some authorities, because it touched the "originality complex."

How large is this project?

It's large but not if you measure it against the American scale. However, we can live on very small amounts of funding, and this is a very nice donation from our government.

Is this funding for fundamental research?

This is for professors for basic research. However, our Ministry is always under a lot of pressure from the Ministry of Finance to demonstrate that important applications come out of fundamental research.

Would you tell us something about your background?

My father came from a farming family, near Tokyo. He was the second son in his family. Japanese villages had this system: the first son was always the most important. In the case of the second son, if there was enough land, he was given part of the father's land, and then he could also settle in the village. However, after one or two generations, there was no land to give to the younger members of the family, and then they had to leave. My father knew that education was important and put all his five children through college. We were given books at a very young age. I liked chemistry for some reason which I don't really understand. It was probably the atmosphere of the time. Before and during World War II, Japan had very high respect for Germany and German technology, including chemistry. Probably I was influenced by something like that, and I decided to major in chemistry already when I was in high school.

Eiji Osawa and his son, Shuichi (b. 1965), also a computational chemist, in Toyohashi, 1994 (photograph by I. Hargittai). Suichi was four years old when Eiji Osawa "noticed a soccer ball he was playing with and recognized the pattern of intricate combination of corannulanes in the surface design, leading immediately to the conjecture of the structure of C_{60}" (private communication from Eiji Osawa, 1998). Shuichi continues doing fullerene chemistry both computationally and experimentally.

What is your main interest now in research?

I have two. One is computational chemistry, and the other is fullerene chemistry. I am very interested in computational methodology, but I am not a theoretician and cannot invent anything better than the Gaussian technique. I just use it as a tool, and what I do is make the concepts of organic chemistry into algorithms. Our latest product is the conformational space search for generating conformations of molecules automatically and determining their stability.

What do you do in your other interest, in fullerene chemistry?

We calculate. Because of my previous involvement with fullerenes, I always think that I must do something original, especially in this field. One nice outcome of this effort is the idea that we should try to coat C_{60} with other elements, such as silicon. People have been interested in putting something *inside* the C_{60} molecule since 1986. Our idea is different. We would like to cover the C_{60} molecule from the outside, coating it with another element. C_{60} is now so cheap and the price is going down quite quickly, so we can use it as a raw material. What I have in mind, for example, is to take the vapor of fullerene and bombard it with silicon atoms. They will react. When sufficient number of Si atoms are attached to the C_{60}, they will bind each other, making a Si_{60} ball outside the C_{60} structure.

Eiji Osawa and Kenichi Fukui (courtesy of Professor Osawa).

But bombarding C_{60} with silicon is experimental work.

Yes, this is a project I'd like to start. So far we have done the computation. Silicon is very nice because we have found that when we put 60 silicon atoms in the same pattern as C_{60} you can make a bond between silicon and carbon. Si_{60} by itself would not be stable and we have also shown this by calculation. This is also in agreement with the results of arc-vaporization experiments on silicon. Once, however, you have succeeded in making Si_{60} bind to the internal C_{60} unit, the substance will possess lots of interesting properties. The stability of the Si_{60} structure with C_{60} inside comes from the strength of the silicone-carbide bond. There are then many other elements, which form strong carbides, such as Ti and Mo, so a lot of experiments can be done. Again, they will be done by our colleagues in other groups.

Where will you publish this paper on the Si_{60} problem?

First we submitted the manuscript to *Nature*. The Editorial Board selects only half of the manuscripts submitted for refereeing, and our paper was rejected in this first stage. So we shall publish this paper in *Fullerene Science and Technology*. I think people in England don't like this kind of imaginative idea. The other thing I have been increasingly interested in is shape. I am now editing a special issue in *Materials Research Society* (MRS) *Bulletin*. I've asked several people for articles, but some of them declined because there is so much demand for articles, and famous people are overloaded. So I have to write an article myself. One of my favorite illustrations from our article is a double helix formed by fullerene coiled tubes. In order to make these coils, and also a torus, you need to have seven-membered rings in some strategic positions in addition to the six-membered rings. This is demanded by geometry.

Did you cooperate with geometers on this?

No. My students do all the calculations, and we have also consulted the appropriate literature in geometry. We have learned a lot from Alan Mackay's works. We can also learn from other analogies. Now people are trying to find five-membered rings among six-membered rings. You can observe this on the soccer ball. You can also recognize this along the cracking pattern of the trunk of a pine tree, and in Japanese basket weaving. The number of non-six-membered rings is always demanded by the geometry. Finding something like this in nature is more interesting. This is why I am interested in the cracking lines on the trunk of the pine tree that are formed when the pine tree is bent. Bending causes strain, and the strain is relieved by making five-membered rings. Similar patterns formed by the cracking lines can be observed in sedimentary rock formation, and I noticed something similar in one of your symmetry books. It is also interesting to observe the cracking of mud in water beds that are drying up. There the pattern will be different. The surface of the pine tree trunk is convex whereas the surface of the dried water bed is concave. But the fundamental effects are probably the same. When the pine tree grows, there is a lot of pressure on the skin of the trunk from the inside, and so cracks develop.

Do you use any of these analogies in teaching your chemistry courses?

It's funny that although I very often showed my students polyhedra in my lectures, I never used the soccer ball. Of course, today it's so easy because people know all about the soccer ball so I can refer to it.

Finally, what question would you like to be asked?

About hobbies. My hobby is my work, and the best part of my work is that I collect literature. I have a large reprint collection. I have this strange inclination that I must cover all the past literature before I make my move.

Elena G. Gal'pern in 1988 (courtesy of Elena Gal'pern) and Ivan V. Stankevich in 1990 (courtesy of Ivan Stankevich).

25

Elena G. Gal'pern and Ivan V. Stankevich

Elena G. Gal'pern (b. 1935 in Moscow, then Soviet Union) and Ivan V. Stankevich (b. 1933 in Moscow, then Soviet Union) are scientists at the A. N. Nesmeyanov Institute of Organoelement Compounds of the Russian Academy of Sciences in Moscow, Russia. To this date they are active in carrying out and publishing research papers in the field of fullerenes.

Elena Gal'pern graduated from the Faculty of Physics and Mathematics of the Moscow Pedagogical Institute (V. B. Potemkin Institute) in 1958. Following a brief spell of teaching physics in high school, she joined the Institute of Organoelement Compounds of the Soviet Academy of Sciences. First she worked for eight years in the theoretical spectroscopy group and since 1968, she has worked in the quantum chemistry department of the Institute. Her research interests include simulation of molecular and electronic structure of carbon clusters by quantum chemical methods and development of computer programs.

Ivan Stankevich graduated from the Department of Probability and Mathematical Statistics of the Faculty of Mechanics and Mathematics of Moscow State University in 1956. Ever since, he has been with the Institute of Organoelement Compounds except for the period of his doctoral studies between 1958 and 1962, which he spent at the Department of Function Theory and Functional Analysis at Moscow State

University. He has been in charge of the quantum chemistry department of the Institute since 1994. His research interests include development of quantum theory of chemical structure and reactivity of molecules and mathematical chemistry.

The first calculations predicting the stable truncated icosahedral structure of C_{60} were reported by D. A. Bochvar and E. G. Gal'pern in 1973.[1] In spite of the feverish international activity in fullerene chemistry by the early-mid-1990s, there was hardly any information about these authors. Thus, following my interview with Eiji Osawa in July, 1994 and prior to the interview with Harold Kroto for which an October, 1994, date was set, I decided to contact the Russian scientists. I had learned the sad news that the senior author, Professor Bochvar, died in 1990 but succeeded in calling Dr. Gal'pern. She told me at once that I should rather talk with her boss, I. V. Stankevich, about my questions. To my surprise she explained that the C_{60} calculations were originally Stankevich's idea and his name was inadvertently omitted from the authors of the paper. Then she immediately asked Dr. Stankevich to the phone, and we agreed that I should send a set of questions in English (the conversation was conducted in Russian). Drs. Stankevich and Gal'pern would send me their joint responses in Russian.

I prepared an extensive set of questions and received answers to many of them soon. I then sent an edited English version of the interview to Drs. Stankevich and Gal'pern for checking and corrections. In the following, the person giving the response is identified only if the response is given individually, the rest of the responses were formulated jointly. The interview appeared in *The Chemical Intelligencer*. We interacted again during the preparation of the present volume.*

Please, tell us about the background of the work that led to the publication of your 1973 paper in Doklady.

At the end of the 1960's, we were working on a project that, at the beginning, did not have any direct connection to the modeling of carbon materials. Our research institute, the Institute of Organoelement Compounds (INEOS) of the (then) Soviet Academy of Sciences (today, Russian Academy of Sciences), Moscow, was busy with the chemistry of π-complexes of transition metals, and in particular, with ferrocene. The director of INEOS was Alexandr Nikolayevich Nesmeyanov who was also the President of the Soviet Academy

*This is a slightly modified version of the interview originally published in *The Chemical Intelligencer* **1995**, *1*(3), 11–13 © 1995, Springer-Verlag, New York, Inc.

Academician A. N. Nesmeyanov (1899–1980) on Soviet stamp. He was Professor of Organic Chemistry, Moscow State University; President of the Soviet Academy of Sciences (1951–1961); Director of the Institute of Organic Chemistry, then Director of what is today the A. N. Nesmeyanov Institute of Organoelement Compounds of the Russian Academy of Sciences.

of Sciences at the time. He had repeatedly told us all at the Institute to look for the synthesis of new heteroorganic compounds in the form of endohedral polyhedral clusters $M@C_nH_n$ with saturated carbon skeletons. One or a few heteroatoms would be included within the carbon cage.

The simplest of such systems might be envisaged by linking the two rings in ferrocene or dibenzenechromium or cyclophanes with polyene or polyyne groups. Thus sewing the rings together would create the cage structure. The stability of such inclusion cage complexes and the nature of their chemical bonds were very intriguing questions. These questions constituted a research project of the Quantum Chemistry Laboratory of INEOS in the late 1960s.

At the time, however, there were great difficulties for computational work in chemistry in the Soviet Union. Therefore, it was decided to restrict the computational work to the investigation of the stability of the carbon cage. This was considered to be the first step toward the investigation of the polyhedral inclusion complexes. The work was started with the C_{20} molecule of dodecahedral shape, called carbododecahedron. Since it was difficult to predict its stability in the valence approximation, we decided to use the Hückel method, which had proved to be successful for classical conjugated hydrocarbons.

The structure of C_{20} carbododecahedron lends itself conveniently to a hybrid orbital description. For each carbon atom it is possible to construct four hybrid orbitals from one 2s and three 2p atomic orbitals (AO). Three of these four orbitals are directed along C–C bonds and are used to form localized two-center, two-electron σ-bonds. The fourth orbital, called ρ-orbital, is directed along a radius of the sphere enveloping the nuclear skeleton of C_{20}. This orbital is occupied by one of the four valence electrons. These ρ-electrons form a system which is analogous to the π-electron system of planar conjugated molecules. However, there is less overlap between the ρ-orbitals of adjacent carbon atoms in C_{20} than between the π-orbitals, for example, in benzene.

It was a general assumption at that time, based on Hückel calculations, that the stability of a conjugated hydrocarbon was related to the closed π-electron system. It was considered closed, in the Hückel sense, if all bonding orbitals were doubly occupied and all nonbonding and antibonding orbitals were vacant.

The presence of a closed electron system was connected with stability as a whole. Kinetic stability was characterized in terms of the energy gap between the occupied ground-state levels and the vacant ones. The relative thermodynamic stability was estimated by various characteristics such as delocalization energy, resonance energy, and other topological indices. The calculations, however, showed the lack of a closed electron system for C_{20}. It was found that the ground state of C_{20} is either a triplet or it can exist only as a dication. The possibility of a lower symmetry was also considered for C_{20}. This would mean a decrease in the number of the degenerate energy levels, in accordance with the Jahn-Teller effect. Subsequent calculations with the extended Hückel method have supported these conclusions.

The small size of the carbon cage was a limitation of a different kind. It greatly restricted the choice of atoms to be placed within the cage. Thus larger systems were sought for further calculations, and eventually the truncated icosahedron was selected. The corresponding carbon cluster of 60 atoms was called carbo-s-icosahedron. The C_{60} cluster was found to have a closed ρ-electron system, and a large enough energy gap to separate the occupied and vacant energy levels to provide kinetic stability. Besides, the size of C_{60} was sufficient for inclusion complexes of a wide range of atoms; the diameter of the inscribed sphere is 6.35 Å, assuming the

C–C bond length to be 1.40 Å. Thus, the theoretical aspects of Nesmeyanov's question about the possibility of inclusion complexes of polyhedral hydrocarbons were reduced to finding suitable models of new allotropic modifications of molecular carbon.

We have also performed studies on heterofullerenes and endohedral complexes $M^+@C_{24}$, (M = Li, Na) in which the C_{24} cluster had the shape of a truncated octahedron. These studies were reported at meetings in 1980 in Kishinev and in 1981 in Baku. The Baku meeting was the most prestigious in the Soviet Union, called Mendeleev Congress.

How did it happen that Dr. Stankevich's name was omitted from the list of authors?

Gal'pern: I was a research associate of Professor D. A. Bochvar, working on my dissertation in the Laboratory of Quantum Chemistry of INEOS. This laboratory was established in 1955 by Professor Bochvar, who had remained in its charge until 1988. He received his title of Professor in 1950 at the Department of General and Inorganic Chemistry of the Moscow Textile Institute.

Stankevich: I had completed my graduate studies at the Department of Function Theory and Functional Analysis of the Faculty of Mechanics and Mathematics of Moscow State University.

I used to play soccer and had thus often been in contact with a ball of the shape of truncated icosahedron. One day, three events coincided. The first was a soccer match in which I participated. The second was a laboratory seminar about the quantum-chemical calculations on C_{20} in which it was concluded that C_{20} was not stable. The third event was a major soccer match televised by Moscow Central Television. In those days, the soccer telecast began with an image of a soccer ball similar to a Schlegel graph of a truncated icosahedron. It was this image that prompted me to suggest probing into the stability of C_{60}.

The "crazy idea" of a C_{20}-molecule in a polyhedral shape belonged entirely to Professor Bochvar. It is almost irrelevant that such a cluster proved unstable. The potential for extending this idea was obvious. One day I brought a soccer ball into the Laboratory and told Gal'pern, "Lena, 22 healthy men are kicking this ball for hours and it is not destroyed. A molecule of such a shape must also be very sturdy."

Gal'pern: I was at first rather skeptical about this suggestion, the more so because of the great difficulties involved in computing the electronic structures and estimating the stability of molecules with more than 20 atoms. The cluster C_{60} was at the limit of the computational facilities available to us at that time. However, I succeeded in the C_{60} calculations. Stankevich participated in discussions of some details and results of the calculations. He declined authorship of the paper though and agreed to appear in the acknowledgments only. The report about these calculations was published in 1973 in *Dokladi Akademii Nauk S.S.S.R.*

Were you convinced that C_{60} should be a stable molecule and did you try to get somebody to make it?

Results of quantum chemical calculations always leave an unsatisfactory feeling behind. The polyhedral C_{60} molecule seemed too fantastic if not incredible. It was not obvious at all that we used adequate computational techniques to establish its stability.

We have been unsuccessful in inducing our chemist friends to synthesize C_{60}, or related molecules. Incidentally, nobody has yet really synthesized such a molecule by chemical means. We ourselves considered C_{60} and even more so its possible inclusion complexes, as "soap bubbles." We could not, at that time, seriously discuss the realization of such "soap bubbles."

How did Professor Nesmeyanov view your results?

Our paper on C_{60} was presented to *Doklady* by Professor Nesmeyanov. We don't know what he really thought of its scientific merits. We don't think, though, that he strongly believed in the possibility of such fantastic carbon molecules. However, he had always supported theoreticians and encouraged the introduction of new techniques to chemical research, including the application of computers.

How did you feel about Kroto et al.'s reporting their discovery?

In the beginning, we reacted quite calmly to the first reports by Kroto *et al.* about the formation of C_{60} and C_{70}. There were no straightforward proofs of the polyhedral structure at the beginning. Initially, there were only hypotheses based merely on the mass spectra of carbon vapors. These hypotheses became reality only following the production of gram quantities of C_{60} and C_{70} by Krätschmer *et al.* The ensuing studies by ^{13}C-NMR, X-ray crystallography, and gas-phase electron diffraction have then eliminated all doubts about the polyhedral structure of these surprising carbon molecules. Of course, we were happy to see that quantum chemistry proved once again to have great predicting potentials.

Gal'pern: When Dr. R. Taylor from the University of Sussex gave us a gift of an ampoule containing powdered C_{60}, I was overwhelmed with delight. It seemed like a fairy tale to me. I didn't think I would live long enough to see the realization of the molecule over whose calculations I'd suffered so much!

How did the world learn about your publication of 1973?

It seems to us that Kroto and his colleagues had not been aware of our work on the carbon clusters. As far as we know, they learned about the works in the Soviet Union from our review published in 1984 in *Uspekhi Khimii.*

Did you know about Eiji Osawa's prediction?

We learned about Osawa's work in 1986 or 1987 from the literature as it was cited in some papers on fullerenes. Dr. Taylor then told us more about Osawa's work a few years ago when he visited our institute. Then Professor Osawa graciously sent us the English translation of some parts of his publications in which he described the theory of C_{60} clusters.

What were Professor Bochvar's research interests?

Professor D. A. Bochvar was a specialist in inorganic chemistry, theoretical chemistry, and mathematical logic. He was interested in the theory of molecular structure, the mathematical apparatus of quantum chemistry, and in the prognostication of physicochemical properties of molecules. His most important result, in theoretical chemistry, was the prediction of possible existence of a polyhedral C_{60} molecule. He had also suggested a new method for the analysis of molecular wave functions. This method was based on the notion of information entropies of quantum mechanical distributions. He also refined the uncertainty principle through information entropy distribution.

D. A. Bochvar (1903–1990) in the late 1960s (courtesy of Elena Gal'pern and Ivan Stankevich).

What are your fields of interest?

Gal'pern: My scientific interests include prognostication of the physico-chemical properties of molecules by quantum chemical calculations and programming. I am currently a Research Associate at the Laboratory of Quantum Chemistry.

Stankevich: My research fields embrace the mathematical foundations of quantum chemistry, the theory of the Schrödinger equation, investigation of spectral properties of periodic quantum structures, and mathematical chemistry. I was co-author of the majority of D. A. Bochvar's theoretical works on molecular structure. I am currently in charge of the Laboratory of Quantum Chemistry.

Are you doing any work on fullerenes currently?

The fullerene studies used to constitute only a small portion of our work. Currently, however, this topic is being pursued with increased intensity.

Do you have children, and what are you telling them about the fullerene story?

Stankevich: There is a daughter, two grandchildren and a great grandchild of D. A. Bochvar. I have a daughter and a grandchild. My daughter works in mathematical chemistry. The children are aware of the fullerene story but have not shown too much interest in it.

Gal'pern: I have two kittens and a dog and when I try to talk to them about fullerenes, they stare at me with great bewilderment.

References

1. Bochvar, D. A., Gal'pern, E.G. *Dokladi Akademii Nauk SSSR* **1973**, *209*, 610–612.

Harold W. Kroto, 1994 (photograph by I. Hargittai).

26

Harold W. Kroto

Sir Harold W. Kroto (b. 1939 in Wisbech, Cambridgeshire, U.K.) is Royal Society Research Professor at the School of Chemistry, Physics and Environmental Science of the University of Sussex in Falmer, Brighton, East Sussex, U.K. He shared the Nobel Prize in chemistry in 1996 with Robert F. Curl and Richard E. Smalley, both of Rice University in Houston, Texas, "for their discovery of fullerenes."

Harold Kroto graduated from the University of Sheffield in 1961 and received there his Ph.D. in 1964. He started his academic career at Sussex University in 1967 as a Tutorial Fellow and went through the ranks of Lecturer, Reader, and Professor. Even prior to the Nobel Prize, he had received many awards and distinctions. He was elected Fellow of the Royal Society (London, 1990) and Member of the Academia Europaea (1992), shared the International Prize for New Materials of the American Physical Society (with R. Curl and R. Smalley, 1992) and the Hewlett-Packard Europhysics Prize (with D. R. Huffman, W. Krätschmer, and R. Smalley, 1994), among many others.

Our conversation was recorded in Harry Kroto's office on an October day in 1994. That fall I was spending a month at Birkbeck College in London as guest of the Royal Society. During that stay I was invited to give a seminar at Sussex University. I took a very early train to Brighton to attend Harry's early morning class in spectroscopy for third year students. His enthusiasm shone through every sentence he uttered, every formula he drew on the board, and every mode of molecular motion he demonstrated with his body.

Our conversation reproduced below appeared in a somewhat different version in *The Chemical Intelligencer.**

Your 1974 book Molecular Rotational Spectra *was one of the texts I used for learning about microwave spectroscopy. However, the* C_{60} *molecule could not be investigated by microwave spectroscopy because it does not have a permanent electric dipole moment.*

That's one of my great misfortunes, that the only thing I really understand is microwave spectroscopy for which I need a molecule with a dipole moment and I have ended up with this amazing molecule which has none.

Harold Kroto at Sussex University in 1994 (photograph by I. Hargittai).

So how did you end up with this molecule?

For me the origins of the discovery go back to Sheffield when I first discovered spectroscopy at the University and I realised that this was my field. Then I did a Ph.D. in spectroscopy and subsequently became interested in microwave spectroscopy. The research I am most proud of is the detection

---

*This is a modified version of the interview originally published in *The Chemical Intelligencer* **1995**, *1*(3), 14–23 © 1995, Springer-Verlag, New York, Inc.

of molecules with multiple bonds; phosphorus-carbon double and triple bonds, sulfur-carbon double bonds. But then around the early seventies the tremendous breakthrough by Townes and colleagues occurred and that was the detection of interstellar molecules. I think that many microwave spectroscopists immediately felt that they could dive into this field. As professional spectroscopists, we knew more about the radio spectra of these molecules than anybody else did and so a number of microwavers immediately became radioastronomers or worked with radioastronomers.

In 1972–73, cyanoacetylene was detected in space and together with my friend David Walton who is a synthetic chemist and an undergraduate student, Anthony Alexander, we started working on polyynes. David is the world expert of the synthesis of polyynes and Anthony was on our Sussex B.Sc. by Thesis course. This was a wonderful course which was designed to involve two supervisors, one from a synthetic chemistry discipline and the other from the physical chemistry discipline and the undergraduate got his degree by research work. We had this fantastic student who insisted on working with us. Alex synthesized cyanodiacetylene and just at that moment the microwave spectrometer that I have now arrived in 1974. The first molecule observed with my own microwave spectrometer was cyanodiacetylene. At that same time there was this tremendous enthusiasm by microwave spectroscopists to do some work in radioastronomy. We were perfectly placed. Everybody loves astronomy. If you are a scientist and if you have a technique that is absolutely perfect, and one that is as complicated as microwave spectroscopy is, then you could really make a synergistic complementary contribution if you worked with a radioastronomer, who knew where the sky was. They could not overnight learn the theory that was required to understand spectroscopy.

My great fortune was to know Dave Walton with whom Alex and I synthesized cyanodiacetylene, and also to know Takeshi Oka who was a postdoc in Ottawa when I was a postdoc there. He was a good friend and I noticed that he had done some work in radioastronomy with some Canadian astronomers. I contacted him and I told him that we had this molecule, cyanodiacetylene, with five carbon atoms and I asked him if we could try to detect it in interstellar space by radioastronomy. Takeshi and his astronomy colleagues actually were able to detect it. I really was very fortunate, first of all in wanting to do this experiment, and then picking the right guy — thus we were able to see the first polyyne in space. Then, of course, if you could do that, you might be able to see the next one — what about the HC_7N? Dave said that we could synthesize this too.

Lady Margaret and Sir Harold Kroto in Austin, Texas, 1998 (photograph by I. Hargittai).

Wasn't it surprising to see such large molecules in space?

Yes, I remember, Takeshi's letter to my question about HC_5N, when I asked, are you interested in this, he wrote, yes, yes, yes, yes, yes, — five yes's. He was obviously very interested. When I discussed it with him he said it was a long shot because no one would expect to see a molecule with so many heavy (C/N/O) atoms. But HC_5N had a very large dipole moment, so there was a possibility and as it turned out, the radio lines were actually much stronger than expected.

When we had made the HC_7N molecule, I went to the Canadian observatory in Algonquin Park in 1977 to work on the telescope to try to detect it. That was tremendous. I don't think that I have had a more exciting and cathartic moment than that when HC_7N came on the screen. The C_{60} discovery was more gradual. We saw an interesting signals over Wednesday, Thursday, Friday, and began to realise that this is very, very important and thus it took about four or five days to really discover this beautiful molecule. In the case of HC_7N, we worked very hard with the synthesis, then we thought that we were not going to see it in space and then it came out on the screen and it was unbelievable, it appeared on the screen just like that.

We were looking for a particular line of the spectrum according to our measurement and we knew which line we were looking for. At Algonquin at the time there was not enough computing power to see the line grow on line. It was a most amazing evening. What happened is that when I left the U.K. my graduate student Colin Kirby had not yet got the spectrum, so we discussed various contingency plans. I got out on the telescope on a Thursday, and during Thursday and Friday we were working on the telescope — on an alternative project. On the Saturday Colin managed to get the frequency to us and then on the Saturday evening, we started looking and concentrated on the channels where the signal was expected to appear. We were sitting there for six hours and every time the channel signal was high there were cheers, and every time it was low, there were groans. However, the statistics were good and we became convinced that the line we knew was there, and — then it came out of the computer and it was a tremendous moment. I have never experienced anything like it since. This was the longest and heaviest molecule detected in the ISM and it presented problems to the interstellar chemistry community. Just about the same time as the seven-carbon chain was found, a fascinating star was detected which has become very famous as IRC+10216. It's a carbon-rich Red Giant, and this star has ever since been a focus of great deal of attention mainly because it's closer than any other star so it can be observed in more detail. As time went on, HC_7N, HC_9N, and $HC_{11}N$ were all discovered in this star. It was clear that this star was just pumping the carbon chains out.

How did you get to know the Houston group?

It was after one of the Gas-phase Molecular Structure meetings in Austin, Texas, that Bob Curl invited me for a visit to Houston in 1984. I had known Bob previously and he had also visited Sussex. During the visit to Rice he suggested that I meet Rick Smalley — Bob was very excited about SiC_2, which turned out unexpectedly to be a triangular molecule, whereas the analogue C_3 is more or less linear. When I saw this, I thought, "aha, click, click, click, now I understand a lot more about Si=C double bonds than I understood before." That goes back to my earlier work on carbon-phosphorus and carbon-sulfur multiple bonds when I was considering going to go on to detect carbon-silicon double bonded species but that could not be done without completely redesigning my microwave apparatus and I really did not want to do that. I wanted to do spectroscopy and I wanted to do chemistry, but I didn't want to build equipment. I could have done it but I preferred to focus on using spectroscopy for chemical purposes. The microwave spectrometer from Hewlett-Packard turned up just in the nick of time. Without that spectrometer I couldn't have done the work on the carbon chain molecules and multiply bonded carbon to phosphorus and sulfur species during the period 1970–1980.

Coming back to your visit to Houston.

Yes, this was in the Spring of 1984, and it was my first visit to Rice University. The previous year, Bob Curl had visited us and stayed in our home and now he invited me to stay with his family. At about that time Rick had developed the cluster beam apparatus. I had seen his papers before, but I had not looked at them very carefully. He was studying clusters and one could not do very much spectroscopy with them, and specifically not microwave spectroscopy very easily at all.

So I got acquainted with Rick who was very enthusiastic and was jumping all over the apparatus. He is very good at describing his work and he was very excited. I was very excited too by the SiC_2 result because it explained why it was so difficult to get a C=Si double bond. Rick had looked at silicon carbide and was also looking at aluminium clusters and other metal clusters as well as semiconductor clusters such as gallium arsenide.

I did not say anything immediately but I thought that if I could put graphite into this apparatus we could make a plasma similar to that in the atmosphere of a carbon star. I don't remember it fully but I remember clearly thinking about it during the discussion but I didn't tell Rick. I was very keen to do this experiment and I thought, how can I ensure that my experiment would get done. When I got back to Bob's home I was quite excited, and decided I must convey my enthusiasm to Bob. I wanted to make sure that they would drag me over for this experiment. It turned out to be no difficulty because Bob was also very enthusiastic to do this experiment.

I had two experiments in mind. The first experiment was a very simple one; to use graphite to produce a carbon plasma and that should produce carbon chains. I knew other peoples' work, like that of Krätschmer and Huffman because they'd done work on carbon chain molecules by matrix isolation spectroscopy, for example. I had a discussion with Alec Douglas in 1977 when our carbon chain papers came out. He was developing his ideas of carbon chains being responsible for the diffuse interstellar bands. The simple experiment however was to make carbon chains and react them with hydrogen. It was just to show that carbon clusters could react to form polyacetylene. That was the basis of the experiment.

The second experiment was to try resonance-enhanced two-photon ionization (R2PI) spectroscopy — the same thing that Rick had done with silicon carbide. So if Alec Douglas's idea was right, you could put graphite into the cluster beam apparatus and we might be able to see C_5, C_6, C_7, ... etc., by R2PI. So those were the two ideas. Bob was more keen on the second and I was keen on both. I knew, of course, how important the second was but there were also major experimental difficulties.

I had a personal attachment to the reaction experiment because I'd spent a lot of time with people like Bill Klemperer, Alex Dalgarno and Eric Herbst, discussing whether they could produce the carbon chains by ion-molecule reactions in the cold tenuous interstellar molecular clouds where they were observed and I felt that they couldn't. My feeling was that if we could do this experiment, we would establish the role of carbon stars in the genesis of interstellar molecules. This was very important at least to me. At the same time, nothing could be more important than the second experiment on the diffuse bands. Subsequent letters between Bob and myself were discussions on how hard that experiment would be. However, I felt that in any case we could do the first experiment and we could get mass spec and that would be an easy experiment. Bob agreed to work on Rick.

How did things continue?

My visit to Rice was in the spring of 1984. Then one day in July '84, I was sitting in the coffee room and Tony Stace, one of my colleagues at Sussex, gave me a copy of the Exxon paper by Rohlfing, Cox, and Kaldor, basically it was the same experiment. As I read the paper I thought "Shit! Why didn't we do this carbon experiment at once while I was there at Rice? Then we should have discovered all the new big carbon clusters with more than 30 atoms!"

I looked at this paper and I saw the distribution of clusters with less than 30 or so carbon atoms — the ones that everybody knew about, but there was this new set, from 30 to 100. They suggested that the new clusters might be carbynes, and I thought that this was nonsense because I was sure that carbyne is impossible material on the basis of the chemistry we know. You surely cannot get highly condensed polyynes {acetylenes} because they would certainly blow up. We know that when polyyne chains lie close together they cross-link — violently. I realised that they probably had already had done it before I suggested the experiment when I was at Rice, but my first thought was that we should have done the experiment there and then and we would also have found this new set.

I also thought that the new set might have been arrays of hexagons because simplistic ideas suggested that one might expect to see even numbered graphitic sheets. So I did know of the paper. Then almost out of the blue in August '85, Bob Curl rang me up.

So nothing happened between the Spring of '84 and the Summer of '85?

Nothing happened for seventeen months other than one or two letters from Bob, the paper from the Exxon group and another paper from Bell Telephone. The Bell group also worked on the carbon chains and they actually looked at C_{60} *specifically* and they fragmented it — but they didn't appear to ask anything about the structure of the species. I didn't know of this paper and only saw a preprint when I got to Rice. I don't remember exactly when that paper came out. Neither of these papers figured in my analysis because my first experiment was to show that carbon stars were important sources of interstellar molecules. The second was to see whether Alec's idea was right. The newly discovered larger clusters certainly weren't carbynes, it seemed to me, and we could look at them again. I don't remember much about my thoughts on these species, you see, I wasn't in the cluster field and I wasn't focusing on them, I was focusing on my experiments. So it was seventeen months, and it was August, 1985, when Bob rang up and said we could do the experiment. I said, I am coming and I immediately rang Continental to get the cheapest ticket and I was there within three days on a Thursday (August 29).

I don't need much incentive to go to Texas, and I actually liked, in particular, a chain of secondhand book stores known as Half-Priced Books. I collect magazines and books — I have a huge collection and my main passion is graphics, design and art — and have millions of images.

Coming back to the story, when I arrived in Houston and I told everything that I thought the Rice group ought to know about carbon chains, radioastronomy, carbon stars, etc, ... to encourage them to really dive into this project. I must have talked for at least two or three hours.

The experiments started in earnest on the Sunday or on Monday, and carried on through Tuesday, Wednesday, and during this period we concentrated mainly on the small species — but we also could look at the large ones.

Tell me about the actual experiment.

It's a big apparatus, a huge vacuum chamber. In one wall there is a small nozzle with a solenoid valve which lets gas through in pulses, so that it passes over a rotating disk of 2.5 cm diameter which was changed to graphite. There is a hole for the pulsed laser beam to focus on the disk and produce plasma pulses. The helium blows the plasma vapor into and across the chamber where it passes through a skimmer. The helium-entrained plasma then is skimmed into a beam which then goes into the mass spectrometer. At this point, the beam is crossed by a pulse from an excimer laser which ionizes the clusters in between two electric field plates and the clusters then are flicked up a time of flight tube. It sounds simple but it is not and it's Rick's genius to be able to create this sort of technically very complex apparatus and get it to work. The timing of all the lasers, etc. has to be done perfectly — and that's Rick. Rick created this apparatus which is so exciting to look at and so effective.

When I first saw the apparatus I thought, let's change the disk for graphite and that was my first contribution to the story. Then we did the experiment, and the students and I worked on them together. There was Jim Heath and Sean O'Brien and also Yuan Liu as well as another student Qing-Ling Zhang who was just starting. All this was nine years ago, and as far as I can remember, during the experiments there were problems in printing out the data. The whole thing was stored on a PC. Yuan sorted this out. For me it was wonderful because I could just sit in front of the display — I didn't even need to type. I could just watch the results as soon as they appeared.

I worked mainly with Jim and Sean — two students with whom I became very friendly indeed — and we would work at all times and go on till late at night and then go to a 24-hour coffee shop late at night — so I was a student again, but it was very easy for me because I didn't have to do the hard work. I just could concentrate on the results as they appeared and assess them with Jim, Sean and Yuan and then decide what to do next. If I said I wanted to land here, the students might say we can't do that but we can do this. It was complicated — it was like flying a 747 without needing to know how to fly myself.

The actual memories of what happened are vague. We started to get interested in the C_{60} signal, a signal which we and apparently no-one else had really noticed in the Exxon data. The Exxon work played very little role, because initially it was the low mass clusters which I was most interested in. Then, I think — gradually — we started to look at the large clusters with more than 30 atoms as well.

At some point we just sort of realised that "Oh, this is big — this thing" and we noticed that it started to vary in relative abundance. In the Exxon data, it was not that strong. Somehow it did not hit anyone in the eye in their study. With hindsight it does but at that time C_{11} and C_{15} and C_{19} were much clearer magic numbers than C_{60} was in the Exxon data and they were also well-known to be magic numbers.

In our case the experimental conditions were being varied, while in the Exxon data they don't seem to have varied the conditions very much. They seem to have done this experiment under one set of conditions and that's it. We were not focusing on C_{60} and already were changing the conditions originally because of my hydrogenation suggestion, and so we started to do reactions with various gases, and hydrogen was one. At the same time we started to look at C_{60} and we discovered that it was sometimes off-scale. When it's off-scale, this is interesting and you start to look at this because it is a new phenomenon and you could go back to the hydrogenation. I think it was partly because we were putting gases into the system and changing the pressure that the C_{60} signal became erratic and started to show characteristics that the group at Exxon had not seen. It was critical.

At which point did you come to the conclusion that C_{60} and, of course, C_{70} were more than just a stronger peak, and at which point did you start feeling that you must have an explanation for it?

It's difficult to say. When you work on something like this there is a gradual realisation that C_{60} is very strong, and you want to find out under what conditions this occurs. One has to look at the notebooks. I had a printout and it indicated that we noticed the signal on Wednesday, 4th September 1984 — and the student's lab book indicated the same day. The actual moment is difficult to assess because what you are in fact doing is optimizing the experimental conditions. The way that you do this is by looking at the display and adjusting the conditions. To do this sort of adjustment, you pick out the strongest signal on the block — and maximise. We thus decided to pick on the C_{60} signal which was the strongest of the big guys and adjust it, and then we saw that it was sometimes ten or more times bigger than the rest — and that's what we saw. You do not necessarily remember exactly when this happened and then you look at the notebooks. When you do this, you find that I annotated my printout on 4th of September which is what I published in *Angewandte Chemie* — and we find that a student made a note in the lab notebook on the same day. Then you start to ask what are the conditions necessary to optimize C_{60} and C_{70}. We know that we had such a discussion on Friday 6th during a group meeting. We do not know much more than that as memories after so long are vague — and memories also differ.

How did the question, what makes C_{60} so stable, come up?

Those discussions we know certainly took place in a group meeting on Friday (September 6th). There is some evidence which indicates that Rick did not learn about C_{60} until Thursday and certainly Friday was the first group meeting during which we assessed the findings. I do not remember much of the details of the discussion during that group meeting other than it took place and that a decision was made that the students should work over the weekend to try to optimize the conditions. This suggestion probably was made by Bob Curl. Bob was a key player in the whole programme — he was an equal co-worker — involved in all the discussions during this period. I feel his role has not been recognized for various reasons. He was an equal co-author with Rick on almost all the C_{60} papers emanating from Rice for three or four years after the discovery.

After that meeting on Friday, I went to Dallas. Sean O'Brien started the experiments which focused on the conditions on the Friday evening and they were continued by Jim Heath on Saturday and Sunday. To the best of my memory — I came back to the lab on Sunday afternoon from Dallas and saw Jim with the data — the C_{60} signal was about 40 times stronger than any other cluster! Sean had got 30 to 1 on Friday but the actual conditions were not clear. So this was between the two students and it was finished off by Jim Heath on Saturday and Sunday when Jim carefully adjusted conditions and rebuilt the nozzle. Basically by Sunday night, the pressure conditions were such that we knew that the C_{60} signal was 40 times stronger than C_{58}, etc., and maybe about ten times stronger than C_{70}. Thus by Wednesday, the 4th of September, we knew that C_{60} and C_{70} were very special and by Sunday night/Monday morning, September 9th, we knew the conditions which would make them special. I do not remember a lot more than that.

How about the question of the structure of C_{60}?

I don't remember any discussions — Rick remembers discussions earlier — but I do not remember any discussions other than the Friday discussion —

and that only peripherally. On Friday I went to Dallas — I do a lot of my thinking in the car, so whatever thoughts I might have had prior to Monday, were mainly with myself on this very long, 400-mile drive.

When did you first remember your visit to the 1967 Montreal Expo and about Buckminster Fuller?

I think that was certainly during the discussions probably on the Friday, but I don't remember. The image that was on my mind was from the magazine *Graphis*. I remember it and Bob Curl remembers Rick suggesting that I should get a book on Buckminster Fuller from the Library. I don't remember that but I do remember Rick getting the book — I didn't have a library ticket so I couldn't get it. You have to remember that I had looked a lot at a particular *Graphis* issue, and there is one particular image that I remember. I have subscribed to *Graphis* ever since I came across it in the 1960's or 50's. It is a graphic art and design magazine and there was a whole issue on the Expo67 and there was a photograph of the US Pavilion that I remember which was of a dome almost entirely of hexagons. This was a very strong memory. I visited the Expo in 1967 while I was at Bell Telephone, NJ, we went up to Montreal and it was our only holiday in the year before I came to Sussex. I had been at NRC Canada in 1964–66 and with Bell Telephone in 1966–67. At some point — probably the summer we went to Montreal and went to see the Expo.

Buckminster Fuller's truncated icosahedron model of the sky, made of cardboard (courtesy of Harold Kroto).

So I visited the US Pavilion at the Expo and over the years the memories were refreshed by *Graphis*. This magazine is a sort of *Angewandte Chemie* of Graphic Design — out of 293 issues I am only missing 6 — it goes back to 1945. This is more important to me than Science in the sense that Graphics fascinates me. In fact when I came back to Sussex, after Bell Telephone, as a postdoc, I started to look at the possibility of working with Buckminster Fuller. Not on these domes but he had some ideas about urban growth and development and how they should be very carefully thought out so that as a city grows you should build in various aspects of the infrastructure which can be reproduced for the good of the overall macroscopic system. I was thinking of writing to Buckminster Fuller to see whether I could work on this project and even got his address but just at the point that I was thinking that I ought to write to him — and other people — for a job, I was offered a position at Sussex.

Coming back to your work at Rice, when did the suggestion of the truncated icosahedron structure come up for C_{60}?

That's a bone of contention. My memory is the following and I have written about it in *Angewandte Chemie* as accurately as I can remember, though for various reasons I did not detail everything that occurred. During the Monday (September 9th), there were discussions. I certainly discussed with Bob Curl, while we were going to lunch at mid-day, the fact that I had a polyhedral model of a star-dome back home. You should realise that there were lots of ideas floating round — I only really recall mine. I remembered that I'd built the stardome model for my kids and it was in my home in a xerox box. I remember discussing Buckminster Fuller's domes as well. I also remember, if I remember anything, that as we were walking home for lunch on the road between Rice and Bob Curl's home on Monday at lunchtime, that I described this model to him and Bob remembers a discussion.

I don't remember any specific discussion with Rick, prior to that during a meal in the evening, except when Rick came back with the Buckminster Fuller book in the afternoon, and we looked at that. I had a third idea; that a four-deck sandwich of a hexagon/coronene/coronene/hexagon would make a little round ball-like cluster. Then because of symmetry, there might somehow be stabilization of the dangling bonds on the edges of this thing. It would have $6 + 24 + 24 + 6 = 60$ carbon atoms. So I'd thought of this as well as about Buckminster Fuller and the stardome and actually suggested them all as possibilities.

Do you remember how the idea of looking for a structure at all came up? I'm asking this because in the paper of Rohlfing et al. there are strong peaks of C_{60} *and* C_{70}*, but they were not suggesting a structure for them or a structural reason for their relative stability.*

But they did suggest that they were carbyne.

So do you remember how the idea of structure came up?

No, I don't. I would say that that is an automatic reaction. I doubt that it was spelled out.

Osawa and Bochvar and Gal'pern had also been looking at these constructions as of very high stability, whether in chemistry or elsewhere.

My guess is that we are all spectroscopists and when you see a signal like this, you automatically try to think what the structure might be. In the case of this one, I don't remember it. Maybe others remember it — maybe they do, maybe they don't. But I'll answer a few aspects of this because they are important. I do not remember very much other than general discussion about Buckminster Fuller, but I do remember Rick saying something about a chicken wire cage at the Mexican Restaurant on Monday evening. That was probably, though I cannot remember, not the first time that the chicken wire cage concept was mentioned in a discussion, and, of course, earlier on in the day we had got the Buckminster Fuller book.

I think there was a consensus at that time that something like a closed cage could explain it. After all we had the book and there was a load of hexagons in the geodesic domes so it was possible to close it off. The picture was that you could get rid of the dangling bonds by closure. So the stability criterion was clearly a major part of the discussion on Monday. There was a lot of discussion on Monday. I went home for lunch with Bob Curl and described this stardome, so I know that at midday I was thinking in terms of this stardome. I don't know specifically what was in the other people's minds.

I remember, at the restaurant in the evening, describing the stardome again and mentioning that there were pentagons in it and in Rick's writing he remembers that I mentioned the pentagons. But I don't remember very much more. I remember the general feelings, but any more specific things? No. The only reason I remember specific things is that specific incidents have refreshed those memories.

I also remember that Rick got very interested in the pentagons and he certainly mentioned the concept of the chicken wire cage at the restaurant — whether he was the first or not or whether that was a consensus I do not know. I would say that the closed cage concept was the general consensus that, in the discussion, we were all coming to. I certainly had the stardome on my mind, as Bob has attested, ideas about Buckminster Fuller — I believe that I was the first, but others could have had the idea too, and I also suggested a third sandwich possibility. I thus had three different models of which two were right and I do not remember any other specific model other than the chicken wire cage. At mid-day I thought of ringing my wife about the stardome model — it was 6 pm in England but for various reasons I unfortunately did not ring her.

The next day the phone rang — it was Bob. He always went in early. I do not remember the exact words but he said effectively "You have got to come in because Rick's been playing around with something to do with that thing you were talking about yesterday and has come up with something bigger than any of us have ever been involved with before."

I went in and then Rick walked in and threw the little paper model that he had built during the night down. I saw this and thought "Oh-Oh, I've got something a bit similar to that at home" — but I was ecstatic. My actual thought was that I should have rung home, but then I thought that it was nice that we could have got the solution as a team. I felt that I had played a part in the quest by suggesting the stardome, the pentagons and Buckminster Fuller's domes. That's my picture. People have different views and different versions and I know that Rick feels that I was keener on the 6:24:24:6 model than I believe I was — but our memories are different.

Who funded this research?

I paid my trip to the U.S. myself. Concerning the experiment itself, of course, you can't build a multimillion dollar equipment without outside funding. The students who were working on this project were students working with Bob Curl, Rick Smalley and Frank Tittel.

There was a project on laser vaporization of semiconductors. Gallium arsenide and silicon clusters were the aim, and Rick and Bob thought that the graphite experiment could be worked into that program. This was expected to last a week or two. All this was part of the semiconductor project.

I understand that you were unaware of the previous works.

We were totally unaware of Osawa's as well as Bochvar and Gal'pern's work. The only work on C_{60} between those and our own work was that of Orville Chapman at UCLA who was thinking about synthetic routes to C_{60} — and we were not aware of this either. I think that it is important to realise that the real surprise is that C_{60} forms spontaneously in such high yield. I think that this is the most important thing and that observation changes the whole perception of what graphite is and the behaviour of carbon.

It is expected that there should be a Nobel Prize sooner or later for the discovery of Buckminsterfullerene. If I were the Nobel committee I would find it very difficult — on the basis of your description — to decide who the one or two or three most important players are in this work.

That is a problem for the Nobel Prize Committee and not me. People have said that a Nobel Prize may or may not be awarded for this work; however, there are more than three players. I don't know the answer to this question. If it went to, say, Krätschmer and Huffman, I think that some people would feel disappointed. If it went to Rick, me perhaps, and Bob, I think Krätschmer and Huffman could feel disappointed — I think Osawa could possibly feel disappointed too.

The Nobel prize is the Nobel prize and it is the decision made by people and, of course, it is very hard to win. Most scientists would be, as I am, absolutely overjoyed to have done something which can even be considered as worth it. That is the first thing — I am not a scientist who has that sort of ambition. I am ambitious to do as well as I can. Then when I look round and assess my work, I feel pretty happy with my microwave work and then also that I went on into another field such as radioastronomy and made a nice contribution. I am not the sort of guy who focuses on one thing, building up a consolidated research programme in a particular field for a particular goal. Then I went on to something else that interested me, clusters, and we discovered C_{60}.

It was not only the original discovery that was important, but also the work we did together with the Rice group until 1987 and further work that Rick did, such as the photoelectron spectroscopy as well as work done here at Sussex also. In fact, one of the things that I am most proud of is the discovery of the reason why C_{70} is the second magic number as well as the existence of other fullerene magic numbers — that proved unequivocally that C_{60} was a cage — to me. The C_{70} result was also discovered and proven by Schmalz, Klein and Hite in Galveston at the same time.

The real problem is that the Nobel Prize is so much more important than all the other prizes in science — and I think it is unfortunate. In tennis, for example, you can win Wimbledon or the US Open or other major championships in a given year and they are comparable. No scientist who is in the position that I am in, can feel other than that it's nice that I am here, great regrets if it goes to some other worthy contenders, and tremendous elation if we were to be so fortunate. That's the way I feel and I can't feel any other way.

I actually think that Krätschmer and Huffman are often not given credit for what they did. One thing that is not fully appreciated is that they detected the UV spectrum of C_{60} in 1983 and their suspicion that this blip might be C_{60} is a real piece of *real* science. To actually have thought that it might be C_{60} — as it was — and then go on to prove that it was, is, I think, very important as a piece of *real* science. I think it was a great piece of science, nice and elegant. What I really like is the fact that the discovery of C_{60} came out of astrophysical work.

You were also trying to extract C_{60}.

Yes, we were actually pipped at the post by Krätschmer and Huffman. My student, Jonathan Hare, had extracted it and had actually put a red solution on my desk and we also had a mass spectrum and were trying to obtain a mass spectrum of the extracted material when I got the Krätschmer-Huffman paper to referee. I was shattered. It was definitely a very bad day — the worst moment — because I realised that we had been so close to extracting C_{60} ourselves.

Was it the worst day of your life?

I think so.

Eiji Osawa used this expression, describing what he felt when he noticed your paper in Nature.

Is that right, I didn't realise that he felt that way. He never told me that.

Coming back to the buckminsterfullerene discovery, you seem to have rather detailed recollections of what happened. Is there a consensus about the story among the participants?

There are some clear differences. Rick feels that when I wrote up the story of the run-up to the structure, I described my role as more important than he feels it was. This is a disagreement that we shall just have to live with.

There was a description by Rick in a personal account of how the discovery was made, with which I did not agree. I felt that, for instance, more than one person was present and directly involved in the actual discovery of C_{60} — it was not the achievement of a single individual and it seemed to me important to clarify this. A number of people noticed the C_{60} signal. In fact, the first student to write down in the lab notebook that C_{60} is very strong was a student who is not on the paper. The date coincided with my own record. Yuan Liu, a Chinese student went for holiday, half-way through the program, before we knew anything specifically about C_{60}. This was only discovered five or six years later when I asked for the laboratory notebook. She is now with Hoechst in Japan.

My view is that this was a team discovery. If someone goes to the lavatory and returns to find C_{60} on the screen, this is irrelevant. One might think that as soon as you see such a signal someone must have said "Oh, this is big!" The answer is no, such recognition comes much more gradually — one gradually realises that there is something interesting here. You notice it and you want to optimize the conditions, so it is very gradual. Ultimately what is important is who writes what down. I have a record — and so do the students. I feel that the discovery was just as much of Rick's — though he was not present in the lab — as it was mine and I don't think one should differentiate.

There is another difference in our recollections and that is over the origin of name of the molecule. In a number of places it is written that Rick had invented the name. This is incorrect. I invented it and I remember the moment clearly. I feel rather fond of the name, and fond of the sort of persona that I have to think of it. So I just wrote that up — but Rick remembers it differently. As in many, many cases, memories differ. I feel that if I am writing a personal account, I am writing my memories, and if someone else wants to write a personal account they should write theirs — and if they are different then they are different and that will always be the case. These are minor differences in some ways, but they have caused some problems. Memories are not perfect but I have written my account to the best of my ability and I stick by it.

You never sat down with Rick Smalley and discuss these differences?

We did sit down and discuss things during a conference in New England and — well — I suppose that we agreed to differ.

That is something already.

Oh, well, I mean "agreement" in inverted commas!

Is this difference very important?

I generally invoke "Rashomon" related arguments here in that the "true" story is only to be found in the totality of subjective accounts of the participants and though they may not on the face appear always to coincide on important issues, they need not necessarily be in conflict with what happened — basically each experiences events differently.

As far as names are concerned — one feels very sensitive about them. There is something in Shakespeare's Othello about one's attitude to names. It's a very personal thing. There is an interesting aspect to this type of issue which I've often thought about because of my interest in graphics, design and art which is the difference between Science and Art. If we had not discovered C_{60}, it would have been discovered within a year. In fact, it should have been discovered already in the sixties. The point is that if I paint a picture, it is the creation of an individual — so there is a certain individuality quality.

That is true for any scientific discovery — the individuality is not very high.

However, in the case of C_{60} the name is such an individual thing. The one difference that I made in the C_{60} story is that, had I not named it, it would have been called footballene or soccerene. At the time we discovered C_{60}, almost everybody who wrote to us had a different suggestion as to what the name should be — and only a small number of people liked buckminsterfullerene at the start and these people usually recognised the humour involved. My view is that this is the only little bit of individual immortality in the discovery. One thing that was very important for me was that Rick and Bob thought that I should be the first author on the paper. I was very pleased with this.

I was invited to write a personal account for *Angewandte Chemie* and I wrote what I remembered from my own personal perspective and I stand by the account in every essential detail. The way that other people remember events may be very different because each person is egocentric and Rick and I are both egocentric to very great degrees. No one can know exactly how another contributed to the discovery or what original thoughts were in another's mind, what they did, what they can remember, etc. I suspect that Rick's view is that this discovery came primarily out of the cluster science advances that he'd made. My view is that it was a grand synthesis of ideas from astrophysics welding together in a very beautiful and elegant way with the incredible achievements that Rick had made in developing cluster techniques — they are both perfectly valid views. Then there was a further synthesis by Krätschmer and Huffman from their ideas on interstellar dust. To be aware, as physicists, of this work was interesting by itself. I think the whole advance is a shining example of how fundamental, non-strategic, non-applied science can lead to an important discovery.

It would have been difficult to write a proposal for the discovery of C_{60}. What I mean is that for the most interesting and unexpected discoveries, you cannot write proposals.

Of course, the most important discoveries are serendipities, because they are things you don't expect. That is why the discovery of the first carbon-phosphorus double bond is different, in some ways. In fact I am more proud, in some ways, of that discovery than of C_{60} because I thought about it, thought that this bond could be formed, worked out how to do it, and then we did it with my student. Then, however, I didn't stay in the field. I'll make another observation; my role in the creation of the field of carbon-phosphorus double and triple bonds and their chemistry has, all but, been forgotten. The reason is that once having shown how the molecules can be made and used as a synthon — with my colleague, John Nixon — I left the field and went on to other things. Unless you stay in the field and keep reminding people of how the field originated, then that aspect is actually lost. That happened to me, and I feel — not hurt — but put out. So when C_{60} broke, it's such an important discovery, it seemed to me that I should stick with it for five years. I first collaborated with the Rice group and then I set up a C_{60}-programme here at Sussex. I discovered, for instance, that I could explain C_{70} and other magic numbers; this still remains in my mind the most convincing evidence that our cage proposal was correct and that existed prior to the extraction which was achieved in 1990. We also discovered that the giant fullerenes were icosahedral and that they would have corners — I am also proud of these findings.

Then when the Krätschmer and Huffman manuscript came through and we were just two or three weeks behind I felt that we should try to get the one-line NMR — that was wonderful to do that. It was wonderful to get the third paper in the field. The first paper is the Rice-Sussex paper, the second is the Krätschmer/Huffman paper, and the third paper is our Sussex paper with the NMR proof of the structure. Roger Taylor helped Jonathan Hare, Ala'a Abdul-Sada and me at the end and he found that C_{60} and C_{70} could be chromatographically separated.

Are you a local celebrity at the University now?

At the University of course but I was known here beforehand, because I had done a lot of work here.

You have contacted Eiji Osawa and invited him over for a meeting. Have you ever tried to contact the Russian colleagues?

Roger Taylor has had most contact with them. He is a physical organic chemist and when the Krätschmer/Huffman paper arrived, he said he would help us to get the one-line NMR. You see that when the Krätschmer manuscript arrived for me to referee, we already had the material; we already had the red solution and I realised how close I had been — my student Jonathan and I, in particular, had been doing the whole thing here at Sussex. Then the question was what to do. As that day progressed, I thought we should go for the NMR, because when I read the paper I realised "Gee they haven't got the NMR in here" — it took me a while to realise that as I was so shell-shocked — I felt that we had lost everything. We already had the red solution and it was the red solution that they'd seen; we already had a mass spectrum but there was no mass spectrum in that original manuscript, so I knew that everything in their paper was right — and they had the crystal structure. We had done almost everything.

So that was a bad, bad, bad day and then it took a little bit of time to recover. But I thought OK resilience — Middle European — my parents were refugees.

From where?

From Berlin.

Jewish?

My father was but my mother was not. They came to England in 1937 and I was born in England in 1939. I was taught resilience, survival from them. My father made balloons — he actually printed faces and pictures on children's toy balloons — but he lost all that when he came to England and the war started. He got a job as an engineer and various other things.

Stockholm City Hall drawn by Harold Kroto as a child of 13 (courtesy of Harold Kroto).

Did your parents witness your success?

No, they knew something about the carbon chains in space but not the great stuff about C_{60} of course. It's a bit sad. My father would have been, perhaps for the first time, really proud of me. Of course, he was proud of my being an academic at the University. He died in 1977.

He had a terrible life. Anyone born in 1900 in Berlin must have had — in 1914–18, there was the war when he would have been a teenager and then the bad times in Germany in the 1930's. When he escaped in 1937, he only just did it. He had a passport as a businessman to travel, but the police had even tried to get him off the train but apparently they were too late by about half-an-hour. I don't know the full details — somehow I wanted to be orthogonal to his world in some strange way and did not discuss it with him very much. I regret it as I would now like to know more. Maybe that is what he didn't like either. We got on very well, but we also had our differences of opinion. My father was a very generous and honest man but he was Jewish and I would certainly be prepared to discuss the Palestinian problem with him — what a rotten problem the Palestinians have — which I hope some day can be solved. I would try to discuss them in a rational way because I had Arab students and could see both points of view — there were difficulties.

My father, like many Jews, would go back to the historical factors. He felt that I was too strong in my discussion of the injustice — perhaps that I was too detached in my view. He recognised the view but he felt that Israel was under such threat, and I think that's true. On the other hand, I'm not going to go back two thousand years to discuss who happened to be standing on this land or that land. We can't go back like that, we have to recognise that millions of kiddies are living in really bad and impoverished conditions today.

He was a very honest man, and I liked him a lot, but he was never really able to understand me. That was the main problem.

Little Harry and his father in a boat on an outing around 1949 (courtesy of Harold Kroto).

Your mother was very courageous too in coming to England with him.

Yes, my mother was an incredible woman — both of them were incredible people. I would have really liked my father to see me on TV. He was proud of what I did, but there were things that were very important to him and he never fully understood me — like all parents.

In what sense?

Well, I am a devout atheist and I don't think he was very happy about it. I'm also a socialist or more accurately a mixture of socialist, democrat and republican (in the sense that France is a republic). He was a socialist once but I think he'd forgotten those things. So there were lots of points on which he had strong opinions when we debated and he was emotional — we both were — we discussed many things.

Self-portrait of Harry Kroto drawn after a picture taken much earlier and painted during college years. The picture is about 8–10' high (courtesy of Harold Kroto).

Isn't it wonderful that you had debates?

Oh, yes, I only wish that he could have seen them my way. I loved him but there were disagreements which went somewhat deeper — discuss it in this way overplays the actual situation. I think as time would have gone on, I would have been able to handle these debates in a much more sensitive manner and then tease him round to understanding me.

Of course, he had a terrible life and even in Britain he was interned on the Isle of Man. Then he got a job as an engineer from 1945 to 1955 and then he started making balloons again, from 1955 to 1966. And then when he was 66, he had this business which he built for me, and then he realised that I was not going to ever take it on.

How did he view this very different line of career?

He never said very much but I suspect that deep down he was probably upset, because he had worked so hard and built this factory — probably for me and it became very clear that I was not going to make balloons for the rest of my life, in fact. At the time I never wanted to see another balloon as long as I lived — I am speaking facetiously, of course. I had seen and weighed millions of balloons in stocktaking by that time. When you are in this sort of a family, you do the work for family reasons, and at that time I wanted to play tennis and get away from the factory.

R. Buckminster Fuller (photograph courtesy of Lloyd Kahn, Bolinas, California).

27

The Fuller Connection

R. Buckminster Fuller's name has become a household name in chemistry since 1985 when the name of "buckminsterfullerene" was introduced for the C_{60} molecule, followed by the name of "fullerenes" for a whole class of molecules and materials. Both Kroto and Smalley had recalled a visit to the Montreal Geodesic Dome during the 1967 Expo in Montreal where this construction housed the exhibition of the United States. Fuller's physical geometry was an important influence on others as well, among them being Aaron Klug and Donald Caspar in their pioneering studies of the icosahedral virus structures:[1]

"The solution we have found was, in fact, inspired by the geometrical principles applied by Buckminster Fuller in the construction of geodesic domes. The resemblance of the designs of geodesic domes to icosahedral viruses had attracted our attention at the time of the poliovirus work. Fuller has pioneered in the development of physically orientated geometry based on the principles of efficient design. Considering the structure of the virus shells in terms of these principles, we have found that with plausible assumptions on the degree of quasi-equivalence required, there is only one general way in which iso-dimensional shells may be constructed from a large number of identical protein subunits, and this necessarily leads to icosahedral symmetry. Moreover, virus subunits organized on this scheme would have the property of self-assembly into a shell of definite size.

"The basic assumption is that the shell is held together by the same type of bonds throughout, but that these bonds may be deformed in slightly

The Montreal Geodesic Dome in 1995 and close-up with a discernible pentagon among the hexagons (photographs by I. Hargittai).

different ways in the different, non-symmetry related environments. Molecular structures are not built to conform to exact mathematical concepts but, rather, to satisfy the condition that the system be in minimum energy configuration ..."

Fuller's opus magnum is his *Synergetics*[2] which does not reveal much connection to chemistry but he gives conspicuous emphasis to Avogadro's Law in it: "Equal volumes of all gases at the same temperature and pressure contain the same number of molecules." In Avogadro's Law, Fuller saw a proof that chemists considered volumes as material domains and not merely as abstractions. In producing *Synergetics*, Fuller relied a great deal on the collaboration with his long-time associate, E. J. Applewhite. This is what he told me about Fuller's connections with chemistry:[3]

"I often heard Fuller say that the chemists draw their stick figures and balls of different sizes to describe molecules and their relations. They find this a convenient diagramatic way to handle the data they have, but they don't realize that that's actually the physical truth. I'm just approximating what Bucky was trying to get across, but just as he regarded these volumes as real in Avogadro, he regarded the chemist's diagrams as much more than a device of convenience. He regarded them as a pattern of validity and reality, and found them very dramatic, something to focus on. He hardly gave a speech without mentioning Avogadro, but not much beyond that. I know this because everything what I could I put into *Synergetics,* one way or another and then have him go over it. I had these pages made up of Pythagoras, of Democritos, of the various things, and I said, this is what I think you said, and I typed it triple-spaced and put it in front of him, and he didn't care about the exercise that you and I embarked on of how he got there and he wasn't that interested in documenting them. Whether he was shunning it or bored by it, it was very hard to focus his attention on that. He had reference to Democritos and Avogadro because he thought that they were doing what he's doing which is starting absolutely from scratch to think for themselves. He would argue that they did start from scratch, each of them. He didn't see them in the context of the development of science and Western civilization, and he didn't see himself in that context. So it's very elusive."

References

1. Caspar, D. L. D.; Klug, A. "Physical Principles in the Construction of Regular Viruses." *Cold Spring Harbor Symposia on Quantitative Biology* **1962**, *27*, 1–24.
2. R. B. Fuller, *Synergetics: Explorations in the Geometry of Thinking.* Macmillan, New York, 1975.
3. From a conversation with E. J. Applewhite in 1996 in Washington, D.C. E. J. Applewhite (b. 1921) likes to describe himself as a layman. He studied at Yale University and at the Business School of Harvard University. After Navy service on an aircraft carrier, he worked with Buckminster Fuller in a housing project in Wichita, Kansas. He then spent 25 years with the CIA. In the 1970s, he spent nine years of collaboration with Fuller on his *Synergetics* books. Being aware of the fact that Fuller did not like to attribute, I asked Ed Applewhite whether lacking adequate recognition for his work with Fuller has ever bothered him. Alluding to his long association with the CIA, his answer was: "If you crave public recognition, you don't belong in intelligence."

Richard E. Smalley (photograph courtesy of Richard Smalley).

28

Richard E. Smalley

Richard E. Smalley (b. 1943) is Gene and Norman Hackerman Professor of Chemistry and Professor of Physics at Rice University in Houston, Texas. He shared the Nobel Prize in chemistry in 1996 with Harold W. Kroto, of the University of Sussex in Falmer, Brighton, East Sussex, U.K. and Robert F. Curl, of Rice University "for their discovery of fullerenes."

Rick Smalley started his undergraduate studies at Hope College in Holland, Michigan, received his B.S. from the University of Michigan at Ann Arbor (1965), M.A. (1971) and Ph.D. (1973) from Princeton University. Before his graduate studies, he worked as Research Chemist between 1965 and 1969 at the Shell Chemical Company. At Princeton, he was Graduate Research Assistant with E. R. Bernstein between 1969 and 1973 and did postdoctoral work with D. H. Levy at The James Franck Institute of the University of Chicago between 1973 and 1976. He has been at Rice University since 1976 where he is now also Director of the Rice Center for Nanoscale Science and Technology. A sampler of his numerous awards and prizes: Irving Langmuir Prize of the American Physical Society (1991); Ernest O. Lawrence Memorial Award of the U.S. Department of Energy (1992); International Prize for New Materials from the American Physical Society (shared with H. Kroto and R. Curl, 1992); Welch Award in Chemistry (1992); Hewlett-Packard Europhysics Prize (shared with D. R. Huffman, H. Kroto and W. Krätschmer, 1994); and the Franklin Medal (1996). He is member of the National Academy of Sciences, the American Academy of Arts and Sciences, and many others.

We did an interview by correspondence in the fall of 1994 and it first appeared in *The Chemical Intelligencer.**

Your family background, schooling, teachers, if there was any one who was a great influence on you in your studies or in the direction of your interests?

I came from a fine, nurturing family of what in the U.S. are known as wholesome midwestern values. The youngest of four children, I spent all but the first three years of my life in a beautiful old home in Kansas City, Missouri. I went to the same public grade school and high school as my siblings — all of whom were better in their classes than I by the time I took them a few years later, often with the same teacher. I tended to be pretty erratic in my classes until at the beginning of my junior year in high school I rather suddenly became very serious and "turned" by my studies, particularly chemistry. This was the first course I ever truly mastered, scoring essentially a perfect grade in the class. The next year, I had a similar success with physics, and decided that somehow a career in these fields might be possible. This was the era in the U.S. when Sputnik had made the careers of engineers and scientists the most romantic possible life's mission available for young boys.

I have been helped and mentored along the way by many. My father had a woodworking shop in the basement of our house, and from him I learned to tear things apart, figure them out, and put them back together. We built many things together. I must have spent thousands of hours at these pursuits in the basement shop. My mother went part-time to finish her liberal arts college B.A. degree while I was in high school and I watched her fall in love with knowledge. In hundreds of long conversations we used to call "mommy talks" I fell in love with ideas and knowledge too — and fine arts, and open, questioning discourse. Incidentally, my mother taught me to do mechanical drawing in those early pre-chemistry days when I was certain I was going to be an architect. We designed a castle together, and I still remember her teaching me how to do the final free-hand art work on the front-view perspective. My mother's sister was Dr. Sara Jane Rhoads (Ph.D. research under William Doering at Columbia), one of the first few female professors of chemistry in the U.S. (her career was spent

*This interview was originally published in *The Chemical Intelligencer* **1995**, *1*(3), 23–26

Richard Smalley with his sister, Linda, and their father (courtesy of Richard Smalley).

at the University of Wyoming). She was the only scientist in our extended family, and was one of the most impressive people I have ever met — I used to call her, lovingly, the "Colossus of Rhoads". Her example led me to go into chemistry, rather than physics, and it was at her suggestion that I decided to attend Hope College in Holland, Michigan, which had then (and still has now) one of the finest undergraduate programs in chemistry in the U.S.

Are you a chemist, physicist, physical chemist, etc.? Do these labels mean anything to you?

I have long wanted to understand chemistry as though it were a problem in physics, and be able to manipulate chemistry as though it were a problem in engineering. Where that leaves me in the chemistry/physics/engineering professions is mostly a function of where I came from (Ph.D. in Chemistry) and what I happen to be doing at the moment (which is mostly physics trying to spawn a practical materials science).

How did you get into cluster research? Have you always been interested in building apparatuses?

My entry into cluster research came from spectroscopy. My Ph.D. project with Elliot Bernstein at Princeton was the detailed spectral study of a

Richard Smalley and his historic apparatus (courtesy of Richard Smalley).

symmetrial aromatic molecule, sym-triazine, in single crystals at cryogenic temperatures. For my postdoctoral study, I wanted to get into a more chemically interesting area so I figured needed to get into the gas phase where molecules could move around and react. The closest job I could find was to continue doing detailed chemical physics via high resolution spectroscopy on gas-phase molecules with Donald Levy at the University of Chicago. My problem was that I knew the details of molecular spectroscopy only in rigid crystals at low temperatures, and had a vast amount of learning to do before I could really understand the arcania of rotations/vibrations of even rather simple polyatomic molecules in the gas phase. In those days (1973), the most celebrated problem in such spectroscopy was the spectrum of NO_2. Levy had recently made some major contributions to this field using (then home-built) tunable dye lasers, which were then just beginning to revolutionize molecular spectroscopy.

When I arrived in Chicago in late summer of 1973, Levy was away in Germany for a few months, and I still had to prepare for my final oral examination for the Ph.D. at Princeton, which at that time consisted of presenting and defending three original research proposals. While settling into work with the graduate students in Levy's lab (I was his first postdoc), I found time to read the literature, and think about possible new research proposals, both for my exam in Princeton, and possible new projects to be done in Levy's group in Chicago. While sitting in the chemistry department library, I read a just-released publication of Yuan Lee and Stuart Rice[1] about the reactive scattering of fluorine on benzene in one of Yuan's "universal" molecular beam apparatuses. I was deeply struck by a passage which said that the supersonic expansion used to make the benzene molecular beam was strong enough to cool out essentially all rotational degrees of freedom. That passage lit me up. Since I knew my biggest problem was that I didn't know about the quantum mechanics of freely rotating gas-phase molecules, this capability of supersonic beams to nearly "freeze out" this motion was truly intriguing. So one of my three research proposals I defended in my Ph.D. oral exam a month later in Princeton, was to use a supersonic beam to cool the rotations of gas-phase NO_2, and then use a dye laser to figure out what was going on to make this molecule such a celebrated problem in spectroscopy. When Levy got back from Germany, I talked with him about this, and after a few months we pushed it around enough to go talk with Lennard Wharton about it. He "lit up like a light bulb" when he heard the idea. The result was a stupendously successful series of experiments done by Don Levy, Lennard Wharton, and myself that introduced the power of supersonic beam laser spectroscopy to the world.

The relevance of this story to clusters is that one of the experiments we did was to make the first helium van der Waals complex (HeI_2) and study its chemical physics. We extended this in a rather famous series of papers to such clusters as NaAr, He-tetrazine, and He_2-tetrazine. Spectral studies of such van der Waals clusters in supersonic beams led me to dream of being able to make clusters from any combination of atoms and study them in a supersonic beam and near absolute zero.

After coming to Rice University in 1976, I continued development of this field, introducing a wide variety of new techniques using entirely pulsed nozzles and lasers. In the late 1980s, we managed to get the first laser-vaporization supersonic nozzle device to work. In the subsequent years, we perfected this source, together with time-of-flight mass spectroscopy and two-photon laser ionization (R2PI) of such supersonic beams, ultimately realizing the dream that was hatched in my 1973–76 research work in the Levy-Wharton basement lab in the James Frank Institute of the University of Chicago.

With the new laser-vaporization supersonic cluster beam apparatus [AP2, and its successors: AP3 which was cloned and went to Kaldor's group at Exxon, AP4 (Ion Cyclotron Resonance, or, ICR), AP5 (Ultraviolet

Photoelectron Spectroscopy, or, UPS, etc.)], we were able to examine a whole new world of fundamental problems in chemical physics — a task with which we were merrily engaged in when in 1985, C_{60} popped up its pretty head and said "Howdy." Our lives have not been quite the same since.

Louis Pasteur said that "In the field of observation, chance only favors those minds which have been prepared." Your experiment was not the first experiment ever that showed the presence of C_{60} species. What had prepared you and your colleagues to be favored by this chance in Pasteur's sense?

With Robert Curl, Frank Tittel, and a fine group of students, I had been studying semiconductor clusters [of silicon, germanium, and gallium arsenide] in supersonic beams. We had had many hours of conversation over just what sort of tying-up of surface dangling bonds of these species was going on to explain the "magic number" behavior they often displayed. This turns out to be *the critical intellectual path* one must follow in order to realize, when looking at carbon clusters, that C_{60} must be a soccer ball. Our long and detailed familiarity with these other sp^3-valence-dominated elemental clusters sensitized us to just how utterly weird and special carbon was. In addition, I was deeply involved in metal cluster research in my labs at the same time — doing some of the first studies of the chemical reactivity of the bare metal clusters, and worrying about how their surface looked and how this dominated the chemistry. In clusters of this size, nearly every atom is on the surface anyway. At that time, I was calling it a new form of surface science. Carbon's ability to form a two-dimensional sheet, and wrap it into a hollow ball is revealed by the chemical passivity of its surface. That's principally what "AP2" was designed to study, and that's principally what my research group was about those days. Under these circumstances, it would have been deeply embarrassing if we had not discovered the fullerenes.

When did it occur to you that something fundamentally new was being discovered?

At about 1 a.m., Tuesday morning, September 10, 1985.

How did the question of the origin of the great relative stability come up?

During the week of September 2–6, Harry Kroto and I had many, many long talks in my office during the day. If you look at the times of the

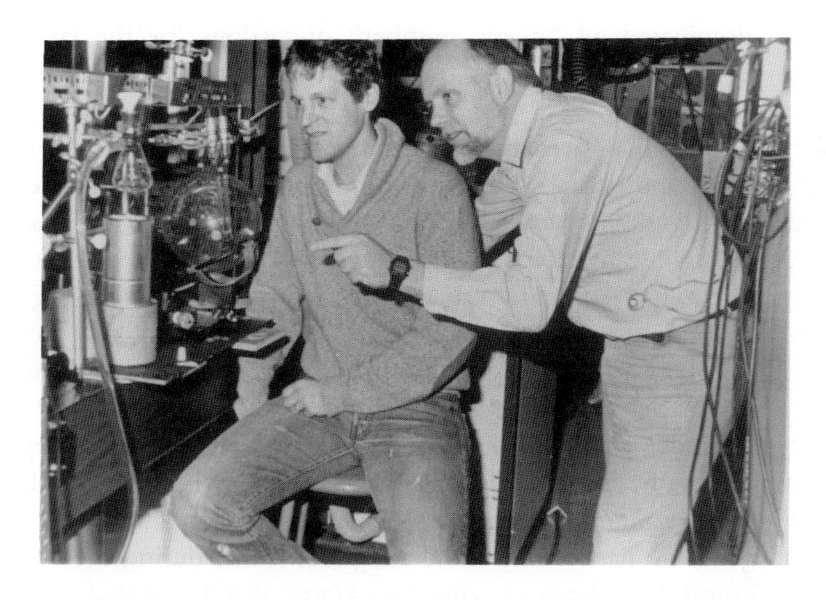

Richard Smalley and Jim Heath (courtesy of Richard Smalley).

data records during this week, you'll find most of them started in the late afternoons, and ran on sometimes late in the evenings. Most of these long discussions were just between the two of us, alone in my office. This is the principal reason no one else is much able to tell firsthand who said what to whom.

The main issue from the start was what the even-numbered cluster distribution was all about. Both of us thought the flat graphene sheet was the most likely form for these clusters, but it wasn't clear at all how one could explain the preference for even numbers. We realized that one property of even-numbered graphene sheets is that you could always find a structure which made every carbon atom part of a double bond, and that odd-numbered sheets will always leave at least one carbon with only single bonds, but I argued that couldn't be enough to explain it: the sp^2 dangling bond energies for such a structure ought to be by far more important energetically than a tiny effect in the pi-system. We realized from the start that there was something about the 60th (and to a lesser extent the 70th) cluster which made it reluctant to engage in the "upclustering" which all metal and semiconductor clusters do in the laser vaporization cluster source as condensation progresses. I argued that somehow, something was being done in the cluster to hide these dangling bonds on the edge to explain the special slowing down of clustering reactions. It was precisely this question that had dominated all "magic number" discussions for the last several years in my group in our work on silicon and other semiconductor and metal clusters.

Harry argued that perhaps two or more sheets were stacked one on top of another, and perhaps there was some sort of stabilization interaction between the edges. But we couldn't imagine how anything like that could have such a significant effect on the chemistry. I argued instead that the graphene sheet had somehow curved up to tie up the dangling bonds of the once-opposite edges — an idea we termed the "closed solution". But neither Harry nor I thought that his flat sandwich, or my closed ball, or any other model we tossed around [we also talked about linear chains, monocyclic rings, and tiny pieces of diamond lattice] were very likely explanations for the increasingly bizarre results from the lab. The data sets taken Wednesday night, September 4, which I saw first Thursday morning, September 5, were particularly stimulating. Sometime during the week, and I think it must have been on this day, Harry and I wrestled again with the question of how any imaginable structure of carbon clusters could explain the evenness of the large carbon clusters and the prominence of 60. To give the "closed solution" possibility a clearer image in the mind, I asked Harry who it was that had designed those domes that I remembered were composed of some sort of (hexagonal?) latticework. He said it was Buckminster Fuller. And I quipped: "Gee. It's Buckminsterfuller — ene!", to go with the sort of Monty Pythonesque humor and word games we were having fun with all through this week.* We kidded around about it some more, and I asked Harry if he would find the time to go over to the Rice Library to check out just what these geodesic domes looked like. Harry said that he would, and I assumed he would look into it. But, in fact, he never did. Note that his failure to do this had nothing to do with the fact he had no library card. All he had to do was walk into the library and look.

---

*(Added by R. E. Smalley) Note here in connection with Hugh Aldersey-Williams' footnote relative to this controversy,[2] it was quite clear to me, as it would be to most chemists, that any sort of wrapped up "closed" graphene sheet with no dangling bonds and no hydrogens would be most appropriately named with an -ene suffix. That's what the "closed solution" structure would have been. At the time I suggested the structure, I was hoping there was some way of doing it all with hexagons. Of course it would be aromatic. The only question about aromaticity in C_{60} has to do with the presence of pentagons, the critical importance of which was to remain in the dark until the paper model was made successfully the next Monday evening.

At which point did you decide that a structure must be suggested for this species?

See above. Structure was discussed essentially from day 1 (Monday, Sept. 2). On Monday, September 9, after Jim Heath and Sean O'Brien showed us their stunning flag-pole data of the weekend and showed under what conditions the flag-pole came and went, we knew we had a paper to write. In a private conversation a little later that morning, Harry suggested we send the flag-pole paper to *Chemical Physics Letters.* I said I agreed, but I thought we had to put at least some discussion in about possible explanations (i.e., the structure of C_{60}). Harry said he agreed, but we just didn't have time to work one out. I then asked him if he had looked into the "closed solution" possibility by going over to look up Buckminster Fuller in the library. He said, flatly, that he hadn't. I looked at my watch and said, "Well, look, I have enough time. I'll go check myself". The rest of the story as portrayed in the Baggott's and Aldersey-Williams' books are pretty close to reality as far as I can recall.

It has bothered me for some time, as another chemist interested in symmetry and shapes and structures that it seems almost agonizing how difficult it was for the discoverers to get to the idea of the truncated icosahedron. I know that everything is easy in hindsight yet I can't help feeling that we, chemists, are revealing some ignorance about some quite basic things in geometry. I understand and like the name of C_{60} but for the truncated icosahedron, the name could have also been derived from Archimedes or Leonardo who is known to have made beautiful drawings of the Archimedean semiregular polyhedra, and there are other examples. It is curious that for such a simple shape, you had to consult Fuller's book and the Chairman of the Mathematics Department. In 1984, I attended a meeting at Smith College, whose title was Shaping Space. The main goal of the organizers was to encourage people to try to get back geometry into the curriculum. I have myself experienced, in teaching general chemistry in the U.S., that my favorite model, the VSEPR, was difficult for some of the students just because of its spatial geometry. So the question is, why was it so difficult to arrive at the truncated icosahedral model once you knew that the number of atoms was 60 and suspected that it was a closed system?

I was simply totally ignorant. I certainly knew about icosahedrons and the fact they had 12 vertices, and I suppose I knew that dodecahedrons would have 20 vertices. We were discussing icosahedral structures at the time for

Richard Smalley's paper model of the truncated icosahedron (courtesy of Richard Smalley).

the outermost layer of metal clusters of 13 atoms. But I think I must have been asleep during the lectures in solid geometry (if I attended them at all) where truncated Platonic solids were introduced.

Would you also care to say anything about how to avoid controversies about looking back to discoveries? Sometimes historians present different versions of the discovery after a hundred years or fifty. Yet we are having here a recent event and there is already controversy.

Keep a tape recorder running at all times [in the U.S. this ought to be called "Nixon's Rule"]. Seriously, I think when you boil this "controversy" down, it's really about two pretty trivial points: who first thought of the structure, and who first named it. As you pointed out in the way you phrased one of your questions, the fact that a truncated icosahedron is the answer to the question of "What's special about a cluster of 60 atoms" is so obvious to someone with a good training in mathematics (and the VSEPR method) that I think we can be sure someone would have pointed this out within a week or so of our flag-pole paper coming out in *Chemical Physics Letters.* And, I frankly don't care about the name.

What is your advice to others how to get best prepared for new discoveries?

Don't worry about it.

Do you anticipate any new discovery of similar importance in the forseeable future?

No.

What is your recent research interest?

Learning how to build stuff that does stuff on the nanometer scale. In particular, I am most interested in learning how to make a truly metallic nanowire, for use in making electrical connections to nanoscale objects, and as a direct information link between the nanometer world and up here in the "real world".

How does anticipation of the Nobel Prize influence your work and your life?

It makes me, and the people around me, a little nervous every year around the middle of October. This gets to be a little bit less of a problem as the years go by and in time it will pass.

Is your family excited about your work and fame?

My son, who is an aspiring musician in New York City, is really proud of his dad. I think he's the greatest.

References

1. Lee, Y., Rice, S. *J. Chem. Phys.* **1973**, *59*, 1427–1434.
2. Aldersey-Williams, H., *The Most Beautiful Molecule: An Adventure in Chemistry.* Aurum Press, London, 1995, p. 297.

Robert F. Curl, 1998 (photograph by I. Hargittai)

29

Robert F. Curl

Robert F. Curl (b. 1933 in Alice, Texas) is Harry C. and Olga K. Wiess Professor of Natural Sciences at the Department of Chemistry of Rice University in Houston, Texas. He shared the Nobel Prize in chemistry in 1996 with Harold W. Kroto of the University of Sussex in Falmer, Brighton, East Sussex, U.K. and Richard E. Smalley of Rice University "for their discovery of fullerenes."

Robert Curl received his B.A. from Rice University in 1954 and his Ph.D. from the University of California at Berkeley under the supervision of Kenneth S. Pitzer in 1957. He was a Research Fellow with E. Bright Wilson at Harvard University in 1957–58. He has been at Rice University since 1958. His numerous awards include the Clayton Prize (Institute of Mechanical Engineers), 1957, the Alexander von Humbolt Senior U.S. Scientist Award, 1984, the International Prize for New Materials of the American Physical Society (shared with H. Kroto and R. Smalley), 1992, and others. He is member of the National Academy of Sciences, Fellow of the American Academy of Arts and Sciences, and many other learned societies.

Our conversation was recorded during the 17th Austin Symposium on Gas-Phase Molecular Structure in Austin, Texas, on March 4, 1998. It was only a few months after Kenneth S. Pitzer's death, who was Dr. Curl's graduate adviser and they also interacted when Pitzer was President of Rice University. So I began our conversation by asking Dr. Curl to tell us about the late Professor Pitzer.

I was a graduate student of Ken Pitzer. I came to Berkeley in September of 1955, intending to work for him. When I was a senior at Rice University,

Kenneth S. Pitzer at Rice University (courtesy of Rice University).

I attended a course in natural products chemistry. I hadn't always liked organic chemistry but Richard Turner, who taught this course, was a fantastic teacher. He spoke enthusiastically and warmly about Pitzer and his discovery of barriers to internal rotation and about the great implications this discovery had for natural products chemistry.

By the time I got to Berkeley though, Pitzer was no longer doing barriers to internal rotation, and I started working with him on extending the theory of corresponding states. It was great working with him. I didn't really want a lot of direction. I didn't want somebody telling me, do this, do that. Pitzer's idea of directing graduate students was to leave them alone, but if the graduate student sought him out and asked him for advice, he always made time for them; he was always willing to discuss the problem, offer advice, suggest a solution, and so on. This was ideal for me. He was real busy then because he was Dean of the College of Chemistry, which included the Chemistry Department and the Chemical Engineering Department. I was always amazed that I never had any trouble getting in to see him. There was another thing he would do for me. I'm one of those people who don't know how to get away when a conversation is over. It was wonderful with Pitzer because we would discuss what I

E. Bright Wilson around 1960 (from the *Proceedings of The Robert A. Welch Foundation Conferences on Chemical Research*, Houston, Texas, 1960).

wanted to talk about, we would reach a conclusion, and somehow or another I would find myself outside the door of his office with a warm feeling. I appreciated this efficiency. I was really happy that he was mentoring me in exactly the way I wanted to be mentored. When I was approaching the end of my time at Berkeley and was getting ready to graduate, there was the question of what I should do and I wanted to go to postdoc with E. Bright Wilson. So Pitzer wrote Wilson on my behalf and Wilson offered me a postdoctoral fellowship.

I spent a year with Wilson and then I went to Rice as Assistant Professor in 1958. Pitzer became President of Rice University in 1962. As President, he wasn't taking any graduate students but he had some postdoctorals working with him. Harry Hopkins, Jerry Kasper, Jürgen Hinze, Stewart Strickler were all postdocs with him. All these guys succeeded academically. Hinze is Professor at Bielefeld (Germany), Strickler is Professor at Colorado, Kasper went to UCLA and got tenure but then decided to leave academia to go into business for himself, and Hopkins went to Georgia State.

Is it common for a University President to have postdocs?

It's not uncommon.

Is it common for a Professor to become University President?

It's fairly common. Remember, Pitzer had a lot of administrative experience. He was Director of the Maryland Laboratory during the War. He was Director of the Atomic Energy Commission, and Dean of the College of Chemistry at Berkeley.

When he came to Rice, we started collaborating on research. We worked on some really interesting problems together; nuclear spin species conversion for molecules other than hydrogen and some matrix isolation infrared spectroscopy. We wrote a paper together on the kinetics of tunneling at low temperatures in solids. Pitzer had relatively little time to spend over in the lab, but when he was over there, we always had a chance to talk with him and work together.

Didn't Rice use to have various restrictions as regards who could enroll in it?

The only restriction Rice used to have was against African-Americans, and Pitzer was instrumental in getting that restriction eliminated. That was one of the conditions he made in order to accept the Presidency. Since the restriction was in the original Charter, the Board of Trustees went to Court and asked the Court to remove this requirement from the Charter. Pitzer came in 1962 so they started this in 1961, well before the big civil rights legislation of the mid-sixties. The other change he helped introduce was that Rice never charged tuition before, and they changed that too.

How long did he stay at Rice?

He left Rice in 1968 to become President of Stanford. He thought at that point that he was never going to do research anymore. He disposed of his old journals and I didn't think he had any postdocs at Stanford.

When he got to Stanford, he ran into the Vietnam war protest. He was caught in the middle between various people, like alumni who wanted him to bring in the National Guard to put down this protest and, on the other hand, there were the students who were not interested in a responsible discussion about anything. Several ugly things happened to Pitzer at Stanford. I'm told, I wasn't there, that somebody poured a bucket of paint on him. He left Stanford in the early seventies. He came back to Berkeley, he was almost 60, and he was uncertain whether the Faculty of the Chemistry Department at Berkeley would want him back. I think one of the things in his life that he was most touched by was that they were eager to have him come back and rejoin the Faculty. He restarted research in a big way, which is hard to do; he had not thought about research for four years. He published many papers after he came back to Berkeley.

The characteristics which he had that I was always amazed by was his ability to see what forces were at work and what was going to happen far in advance, how people were going to behave, how things would fit together. He made a huge difference to Rice. He was its third President. In the early days Rice was a good regional University. In the State of Texas it certainly had the highest academic standards. When Pitzer came, he had a vision of Rice becoming a national university. During the time he was there he made an awful lot of progress and moved Rice in the direction of becoming a national university. He was a leader, not a driver. He was leading people into thinking of themselves in terms like "We can be more; we can do more; Rice can be more than it currently is."

One of Kenneth Pitzer's long-time colleagues at Berkeley told me that he had learned more about him during the memorial service than during decades of their association.

I don't think this was a question of not being open to others. He had really well-centered values. His kids have all turned out great, for example. They are well adjusted, and, as far as I know, they've never had any chaos in their marriages. He and his wife had a very good marriage. When he died, they had been married for over 60 years. He didn't talk about his family life to people whom he was seeing professionally. My relationship with him was primarily professional. I admired him, but I didn't know his family well. He did invite me over when I was a first year student at Berkeley for Thanksgiving dinner in his house where I first met his wife, Jean, and his daughter, Ann, and son, John. Russell was away at CalTech. There are many people at Rice who think the world of him. His main avocations were sailing and designing sailboats, camping, and outdoor activities.

What would you single out as his most important contribution in chemistry?

The measurement of the barrier to internal rotation in ethane. I feel that he should have received the Nobel Prize for the discovery of the barrier to internal rotation in ethane. From what I had learned in my senior natural products class, this had a revolutionary effect on the way people think about organic structures. Other people single out other activities. His laboratory produced more thermodynamic data than any other laboratory.

His main interests other than internal rotation and the determination of thermodynamic quantities, was the theory of corresponding states. The extension of corresponding states that he invented was used extensively by chemical engineers for years for designing systems involving fluids. He worked on the effects of relativistic forces on chemical bonding in his later years. Balasubramanian at Arizona State was one of his students. He was still doing research until his final short illness. His primary interest in his final years was the theory of electrolyte solutions.

You were also a postdoctoral fellow with another distinguished chemist, E. Bright Wilson, Jr. What was he like and how did he shape your career?

In personal styles, these two great men were remarkably different. Pitzer always seemed very relaxed. He approached science in terms of seeing the big picture. He wanted to get at the nub of the matter. His favorite adverb was "essentially." In contrast, Wilson believed in taking life quite seriously. He seemed to me to be the quintessential New Englander, a man of rock-ribbed integrity and deep moral sense, even though I knew he was born in Tennessee. He rarely put his name on the papers that came out of his laboratory: his name went on a paper only if he felt he had made a major intellectual contribution to the work. He abhorred superstition in any form; no student or post-doc would dare mention the gremlins we believed lived in his spectrometer in front of him. From what I've said, you might draw the impression that Wilson was a stern, severe person. In fact, he was one of the kindest people I've ever known. He rarely volunteered advice and counsel, but, if you asked for it, he gently provided great wisdom and insight. My early independent research as a young faculty member was in microwave spectroscopy, which I learned from him. I was at Harvard in 1957–58, and did not leave microwave spectroscopy until 1978. He had a great positive influence on my career.

You gave a lecture today and it was somewhat unexpected for a Nobel laureate to worry about rotational transitions and about the shapes of simple molecules.

There's a whole issue of what do you do after you get the Nobel Prize. There are several options. One is to continue resolved to get another one. I weighed that option and found my prospects highly unlikely considering my age and considering I was very lucky to win the first one. Another option is, once you recognize you are not going to equal, much less to exceed your previous self, you can decide to do something different. Harry, for example, has decided to get interested in education, particularly in elementary school education. The third option is to decide that though I may not be able to do science that equals the science I've done in the past, that's not a reason to quit. That's the position that I took. Speaking frankly, I was not particularly proud of the work I was reporting on today because a lot of it was relatively old. There's this famous report on a paper, "This paper contains many things that are new and interesting. Unfortunately, what's new is not interesting and what's interesting is not new." This comment shouldn't be taken as too literally applying to my talk today, because there were some new and interesting results in it; I was just disturbed about feeling the need to put some things into it that weren't new. This was bothering me when I was giving this talk, but I still feel there is no reason to give up science.

Was your prize-winning research your most important research you've ever done?

Without a doubt.

In my interview with Harry Kroto before the Nobel Prize, he stressed your role in the discovery of buckminsterfullerene, but he also said that many people have under-appreciated it. Did you ever feel that?

In terms of the publicity that was going around, especially in the late eighties, my name was almost never mentioned. It was always Kroto and Smalley. In popular magazines especially, when fullerenes were mentioned, I was very seldom mentioned. Harry in Europe and Rick in this country were doing most of the interviews with the popular press. The effect was that their names were mentioned all the time, but that was not their fault. I've found myself that reporters usually tend to repeat the name of the person they interviewed and ignore the names of others whom the interviewee tries to bring in.

Were you surprised to win the Nobel Prize?

It would be wrong to say that I wasn't worried because I was concerned that I might get left out, but I really wasn't surprised. What was going on in the popular press was not what the Royal Swedish Academy would base its decision on, that was clear to me. Another thing that stopped me worrying about this situation was that two books were written about the story of the fullerenes. They gave a reasonably accurate historic account of what actually had happened. I felt that the members of the Royal Academy

At the Royal Swedish Academy of Sciences on October 10, 1996: announcing the Nobel Prize in chemistry for the discovery of fullerenes (in the middle, Professor Salo Gronowitz, the then Chairman of the Nobel Committee for Chemistry, photograph by I. Hargittai).

probably would read those books. Originally I resisted the writing of these books. This shows you how lacking I am in foresight. The problem was that we got into this very unfortunate period when everybody was trying to remember exactly who did what in this group project. There wasn't anybody with a tape recorder and a VCR there to actually see who was contributing and how they were contributing. When an effort was being made to reconstruct the past, my feeling was, the past is lost, gone forever, we'll never know who did what and when, and furthermore it's not terribly important. I thought the past is never going to be reconstructed and I tried to discourage these guys from writing their books. But what actually happened was that these two books virtually insured that if there was going to be a Nobel Prize, I was going to be on it.

How would you summarize your role in the discovery? And what was the crucial moment in the whole story?

The summary of my role has always been a difficult one. The reason that it's difficult is that there were various tensions going on between Rick and Harry which, I felt, were over nothing. This was happening later, but it made me wonder if I was there at all? The truth of the matter is that the key thing that happened was that we decided that we had to take a closer look of what was happening to this carbon sixty peak. That was a group decision. It was a Friday afternoon, the five of us, Rick, Harry, Sean O'Brien, Jim Heath, and myself, were in Rick's office. Harry was about to leave on Monday, and we were wrapping up the experiments that he'd come to do, associated with the carbon chains in space. We were all aware that the carbon 60 peak was sometimes quite high, seven times higher than its neighbors. We all decided at that meeting that we needed to see what we could do, to make that peak even more intense by varying the conditions. It's somewhat embarrassing to relate that there were these three professors and two graduate students there; we come to this joint decision and turn to the graduate students and say, "Why don't you work over the weekend and see how strong you can make this signal."

What happened was that on Monday morning, Jim Heath came in with this one enormous spike C_{60}. Once you know that you have this signal, then you know that it has to be explained. We were all in agreement from the way the conditions were varied that the only way you could explain it was that it had a specially chemically unreactive structure. Once you get to that stage, finding the soccer ball structure is just a matter of time. We were all thinking of Buckminster Fuller, of how he made his domes, but he wasn't telling us in the books that we saw. Harry had this recollection of having made this object that had pentagons and hexagons in it which might have 60 vertices, and the thing was that none of us had thought of pentagons. Then Rick was able to construct the correct structure by the process of trial and error. When Harry got back to England, he ran into some scientist and told him about the discovery of a beautiful carbon-60 molecule, and the fellow said, "You mean it's a truncated icosahedron?" But we weren't familiar with it.

Speaking about my role in the discovery, as those books say, I think my role was that of the person who kept our feet on the ground. I was the one, for example, who thought immediately that we had better take this paper object and put double bonds on it and we don't wind up with trivalent or pentavalent carbons in it. I was a well-known skeptic. When we put all the double bonds on there, and it worked, I said, "This has got to be it." I think my saying this provided some element of reassurance.

The whole process evolved very rapidly. First there was C_{60}, then we learned about other structures, and then there was this whole concept of the fullerenes. I saw my job to be the one concerned that we weren't screwing up, that it wasn't all wrong. This doesn't mean that the others weren't concerned about not screwing up, but I've had a long career of being the designated worrier, and I think my colleagues liked having a professional on the job.

Once we decided to follow through on this fluctuating C_{60} peak, the rest of it was almost preordained. You know the old story of the discovery of penicillin. You're growing a bacteria culture in a little dish. You come in one morning and you look and a mold is growing in your culture and all your bacteria are dead. There are two ways of reacting to this. One way is to say: "This damn mold has killed my bacteria; I better be much more careful in the future, not to let mold spores get into my bacteria cultures." The other way is to wonder about how has the mold killed bacteria; is it secreting something? This leads into a whole new direction. The same thing was true with the fullerene discovery. Once you see this peak, instead of saying, "This is not what I wanted to study," when you decide that you're going to do something about it, that's the fundamental moment.

Harry and I are very intelligent people, but Rick is a genius. He's so creative. Following his career I was always amazed, he would come up with a new idea for a new instrument. He would imagine a new kind of experiment and he would invent the instrument to do the experiment. Rick had invented this machine to do this job, and Harry wanted to use it. At that time we were doing experiments on silicon and germanium and gallium arsenide clusters. We thought, we were doing great things because people wanted to make circuits and computers smaller and smaller until in the finite size the number of atoms would matter. This is why we were studying the properties of these clusters. The idea of testing the hypothesis of carbon clusters for radioastronomy just didn't seem of that much interest. I started pushing to do this after a while because I was familiar with Alec Douglas's paper about the carbon chain molecules being the carriers of the diffuse stellar bands. I thought, if we get Harry over here doing the experiments that he wanted to do on the formation of chains, then we might move on to resonance two-photon ionization studies of the spectra of carbon chain molecules. As time went on I began to say, "We really ought to do these experiments with Harry." Of course, when the fullerenes came about, we forgot all about the carbon chain spectroscopy.

How did your interest turn to chemistry?

I got a chemistry set from my parents for Christmas when I was 9-years old. I was fascinated. I spent hours and hours mixing up chemicals.

Did you give chemistry sets to your children when they were small?

I have two sons, 40 and 37 now, and I was the most frustrated father because they were not interested in playing with chemistry sets or electronics sets. My interest in chemistry was not very rational and scientific. I was interested in colors and explosions. Later on I would get practical books on chemistry, how to make stain removers, for example. The theoretical interest came later but from age 9, I never had any doubt that I wanted to become a chemist.

The Curls' wedding, in the middle Robert Curl's father (courtesy of Robert Curl).

What was your family background?

My father was a Methodist minister. He was deeply religious which I never was. My father encouraged me a lot to think about the world. He seemed to want to encourage me in anything that I was interested in. He would buy me books to encourage my interest and we would read these books together. I got a lot of attention and I enjoyed the attention.

I would like to get back to the question of how you would summarize your principal contribution to the buckminsterfullerene discovery.

I've meditated about this for years. One issue is, Are anyone's recollections valid? What always used to trouble me about this was that the way scientific interactions go in a group is that somebody will suggest something that draws no reaction, and somebody else will pick it up again later. I've always felt that no one could ever really sort it out what was actually going in the various discussions. This is why I originally objected to people writing books about this discovery, because I thought that intrinsically they wouldn't be able to catch the truth. An example which I'm fairly clear about now, and I've been over this stuff hundreds of times, was when this business about the pentagons came in. The group went out for lunch to this Mexican restaurant and I didn't because I had a meeting. Harry brought up this cardboard model that he had made, from a kit I think, for locating the constellations. It was a polyhedral structure that had hexagons and pentagons. He thought it might have 60 vertices. He didn't get much of a reaction. Then, during the course of the afternoon Harry began to think about this more and more and, evidently, Rick did too. Around 6 o'clock at night, or a little later, Harry wanted to call his wife in England to bring the thing out and get her to count the vertices. Harry and I had a little conversation about it. It was along the lines, something like, "OK, Harry, where is it?" and he said, "I haven't seen it for years." I had this picture in my mind, of waking his wife at 1:30 in the morning, asking her to get up and find this thing and count the corners of it. I thought my wife wouldn't react to it too well, so I told Harry, "It can wait till morning." But it turned out it couldn't wait. Harry would've felt infinitely happier if he had made that phone call, because his cardboard model was the right structure. The point I'm trying to make here is that Harry's comment about remembering pentagons drifted down and came back up with Rick. I don't think that Rick reacted to it when Harry first brought it up. I don't think the idea that there might be pentagons in the structure had any initial effect. Chemists much prefer hexagons to pentagons. I think Rick finally tried the pentagons, and reluctantly, just because he wasn't getting any solution using only hexagons.

If this was the source of future animosity, it seems to me, it was over nothing.

This wasn't the only source of animosity. That built over time. I've always felt the animosity developed over virtually nothing, but I don't want to create the impression that I was above it all; I'm very much afraid I wasn't. Science is done by human beings. But let me go back a little in history. The hypothesis that soccer ball C_{60} forms spontaneously in condensing carbon was put forth in September, 1985. By the end of 1985, all the other major conceptual advances were there; putting an atom inside of C_{60}, the

Wolfgang Krätschmer and Robert Curl in Heidelberg, Germany, in 1998 (courtesy of Wolfgang Krätschmer).

fullerene concept, the structure of C_{70}, the soot business, i.e., that soot is formed in spiraling icosahedral shells, which may be wrong; all those things were done before 1985 ended. We spent 1986, 1987, and 1988 on testing the fullerene hypothesis. It passed all the tests which made us all very happy. By the end of 1988, we were running out of gas. If you gave a talk, the organic chemists said," Let's see your sample." I was sure at that time, and I'm still sure today if the fullerenes had remained in the status of a few molecules in a big, monstrous apparatus, that there would've been no Nobel Prize. It would've remained a scientific curiosity. The work of Krätschmer and Huffman to make macroscopic samples of the fullerenes was crucial to making this important. In my opinion, it should've been a very close call with the Nobel Committee about who was going to get the Prize for the fullerenes. It could've been them as much as it could have been us. The numbers didn't work to make it both. I didn't even remotely consider that we would get the Nobel Prize before their paper was published. After their paper was published, I began to get sincerely nervous about it. What they did was absolutely crucial to the whole fullerene story. I used to say that in late 1990 I went through a very peculiar experience because I'd get up early morning, kick myself in the seat of the pants for not having done what they did, for not having the sense to have tried something as simple and elegant as they did, and I would thank them for having done it in my next breath.

Wolfgang Krätschmer, 1995 (photograph by I. Hargittai).

30

Wolfgang Krätschmer

Wolfgang Krätschmer (b. 1942 in Berlin, Germany) is Research Scientist at the Max Planck Institut für Kernphysik in Heidelberg, Germany, and leads the "Laboratory Astrophysics" Group in the Institute's Kosmophysik Department. He has been at the Max Planck Institute since 1968. His best known contribution is the first report on the production of C_{60}, Krätschmer, W.; Lamb, L. D.; Fostiropoulos, K.; Huffman, D. R. "Solid C_{60}: a new form of carbon." *Nature* **1990**, *347*, 354–8. Wolfgang Krätschmer received his Diploma in physics from the Technical University Berlin in 1968 and his doctorate, also in physics, from the University of Heidelberg in 1971. In 1993, he was appointed Honorary Professor at the University of Heidelberg. His distinctions include the Stern-Gerlach Prize (1992), the Leibniz Prize (1993), The Materials Society Medal (1993), the Carbon Prize (1994), and the Hewlett Packard Europhysics Prize (shared with D. R. Huffman, H. W. Kroto, and R. E. Smalley, 1994). The interview was recorded on August 30, 1995, in Budapest, and it was finalized in November in Heidelberg. It first appeared in *The Chemical Intelligencer*.*

*This interview was originally published in *The Chemical Intelligencer* **1996**, *2*(1), 17–23

My first question was about Dr. Krätschmer's family background and education.

My father was the first in the family who graduated from high school. He wanted to go on to university but the war spoiled all his plans. The roots of our family extend to the former eastern parts of Germany, regions which now are Poland. I think our name is Slavic in origin and means inn-keeper in English. My father traced the family back because in the time of Nazi Germany, you had to collect the family records but I've never been very much interested in my roots. I was born and grew up in Berlin. Incidentally, I went to the same high school in Berlin — we call this Gymnasium — as my father did. Our family lived in the southwestern part of Berlin which used to be the American sector. Living a few kilometers away in the parts occupied by the Soviets would have made my life completely different.

In 1961, I entered the Technical University of Berlin. That was the year when the Wall was erected. I studied physics.

How did you chose physics?

In the first years of high school our physics teacher was rather bad. This challenged me, and I started to study the textbooks and discovered physics on my own. Faith and religion also played a role. In those days we thought that we had to fight communism and counter their materialism with our idealism. I believed that when God speaks to us he does so through the laws of physics. It may sound weird but this is how I felt. Another reason was more materialistic. There was tremendous progress being made in technology, in atomic energy, and it was the dawn of the space age. However, I had to work very hard because things don't come easy for me and physics, at times, seemed very demanding to me.

I graduated in 1968 in Berlin under Hans Bucka with experimental work on atomic hyperfine spectroscopy. Bucka came from Heidelberg and was always praising the beauty of that town, its university, and its science. That's why I went there and joined the Max Planck Institute of Nuclear Physics. I did my doctoral work there. My supervisor was Wolfgang Gentner, who actually had also founded the Institute.

Gentner was a big shot in nuclear physics in the 1930's. Together with Walter Bothe he had discovered the nuclear photoeffect; with γ-radiation, you can release nucleons from the nucleus. Furthermore, he was an anti-Nazi. That helped him in his career in post-war Germany.

How did he survive the Nazi era?

He was regarded as a war-important physicist, was careful enough not to expose himself openly, and finally he had very good luck. Gentner became a rather important and influential person not only in the Max Planck Society in which he was one of the vice-presidents. He helped to establish scientific relations with Israel, became director of CERN, and organized many other things. For this reason, Gentner was mainly on the road and very seldom came to the lab. I could work rather freely. My thesis was on tracks of high-energy ions in solid materials. Individual ions leave tracks of radiation damage in solids which can be made visible in a microscope by a simple chemical etching procedure. I completed my doctoral work (Dr. rer. nat.) in 1971.

What happened then?

I stayed on at the Institute on a temporary basis. This was the time of the Moon landings. We received Lunar samples from NASA, and we studied them for cosmic ray ion tracks. The intriguing thing was that one could study the tracks of cosmic ray ions produced in lunar samples millions of years ago.

Eventually, in 1976, Hugo Fechtig and Heinz Völk, who were then directors of the cosmophysics department, asked me whether I would be interested to work on interstellar dust particles. This topic became more and more exciting, thanks to the progressing infrared astronomy. Since this offer also implied a tenure position, and I wanted to stay in science, I accepted the job.

Interstellar dust is primarily known through spectroscopy because we can't go easily out into space and collect samples. One takes a telescope with a spectrometer, samples light from a dust-obscured star, and compares it with the spectrum of a similar but unobscured star. When you ratio the spectra, you obtain the extinction spectrum of interstellar dust material. This work is done by the astronomers, not by us. We take the spectra measured by the astronomers, go into our laboratory, and try to produce dust or dust-material that reproduces these interstellar spectra.

In the 1960's, the astronomers worked in the ultraviolet domain. Then in the 1970's infrared astronomy became important. From the electronic transitions in the ultraviolet region and the vibrational transitions in the infrared region, you can speculate about the species that make up the dust. This speculation made our work very interesting indeed, because in many cases nobody knows for sure what the interstellar absorbers really are. The so-called diffuse interstellar lines are especially intriguing. At present, roughly 200 lines are known, but none of them has been identified. Beside these, there is a broad and very intense absorption feature in the ultraviolet region at 217-nm wavelength, and it's almost a scandal that nobody knows what the carrier is. However, there are very strong arguments that this absorption comes from something like graphite particles. The disadvantage of solid state spectroscopy is that you don't have distinct vibrational and rotational structures which usually help to identify the carrier.

As I said, the aim of our work was to produce interstellar-like particles. Don Huffman in Tucson already had years of experience in this field and had also written books and articles about it. To become properly educated for my new job, in 1977 I got the opportunity to work in his lab. We studied silicates because the interstellar dust contains silicates too.

A few years later, in 1982, a Humboldt Award enabled Don Huffman to spend a year at our institute in Heidelberg. He suggested that we should try to solve the puzzle of the interstellar 217-nm feature. Several years before, he had already worked on this problem. Now we again produced soot but varied the conditions to a much larger extent than Don had in his previous work. We recorded spectrum after spectrum, but essentially always observed the same very broad absorption feature with a maximum at around 230 nm. The mismatch of the peak position did not worry us too much, but the width of our features was by far too broad compared to the interstellar absorption. This was very frustrating. Occasionally, however, we noticed some additional weak but distinct absorption bands in our soot spectra. These soon caught our attention. To our knowledge, these mysterious bands had never been reported in the literature and also Don hadn't seen them before. From their characteristic shape, we named these absorption "camel" features.

At that time, we had a rather rudimentary equipment for soot production, just a carbon evaporator in a bell jar, evacuable by an oil diffusion pump. The valves of the pump sometimes didn't close properly, and then we saw clouds of pump oil floating around in the bell jar. After such an event we naturally cleaned the whole apparatus. But some oil may have remained inside the bell jar. When we heated the rods to evaporate the carbon, the whole bell jar became hot and residual oil traces may have evaporated as well.

Didn't you use a trap?

Yes, between the pump and the bell jar. But this wouldn't help if some of the oil had already entered the bell jar. Thus I was concerned about

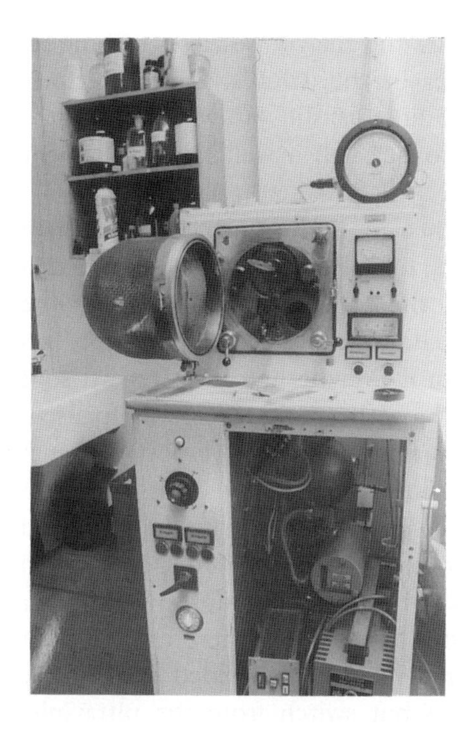

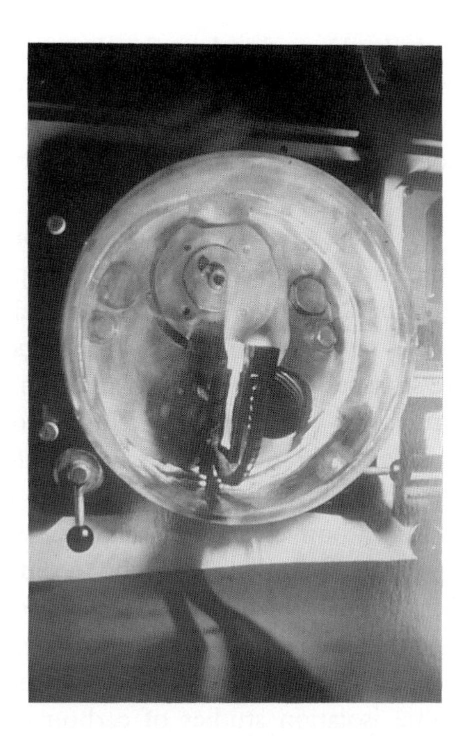

The apparatus in Heidelberg for producing C_{60} (courtesy of Wolfgang Krätschmer).

the possibility that we were just recording junk. But Don was more convinced that these camel humps might have something to do with carbon. He was right — as it turned out years later.

In any case what we had then could have been published only in something like Irreproducible Results Letters.

Did you ever consult with chemists?

No. My idea was really that it was junk. In a few experiments we checked for pump-oil contamination. We tried to extract these contaminants from the soot using basically the same technique that later became so effective to extract C_{60}, namely, sublimation and solvation. However, we sublimed in air and used acetone as a solvent. These were just the wrong conditions. Nothing could be extracted. It really is a pity that we didn't try harder!

In conclusion, we were not very successful in assigning the 217-nm feature to carbon particles — as I mentioned, the true nature of the interstellar carrier is still unknown. We then turned to carbon molecules, another facet of the same field. In 1977, Alec Douglas had claimed that the diffuse interstellar bands are produced by carbon molecules. We set up experiments to study the molecules of carbon vapor by the matrix isolation technique. This field was pioneered by Bill Weltner in the 1960's. Despite efforts by many researchers, it is amazing how little is still known about these molecules. One would suppose carbon to be such a well known element and it's really not so.

In 1984, there was a cluster conference in Berlin and Don Huffman came back to attend it. We had two papers, one about the ultraviolet spectra of carbon clusters and the other about the soot spectra. However, one of the reviewers of the proceedings asked us to withdraw the paper on the soot spectra. Obviously, he didn't like it. Maybe he didn't regard soot particles as the nice clean clusters on which the conference was focused.

This conference was remarkable for us also in another respect: We met Andrew Kaldor for the first time. He was showing the famous mass spectrum of his carbon cluster experiment, which yielded clusters up to C_{100} and larger. I admit that Don and I did not notice that the C_{60} peak was in excess! There just was too much to see — it was very impressive. In our experiments we had gone up to something like C_{10}, and here came Kaldor with such large clusters. After the meeting was over, I decided to continue matrix isolation studies of carbon clusters but switch from the ultraviolet to the infrared. This became the diploma work of Klaus Nachtigall. Don was pursuing similar work with Joe Kurtz in Tucson. Then in the fall of 1985, I learned of the discovery of C_{60}.

How did you learn about it?

Alain Lèger, a colleague from Paris who at this time was also working on interstellar dust, came to give a seminar at our place. He told me about the paper by Harry Kroto and co-workers.

Did you make the connection with your own studies?

No. Again, I was amazed that such a huge molecule could exist at all. I never thought that it had any connection to what we had been doing. I also asked Don for his opinion, and I guess for both of us it was beyond anything we could have thought of. Then there was the name. It gave the molecule an additional fantastic aura. In 1986, I saw Harry Kroto for the first time at a meeting in Les Houches in France where he gave a talk. Harry is Harry, of course; he is extraordinary. He gave one of his intriguing and also rather entertaining lectures. I liked his way of speaking and especially his humor. He told his C_{60} story and explained the work of Buckminster Fuller, of which I had so far no idea, and finally focused all culture and science onto a single point: the closed cage soccer ball structure. Harry convinced me that there is something in C_{60}.

Then it was in 1987 that Don Huffman shocked me with the — as I thought — crazy idea that the camel-hump stuff we had produced is not junk but C_{60}. He even decided to apply for a patent for this. The reason for his claim was that he had gotten hold of a paper by Arne Rosén and co-workers that showed calculations of the ultraviolet spectrum of C_{60}. The predicted spectrum was very close to the spectrum we'd observed. At this time, calculations of the infrared and Raman spectra of soccer-ball C_{60} had also been published. Don had, in fact, made some Raman measurements at the Max Planck Institut für Festkörperforschung in Stuttgart during his sabbatical stay in Germany, and he thought that he could identify at least one characteristic Raman feature of the camel-hump soot spectrum with that calculated for C_{60}.

What was the patent about?

The production of C_{60} by the carbon-evaporation method. The evaporation is done by resistive heating and must be carried out in an inert atmosphere, such as helium or argon. Then you collect the soot and find C_{60} in it.

Were you on the patent?

No, only he was on it. However, he was asked to reproduce the procedure and deliver the material, and he couldn't do it. At this time we didn't know about the importance of the helium pressure for obtaining reproducible results.

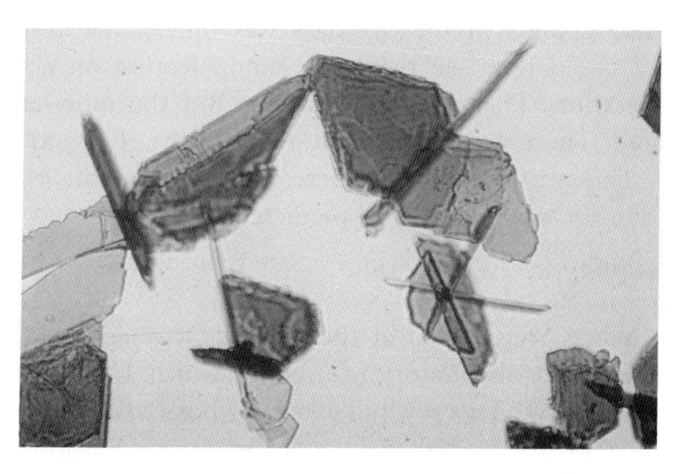

Pure C_{60} crystals of 50 to 100 micrometers (courtesy of Wolfgang Krätschmer).

Where was the original experiment done?

Originally it was done in Heidelberg, in 1983.

Did he invite you to join him in the patent?

No, he filed the application in his name only.

What did you think?

I just thought, why not? After all it was his idea. I was not really interested in this since I didn't believe him. Then, as he could not reproduce the results, he started urging me to do it, again using our apparatus in Heidelberg.

Was his patent accepted?

No, eventually he just withdrew it. But the idea was there. The next event was a conference on interstellar dust in California, and both Don and I were attending it. He again spoke to me, but it was as if he were talking to a deaf person. However, Don can be very, very eloquent and finally I accepted that I should try the experiment again. Incidentally, after my return to Heidelberg a student came to our place. His name was Bernd Wagner. He was from Cologne and wanted to volunteer for one month. I asked him to work on the carbon evaporator and make soot under various conditions. When I had time, I recorded the infrared and ultraviolet spectra of the samples he had produced. In my lab we had a new and very good Fourier transform infrared spectrometer, which was much more sensitive than our old grating instrument. One evening Wagner came to me with a sample produced at 100 and 200 torr Helium pressure. The usual pressure Don and I had applied in the old days was only a few torr of helium. Wagner and I could now see the camel-hump feature on a regular basis in the UV spectrum. That was very exciting. But the most exciting thing was the infrared. There we saw the continuum of the soot and four distinct narrow lines. This was exactly what there should be for the soccerball C_{60}. Also the positions were almost as predicted.

This was in 1988.

That's right. When Wagner left at the end of October, 1988, I was very busy with an infrared observatory project for which I urgently had to do some engineering work. This is why I asked my bosses for a doctoral student to continue the work, and this is how Konstantinos Fostiropoulos joined me. This was in early 1989. I told him the whole story, about the possibility of having C_{60} in the soot, and I also told him about the possibility that our spectra may rather be produced by a junk contamination.

I asked him to check this by infrared spectroscopy in an isotopic replacement experiment. This meant the replacement of all ^{12}C by ^{13}C. Commercially available ^{13}C is a powder. In order to evaporate it in the bell-jar, we had to prepare rods from this powder. I further asked Fostiropoulos to clean the bell-jar and replace the diffusion pump by a turbo pump to minimize possible contamination. After all these cleaning precautions, the camel-hump spectrum was still there. So I presented these findings at a conference on interstellar matter in Capri, Italy. I had the paper for the conference proceedings already with me and gave copies to colleagues who were interested. Mike Jura was one of them. From him the paper went straight to Harry Kroto, so he was aware that we were working in the field and that we might be able to produce C_{60} in bulk.

Was this then the first publication about the production?

Yes. Of course, when you see something in the infrared, it must be something in substantial amounts, not just a trace. We estimated it to be something like a few per cent C_{60} of the soot sample. The conference at Capri was in September 1989.

Was Huffman on this paper too?

Yes. I thought this to be fair. He had pushed us into this direction.

In early 1990, Fostiropoulos completed the ^{13}C replacement experiment and the four distinct features in the infrared shifted by just the amount expected for the isotopic substitution. We also made experiments with partial substitution and everything was consistent. We were dealing with a large molecule. All this was in favor of C_{60} but was not really proving it. We published our findings in *Chemical Physics Letters* of which Rick Smalley is one of the editors, and, of course, he got the smell of it too. For some reason, however, he didn't believe us. This was our good luck. Chemical intuition says that such a complicated and highly ordered molecule like C_{60} cannot be produced under high-temperature conditions. This is, of course, correct in general, but the C_{60} case is an obvious exception to this rule.

I mailed copies of our paper to several colleagues and people I know. Werner Schmidt, a PAH researcher, sent me an encouraging letter and suggested that we separate C_{60} from the soot. He proposed sublimation and also solvent extraction. He recommended that we use trichlorobenzene.

We didn't have this solvent right then so we tried sublimation first. We did it in a rather simple way, with a test tube of soot and a Bunsen burner as heater, floated the whole with argon and it worked! Something brownish was coming out of the soot and coated the walls of the test tube. We collected this brownish condensate and took spectra which duly showed the camel-hump features in the ultraviolet and the four lines in the infrared without any soot continuum. This was very exciting. We did some mass spectroscopy as well. Although the mass spectra were not very good-looking, everything was very convincing that we really did have C_{60}. Then we rediscovered what Schmidt had already suggested to us, namely, that we can dissolve the C_{60} condensate — in benzene. Naturally, the solvent extraction procedure is by far simpler. We were very much intrigued by having crystals of C_{60} that we could see under the microscope. We were, of course, in constant contact with Don and Lowell Lamb, who reproduced these experiments and added new data. Don had some colleagues who produced very nice mass spectra of the soot extracts. These spectra showed a big C_{60} and a small C_{70} line and even the line of the doubly charged C_{60} was visible. Next we had to figure out the structure of these crystals, and Don and I both tried to do some X-ray diffraction.

At this point you still did not publish anything.

No, and it was very risky to wait. At about this time we got a letter from Harry Kroto's group in which one of his students, Jonathan Hare, was asking about some details of the soot production. He showed the infrared

Donald Huffman and Wolfgang Krätschmer with their wives during a conference in St. Petersburg in 1993 (courtesy of Wolfgang Krätschmer).

spectra of C_{60}-soot he had obtained. Fortunately, he didn't mention anything about extraction. But it was clear that we were pursued. If I think back, I wonder why we stayed so cool. I cannot explain this. We all worked steadily step by step. The problem of understanding the crystal structure of the C_{60} solid took us about one month. It was a very shaky time and long enough for us to be scooped by some other group working faster.

Didn't you feel the urge to publish right away?

Not really. Somehow we felt safe and that we still had some time. I think, that among all of us Don had to suffer most. At the beginning of August 1990, after the paper was finally submitted to *Nature*, I went with my family on vacation to Hungary. While I was relaxing in the sun, Don had a very stressful life. He had to contend with various problems that unexpectedly came up, had to argue with reviewers, and so on. By the way, the reviewers were extremely fair to us. On my return, Don told me that he sometimes was close to collapse. For his commitment I will be ever grateful. Without him we definitely never would have started the whole work, and probably also would never have successfully finished it.

Coming back to the structure, Don claimed that it was close hexagonal packing of spheres but our diffraction pattern did not really support this. We had to invoke stacking disorder of the lattice to make the diffraction pattern plausible. By the way, this issue wasn't solved until half a year later. By then we had already published our *Nature* paper. The structure is, in fact, cubic close packing. In a detailed X-ray investigation Paul Heiney, Jack Fischer, and co-workers showed that the spherical shape of C_{60} gives rise to interference effects which extinguish certain lattice diffractions.

You used the term "a new form of carbon" in the title of your Nature paper, but it was the same C_{60} that Kroto and Smalley had reported.

Yes, but they didn't have the material.

When you submitted this paper, did you feel it was a landmark paper?

No. We knew it was the most important paper we had ever written, but we didn't think it would initiate such an avalanche.

It has also bothered us that the whole procedure is so simple. It's almost impossible that nobody had found this before. This also produced a considerable psychological barrier. There were some evaporation experiments of carbon already conducted in the first decade of this century, and given the conditions applied, very likely C_{60} must also have been produced.

Harold Kroto, Donald Huffman, and Wolfgang Krätschmer in St. Petersburg in 1995 (courtesy of Wolfgang Krätschmer).

However, I may also ask the question, What would have happened if we had had the material years before? Probably nothing. We might not even have looked at the region of 720 in the mass spectrum. Kroto and Smalley's discovery was essential for us.

I well remember the 1990 cluster conference at Constance, Germany, where Smalley gave me part of his lecture time so that I could report our work. The night before, there was a public lecture by Joshua Jortner, who made a comparison of the development of art and natural sciences, and he said that, at least according to some deep and funny thinker in this field "the most important thing is for the painter to be there when the picture has to be painted." This was precisely the situation I felt I was in. We had to be there when the C_{60} picture had to be painted. We were just the fulfillers of something that was already floating in the air. It was there indeed. When Harry Kroto got our *Nature* manuscript for refereeing, he also had the material. As we learned later, Donald Bethune and his colleagues from IBM were also extremely close.

How about the patent?

Don Huffman renewed the patent in 1990 and added me as a co-inventor onto the patent. This patent is still pending. As for myself, I prefer publishing to patents. We in Heidelberg are working in an institute of fundamental research. The idea of patenting rarely occurs to us.

What is the patent about?

It's about the production and extraction process of C_{60}.

What's then the recipe for making C_{60}?

You take carbon rods, evaporate them in an inert atmosphere of about 100 torr helium, collect the soot, and wash the C_{60} out of the soot by benzene or toluene, and you get it. The extract contains, of course, admixtures of C_{70} and still larger fullerenes, and you can separate them. Although it is a more difficult problem, it may be done, for example, by chromatography.

How much is one gram of C_{60} worth today?

I think about DM 200.

How about this controversy with your former graduate student Konstantinos Fostiropoulos?

This is one of the bad things of the whole story. During all the fuss about C_{60}, Fostiropoulos started to feel that his contributions were not recognized enough. He demanded to be co-inventor on the patent.

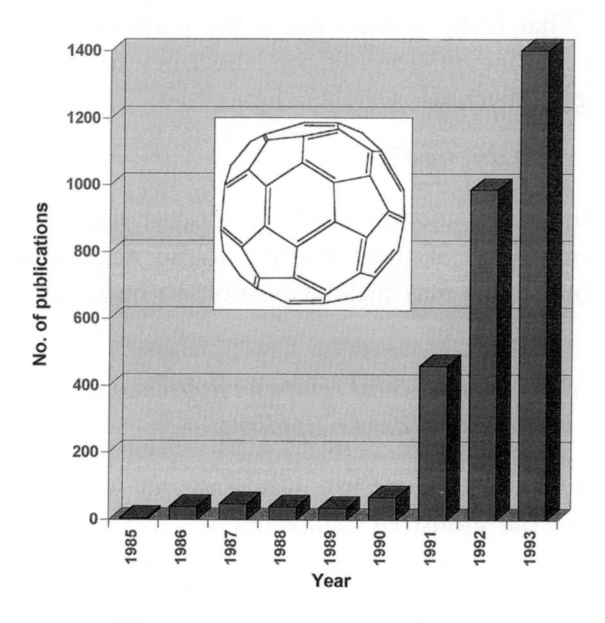

The number of papers on fullerenes during the first few years. Following an initial growth, there is a slight decline in 1988 and 1989. Then, following the paper by Krätschmer *et al.* in 1990, there has been a spectacular increase in the annual number of papers.

He was co-author on your paper in Nature.

Yes. For me it wouldn't have been a big problem to support his becoming a co-inventor if he had not blamed us for scientific robbery. He accused Don and me of being thieves and this made the situation very complicated.

Where is he now?

I don't know. I think he is unemployed. He is somewhere in Germany, and his lawyers are pursuing the suit against the Max Planck Society, which is one of the patent holders.

How about the Nobel prize?

As somebody of my C_{60} colleagues stated, it is very gratifying just to know that one's name is being mentioned in this connection. For me, this is much more than I had ever dreamed of.

Does it make you nervous?

Not very much. I don't do science for the Nobel prize. Also, the more time elapses, the lesser the chance that we'd be getting it. It's also very clear that there should be a breakthrough in application and that hasn't happened yet. What really bothers me is the journalists. In the last years, they have been waiting in front of my laboratory for the announcement. It is disturbing to become a public figure.

What is your research now?

I go back to the roots and investigate carbon molecules. I have a small group of students. One of them recently found an easy way to produce C_{60} dimers in bulk. This may make it possible to make big fullerenes more efficiently.

Has any individual or industrial company approached you with suggestions or asking for advice as regards applications?

I had a contract but the company didn't ask me seriously for advice. It is my impression, the industrial companies are very reluctant to share any information. I think industry prefers to continue with projects that had been proved to work rather than try something new.

How about your research support?

I received the Leibniz prize, which gives me three million German marks for a five-year period. I buy good instruments and I have good students and I just pursue my research, and this is fundamental research.

You are now a celebrity among chemists.

This seems to be the case. I got an honorary professorship in chemistry at the University of Heidelberg. I give a course for chemistry students on carbon molecules. Carbon is a very educational substance.

How about your family?

My wife, Zsuzsa is from Hungary. She keeps telling me that I just had to become involved in C_{60} because she comes from Hatvan. This is a town in the vicinity of Budapest, and Hatvan in Hungarian means sixty! I find this a funny connection, similar to Buckminster Fuller's connection to Carbondale, Illinois. My wife is a teacher by education but now works with me in our Institute. We have no children but my wife has a daughter, 22, from a previous marriage, who lives with us and is preparing for a career in the hotel business.

Any hobbies?

I don't have any really profound hobby. I am interested in many things. One of my favorites is history. The ancient Egyptians, the Greeks, the Romans, and — because of my wife — the ancient Magyars as well! If time permits, I would also like to look at things with my small home telescope or with my microscope.

What is the most important change in your life and work between now and, say, 10 years ago?

Having the freedom to decide how research money should be spent is a good feeling. However, it is a burden too at times, since I feel the pressure to accomplish really good science for the taxpayer's money. Further, I now have many more students and co-workers for whom I feel responsible. This is all a big change.

Do you feel that your most important work has not been done yet?

I guess there will be nothing beyond C_{60}. I think that is why I was there.

Robert L. Whetten, 1995 (photograph by I. Hargittai).

31

Robert L. Whetten

Robert Lloyd Whetten (b. 1959, Mesa, Arizona) is Professor of Physics and Chemistry at the Georgia Institute of Technology in Atlanta. He received his B.S. in chemistry and mathematics in 1980 from Webster State College and his Ph.D. in chemical physics in 1984 from Cornell University. Following a postdoctoral fellowship at Exxon Research & Engineering, he was with the University of California at Los Angeles between 1985 and 1993, quickly rising to full professorship in 1990. He has completed visiting research professorships in Lausanne, Paris, and Berlin, and was von Humboldt Fellow in Munich. His main research interest is in the chemical physics of metal clusters and nanocrystals in molecular and biomolecular forms and their applications. The interview with Dr. Whetten was recorded on June 21–23, 1995, in Erice, Italy, during a NATO Workshop on Large Clusters of Atoms and Molecules. It was an impromptu interview and it is of interest to tell its circumstances. At the workshop I gave a talk, "Lessons of a Discovery" about the buckminsterfullerene story (later it appeared in print[1]). In my talk I mentioned the Exxon work, their beautiful mass spectrum, and their failure to note and suggest a structure for the C_{60} peak. The talk then moved on to the other works in the story. Following the talk, Robert Whetten (yet unknown to me) called the story a sanitized version of what really happened. He stated that the Exxon paper should not have said more than it did and it was the 1985 *Nature* paper that overstepped important limits in its conclusion considering the then available evidence. After the session we continued our conversation and hence this interview. Since the interview was recorded in 1995, and in 1996 the Nobel Prize was awarded to three of the authors of the 1985 *Nature* paper, I asked Robert to reread the transcripts of our original conversation. This second meeting happened at the end of January, 1998, in Wilmington, North Carolina. Dr. Whetten reread the transcripts and told me that his opinion has not changed a bit although many people feel, in hindsight, the inevitability of the discovery on the basis of the then available evidence. In any case, the following conversation took place in 1995.

How did you get involved in the fullerene story?

In late 1983, I was still a graduate student when I talked with Andy Kaldor about the possibility of being a postdoc at Exxon. I was at the Chemistry Department of Cornell. My supervisor was Ed Grant, and Roald Hoffmann was on my Advisory Committee. Although my work was unrelated to clusters, I was interested in the subject. Smalley's work was already well known on metal clusters. The Exxon group was going in a different direction, not spectroscopy, but magnetic, electronic, chemical properties. Half a year later, when I arrived at Exxon, they had moved to a new lab, and it wasn't operating yet so they were looking for a project for me. They gave me a file of mass spectra along with published reports. Kaldor said, look through these, look at our paper, speak with organic chemists who thought about structures a long time, like Hoffmann. Hoffmann has a reputation for studying structures, of course, and he wrote a paper about carbon rings in 1964. So I asked him, what is proposed about the structure of larger pure carbon molecules? Can rings persist? I said about the abundance anomalies, magic numbers, and about the series. I'm sure I said C_{60} was very strong but I'm also sure that I said the same about C_{50}, C_{70}, etc. Hoffmann said that he has never heard of any proposal for a series of stable carbon structures, but he'd take a look.

Did the pile include the mass spectra of the 1984 Rohlfing et al. paper?

Yes, this was in May of 1984, six months before that paper appeared. Dudley Herschbach was very interested in the structure of the carbon clusters from their astrophysical and astrochemical aspects. He was at Exxon three months of the year, and he spent there a good part of the summer of '84. We spoke a lot about experiments that could be done on carbon clusters, that would help resolve mysteries. So it's not as though the Exxon people had not been thinking and working much along the same lines as Kroto. Kroto had the same attitude as Herschbach, in that there was this old proposal that carbon clusters, especially the chains, might help explain the interstellar absorption lines. The problem was that the actual data showed C_{28}, C_{36}, C_{44}, C_{60}, etc. It was obvious that this pattern gave some evidence of structure, and the question was, had anyone predicted such a family of structures? Nowhere did we ever talk about C_{60} as an individual structure.

The hypothesis of the carbon rings was 20-years-old and explained the family of smaller clusters. Then there is a little break, and we clearly would have liked an analogous explanation for the large mass region. Thus the Exxon people said to me, use your time and try to find out if somebody had an idea about this.

So they weren't satisfied with the explanation they had given in the paper.

They wanted more insight into this. It was also a way to keep me occupied. Let me jump ahead for a moment. The pattern of magic numbers was fully explained eventually by Kroto in 1987 in a paper in *Science* where he worked out with a mathematician, a topologist, the consequences minimizing the adjacency of pentagons. Before then nobody predicted these patterns.

To give a wider context, at that time many new experiments were being done on clusters. There were many reports of anomalies for other elements too, equal to what C_{60} was in the Exxon mass spectrum. It is a taboo, however, to go for a single peak and assign it a structure. If somebody working in mass spectrometry manages to go through the referees and publishes a structure, saying that this is the structure for this mass, that person would be excluded from the community and would be looked at as somebody who can't be trusted. Many people in the past, some time in the distant past overinterpreted their results. This created a strong response. It's just like in the legal practice, there are standards of what can be considered as valid evidence. A peak in the mass spectrum, no matter how large, is not considered strong enough for assigning a structure. So the feeling was after the Kroto *et al. Nature* paper that the authors had overstepped an important taboo. Of course, people would not criticize them publicly because public criticism is another taboo, but privately all experienced experts would say that they'd overstepped this taboo. I'm a little bit more tolerant in this; I think they had very exciting new evidence and the peculiarities of the C_{60} peak did demand an explanation indeed. And they gave it. In this way, I feel Kroto and Smalley did the right thing. At the time, though, I felt differently. I felt that Kroto, coming into this field never working with mass spectra, was excitable and had no understanding what's necessary to prove a structure, and Smalley didn't want to be left behind.

How long did you stay with the Exxon group?

I stayed from May 1984 until July of '85. So when the *Nature* paper came out, I was already in Los Angeles.

What did you feel when you saw the Nature *paper?*

Shock, amazement at the audacity and the boldness that they'd put this forward. Everybody felt this, everybody I spoke to. The organic chemists were convinced that Kroto and Smalley must have had some other evidence in addition to the mass spectra, such as NMR and X-ray and so on. It was a structure published on the cover, but it was a hypothesis at that point. The way it was presented, though, gave the feeling as if there were hard evidence.

Didn't you feel that you yourself had missed some opportunity?

No. I know everybody expects that this would be the feeling but I actually sensed more the amazement, that they took the risk like this with their reputation. I felt I definitely wouldn't have wanted to be on that paper.

So what did you do?

We challenged them. I wrote a paper in 1987 and presented new experimental evidence challenging them to go further and prove that their evidence is really what they say is. Some people felt that this was something negative only, but this is the normal way that science progresses in. Every hypothesis has to be challenged. Kroto, especially, preferred just to talk and theorize about it, while other people did hard work in the lab and wrote papers and said maybe it was more complicated.

At one point I came out with an alternative model and told about it at a meeting and Smalley asked me why don't you publish it? And I said it's not my nature to come and draw a little something and then publish it. There was also a pretend game going on. Anytime you'd call Smalley, or Kroto, and ask them, have you tried this, we might try this experiment, they would always say we'd done everything, we'd tried everything. My recollection is that they even said this about approaches that would later become the Krätschmer-Huffman procedure.

What happened to you after you'd left Exxon?

I went to UCLA in 1985 and was working with the chemists, Francois Diederich and Orwil Chapman, who were trying to synthesize C_{60}. We published our first major report on this in 1989.

How was UCLA for you?

I was promoted very quickly, to Associate Professor in '88 with tenure and Full Professor in 1990, when I was 30-years-old. They promoted me very quickly, partly because of this excitement surrounding clusters.

You said that you hadn't felt then as if you missed a big opportunity in 1984. How do you feel about it now?

Well, I've missed many opportunities, and, of course, anybody would feel some jealousy about the attention given to Kroto and Smalley, but I never felt any bitterness about it. Especially then because I was so young, and I felt there were so many opportunities in science. I don't know how to express strongly enough that in mass spectrometry what Kroto and Smalley did is a forbidden act. It must seem crazy. If, after Krätschmer and Huffman, everything would have been explained by a different structure, I really don't think that Kroto and Smalley's reputation would have survived because it would have been exposed that they had deceived themselves for years with this story. It doesn't look like it's conceivable at all now because of hindsight, but I think they lived with this. When you spoke with Smalley at any meeting, he would come up and he would ask you, are you a believer? Do you believe this structure? Believe! Like religious faith. You had to decide to believe or not believe. For five years it sat in that condition.

Coming back to your story, how long did you stay at UCLA?

A total of 8 years, including one year of sabbatical in Orsay and Berlin in '91/92. I was promoted to full Professor so early that people tell me there was only one other person in the history of the whole University of California, and it was Glenn Seaborg.

Why did you leave UCLA?

When I was on sabbatical in 1991/92, the financial conditions of the University of California were under increasing stress, and when I returned I was involved in several efforts to make some changes, to improve things there, and when these failed, I was quite dissatisfied with the situation there, and specifically with the working conditions, the facilities, and the prospects for the future. At that time there were a lot of other places and quite a few other people who said that if you were not happy at UCLA, then why don't you come. I really wanted to be at an institution with stronger engineering colleagues and one that was really building up in these areas. And Georgia Tech in Atlanta is just that institution. I have, naturally, also higher salary, nicer labs, and I have a double affiliation. I am located in the School of Physics but I am also a member of the School of Chemistry with very good connections with the Electron Microscopy Facility. By training I am a chemist, a physical chemist but I started out in mathematics and in graduate school I took mostly physics.

What was your reaction to the Krätschmer-Huffman paper?

Entirely different from the Kroto-Smalley paper. This was a paper, an achievement that I used to dream of all the time. So the Krätschmer-Huffman paper was tremendously exciting to me and it was also the end of that dream.

Synthesis?

Yes, but my responsibility was the analysis side, the final stage, transformation, gas phase, mass spectrometric analysis. In the next two years, 1990–1992, I worked, with my group, on the new fullerene molecules, giant fullerenes, endohydrofullerenes. It was a very rewarding experience. I believe that the fullerenes was the best thing that could happen to me. It was painful at times but it forced me to new developments. Most young Assistant and Associate Professors in America are completely overwhelmed by their research group, teaching, administrative duties, publishing, and have no time to think and learn something fundamentally new.

What fraction of your time goes to writing proposals?

Three per cent.

How much funding do you have?

At the highest point it was close to 300 000 dollars per year, now about half of that. Funding goes to two different things. Instrumentation and then the cost of continuing your research, supporting people and materials and supplies. You only need big funding if you have a very large group, which I don't, or if you have to purchase new instrumentation which I don't need. Currently I have four or five graduate students, a postdoc, occasionally a visitor. At the highest point, in 1990, my group was about 13 or 14 strong and then I decided never again to have such a large group.

Do you still work with your hands?

Yes, when there is a problem in the lab.

What's your present research interest?

It's nanocrystals, that is, nanometer scale crystallites. A nanometer is about three lattice planes in the solid. The upper limit is ten nanometers, i.e., about thirty lattice planes. In terms of the number of atoms they run between 30 to 100 up to about 20 000 to 100 000 atoms. It's a very vast range in terms of volume and mass, and it's a very unknown region.

Our work on gold nanocrystals covered with the self-assembled molecular monolayer to me is the most exciting work I have ever been involved with. We had a three-year period of quietly working at home. Not talking, not publishing, accumulating the skills and the first results. We've shown that perfect nanocrystals can be made in such a form that they can be treated as very pure macromolecules. By their scale and volume, they are similar to biological macromolecules, like proteins. There should be a whole class of molecular materials in which the nucleus is a metal, covered with a molecular monolayer. Then you build a macroscopic crystal with regular packing out of these units. It was a revolutionary discovery about 10 years ago to cover the surface of a metal with an organic self-assembled molecular monolayer. They are also called compact, ordered two-dimensional crystals. The molecules pack themselves into such a structure of a single molecular layer. They are very specific. They have to be matched to the material of interest.

Could this be a protection against corrosion?

Exactly. And there are many other applications, giving the surface the desired property. Take a look at a small crystallite of 1,000 atoms. This is a polyhedron. Each one of the faces of this polyhedron has a definite surface structure which can be taken as a piece of an extended surface. The best thing to protect these surfaces is the self-assembled monolayers. It's remarkable how well these monolayers go over the edges and vertices as well. I am completely consumed with the idea of proving that perfect crystallites can exist and can be handled as molecules, and can be reassembled as solids. The picture I like best when one makes a macroscopic crystal which is composed of identical replicas of nanocrystals. This is a three-dimensional polymer network in which the linkages among the nanocrystals are created by the organic molecules forming the monolayer. Imagine, for example chain molecules with a sulfur at both ends attached to a nanocrystal at each end.

What did attract you to science?

I don't know it myself. This is the honest answer. It's not in my family history. I have relatives, once removed, second cousins, so on, who are scientists. One of them is one of the inventors of the quadrupole mass spectrometer. But no one in my immediate family is a scientist. On the other hand, I was taught practical mathematics at a very early age, from the age I could read.

You might have become a banker or insurer.

From that I was driven away because my father was an investment banker, and he wanted me to be interested in that. It was just teenage rebellion, but I went in the opposite direction as far as I could.

Robert Whetten with his wife, Claire E. Ozin, and their child, Andrew Benjamin in Erice, 1995 (photograph by I. Hargittai).

How about your mother?

She is an elementary school teacher. Once she had three children, she stayed at home and taught us. She told us the Latin names of plants and so on.

What's your ambition?

I really hate all the artificial division among the disciplines, and the disciplines I know best are physics and chemistry. I also don't like people who benefit from this division, by carving out areas as narrowly as possible. I feel unhappy and angry even when I see this kind of boundary. What I would like to do is help. I don't think that any one person can change this all but I would like to see those divisions destroyed. I think the way to destroy them is to provide clear examples of material systems and approaches which can't be defined. Going from this motivation you can probably see why I'd chosen areas between chemistry and physics.

How broadly does Georgia Tech prepare its graduates?

Traditionally, Georgia Tech used to have rather small chemistry and physics and biology departments and it had a very large and highly rated engineering faculty. A few years ago the engineering faculty, which already had very high ranking in the Nation, wanted to find out what to do next to improve their ranking even further. The leader of the engineering school said that every engineering institution rated higher is also very very strong in the sciences, and we are weak in the sciences. So we need to double the sizes of the science departments and cover many more areas than before. This is how they brought me there too.

Who was the most remarkable teacher or colleague whom you've met?

Probably Richard Bernstein and we'd spent about six years together at UCLA. Earlier he was at Columbia University at the time I was a student there. He was the highest-ranked chaired professor possible. At that time in the early '80's corporate labs were building up for the last time. Occidental Petroleum Company, based in Los Angeles, decided to build a beautiful palace for basic research in Orange County, Southern California. They needed someone for Vice President for Research, somebody with a great reputation to recruit many other good people and persuade them to come to this lab. Richard Bernstein, who had spent his entire life in Academia, accepted this position, and his first assignment was to go around and persuade young people that they should move with him and build this place. He was 55 at the time. He relocated from New York to Los Angeles, and many people followed him. They bought everything new, beautiful, and best that money could buy. Then about a year and a half later, when the finances of the petroleum industry stopped expanding, the President of Occidental, Armond Hammer, decided that this operation was a mistake. So Bernstein's next assignment was to tell each one of those people who had followed him, sorry, we have no job for you any more, we are closing the lab.

Following this UCLA invited Bernstein, brought his lab equipment, and this is how he happened to be at UCLA. It's interesting that Don Cram was very skeptical at the Chemistry Department. He said Bernstein is now 58-years-old and he is just looking for a place to retire. So when Bernstein met Cram, Cram challenged him immediately and said, are you going to come here just to retire? And Bernstein was so angry, he was a very intense man, he could hardly speak. He was completely insulted by the question, and then Cram knew that this was a good man to have on the Faculty.

Bernstein really started all over again. He is known for orienting molecules with fields and doing stereochemistry with it, the ultimate stereochemistry. Then he died. He had a cardiac arrest in St. Petersburg, Russia, and a few days later he died in a hospital in Helsinki, Finland. This was three months before the Krätschmer-Huffman paper, the last couple of days of June of 1990. He was more than anyone I remember very upset with Kroto and Smalley and challenged them indirectly to prove their structure and he was very angry that they spoke all the time about it without doing more tests of the structure. I often thought that it's so sad that Bernstein didn't live to see the structure proven because he would have been more excited than anyone by this development.

My personal encounters with Bernstein were at UCLA, but I never had a comfortable conversation with him. We were on the same Faculty, and I was recruited partly by him and partly by El-Sayed. I knew him the last six years of his life. There was never a relaxing moment with him. He was very intense, even in social occasions. He had the attitude that life is short and you shouldn't waste it with frivolous things and that people who had some kind of talent should be working basically all the time to reach the highest level. He never said that this was his philosophy of life, but every way he acted this way and every way he projected this on everyone around him.

During the 85–89 period he was winning every kind of award that the American Chemical Society and the American Physical Society gave and international awards too, for his life-time accomplishment. He would show us the medals that he had won, including the Medal of Science which he received from the President of the United States. But it hurt him that he was passed by by the Nobel Committee.

For personal conversations, we had this pattern, it happened five or six times that he saw me in the library and would start a conversation. It would be about what's new. He looked at every paper we wrote and he was very critical. It wasn't unusual for him to say, this is not a good paper, this is not to your standards, you should really do better. This was his method of encouragement. He had a lot of energy and intensity to monitor many things around him.

Bernstein liked to say, you know, you'd only be recognized in the end for one thing, maximum of one thing. When your career and life is finished, there'll be at the maximum one thing remembered about you. If you're unlucky, nothing will be remembered. Then he'd make a list of his field, saying, I worked on this field and I worked on oriented beams, and I also worked on this and I worked years on multiphoton dissociation and ionization, and nobody will remember this. He said, I'll only be remembered for orientational effects, stereoeffects on chemical reactivity. He said, it will be the same for you, so the point is to find this and stop wasting your time on all the other things. Go as deeply as you can with the one thing which is the best and stop the rest. He repeated this every time we had this conversation. Of course, he never lived his philosophy.

His death was a shock and I don't think people really recovered from it in UCLA chemistry. I know I thought many times later on when I was doing some kind of an experiment, I could almost hear Dick Bernstein's voice saying, "Why do you want to waste your time on this?" or, "If you'd only work like this." I find this was an irritating voice in the back of my head, criticizing the things that I did, for several years after he died.

References

1. Hargittai, I. in Martin, T. P., Ed. *Large Clusters of Atoms and Molecules, NATO ASI Series E: Applied Sciences*, Vol. 313, Kluver, Dordrecht, 1996, pp. 423–435.

Philip Eaton with a sample of cubane in 1995 (photograph by I. Hargittai).

32

Philip E. Eaton

Philip Eugene Eaton (b. 1936) is Professor of Chemistry at the Department of Chemistry of the University of Chicago. He received his B.A. degree in 1957 from Princeton University and his Ph.D. in 1961 from Harvard University. He is best known for the synthesis of new organic compounds, including cubane. He received the Alexander von Humboldt Prize (Germany) in 1985 and an Arthur C. Cope Scholar Award (ACS) in 1997, among many other distinctions. We recorded our conversation on November 10, 1995, in Dr. Eaton's office and the transcripts of the conversation were finalized in June, 1998.

My first question was about cubane.

This story starts in a very strange way. Years ago my mother broke her hip and was taken to the hospital. The woman in the next bed was the wife of the Vice-President of Allied Chemical. So my mother arranged that I'd have a job in Allied Chemical which I started in my second year of college at Princeton University. At that time Allied Chemical was working on kepone, which is a very powerful insecticide; it's banned now because of its chlorine content. Kepone is made by a very strange reaction of hexachlorocyclopentadiene in liquid sulfur trioxide. The cyclopentadienes come together and give a cage molecule. The cage is a bishomocubane. So very early in my career I got interested in cage compounds.

When I went to Harvard, I wanted to continue some of the work I had been doing during the summers in industry, and I did that as part of my Ph.D. research. Peter Yates, for whom I worked at Harvard, was also very interested in photochemical cyclizations. At that time, and this was in the late fifties, every photochemical ring closure was a novel reaction. When I finished my Ph.D. at Harvard, I went to Berkeley as an Assistant Professor and started to look at photocycloaddition reactions in a more sensible way. When I came to the University of Chicago a few years later, I thought that I understood how to make defined compounds photochemically; that it would not be just a matter of accident. I got an idea of how to make cubane from one of those compounds by the Favorskii closure. The idea of the Favorskii closure came from a Harvard cumulative exam I had taken years earlier and never forgotten. The question was wrong, and certainly my answer was also wrong, but I remembered it. I thought if we could make a particular compound, a particular bishomocubanedione with halides in the right positions, it would be worth the try to see if we could make cubane by Favorskii closure. The challenge to us was, since we thought we understood how to make cyclopentadienones with specific substitution and how to do photochemical ring closures, to make this compound on which to try the Favorskii closure. It was a gamble; it might work and it might not work. In any case, we made the compound. The synthesis part went fine, the photochemistry part went fine, and the closure went fine. The entire project took two weeks, from its concept to its execution. At that time we had just gotten our first 60-NMR in this Department. It was one of the early ones. Saul Winstein of UCLA was visiting just when I was taking the spectrum of the dimethyl ester of cubane-diacid, and he came into the NMR room, and there it was, this wonderfully simple spectrum, and I said to him that I think we may have made a cubane. And it turned out to be true.

Is cubane the most important piece of your research? At least it is what you are best known for. Do you resent this?

Yes, it's a little annoying that people think only about the cubane compounds in our synthesis work. They forget about the propellanes and the paddlanes and peristylane, etc. Much of the early work on intermolecular photocycloaddition reactions and medium ring trans-enones was ours. There is a whole list of things that we've done. I think, however, that in the history of chemistry it's not so much cubane itself that will be considered important, but the various things we have been able to get from cubane that have been very important in the development of physical organic chemistry. Cubane was very important in understanding what was happening in metal-catalyzed ring-opening reactions. From cubane we've gotten the most highly twisted olefin, and the most highly pyramidalized olefin — one that rearranges to a carbene — and the fastest rearranging aliphatic radical, and cubane carbonium ion which is certainly the least likely carbonium ion — formed 10^{15} times faster than was originally predicted, and many exceptional intermediates like cubane-1,4-diyl whose existence would never have been dreamed of if we had not been able to make cubane and with a synthesis that was good enough to make many, many grams. Cubane is now being made commercially, ten kilos of the diacid in a run. So we had lots of compounds to work with.

The cubane synthesis was a rational synthesis. It was not an accident. Up until that time, most highly strained compounds had been made essentially accidentally, or shall we say, incidentally, by some bolt of lightening approach. Our cubane synthesis was rational, and it fit in very importantly into the development of synthetic methodology. We proved that you could make strange compounds, things that were very foreign to Nature, in a rational way.

Foreign to Nature?

Cubane is certainly something that Nature has not made.

Are we sure of that?

No, we aren't sure; it's possible that Nature made it a long time ago and then decided it was useless.

Was there any consideration of aesthetics in your search for cubane?

I always appreciated the idea of symmetry. Symmetry is something very attractive to everyone, except, perhaps, to abstract painters, but it was not the driving force behind our work. Once we made cubane, we started thinking of making other symmetrical compounds. The symmetry is there and symmetry is very nice from the aesthetic point of view, but symmetry also has a very great chemical significance. It simplifies spectra tremendously, and it often makes interpretation, of what's happening, easier. Symmetry simplifies things for zero-order approximations. It also hides information. If you look at the NMR of cubane you have one single peak. It's very nice for identification, but all sorts of information is hidden because you only have one peak. So symmetry is a help at a certain level of chemistry and then it becomes a mask.

How about compounds that you could not make yet?

Azacubane is an interesting example. There are even calculations in the literature of octaazacubane, which is an allotrope of nitrogen.

Do you do calculations?

Only in the very simple sense of MM2-type things. For the most part, we don't believe the calculations. Of course, it depends very much on the person doing them. If you do calculations, you must have a good intuitive feeling about the experiment, what I call the real world. There are some who just do calculations and have no sense of reality.

How about buckminsterfullerene? Nature definitely makes this substance but its synthesis is not yet solved.

It's not solved and I do not think it should be solved. I have very strong feelings about this. One has a new kind of synthesis with the flame. I don't mean the laser blasting; that's useless. I mean using the arc, and it's wonderful, and I don't see any purpose at all in synthetic organic chemists spending years of peoples' time to create this molecule by stepwise, old-time chemistry. Synthetically to do this stepwise is a very difficult process, and I see no purpose in it at all. There are some very nice works by people who have made part structures of buckminsterfullerene. There are certain things that you want to know about the part structure; that's important in understanding fullerenes, and that's a very good reason to make these compounds but not to take them on to buckminsterfullerene itself. It would be much more useful and exciting to take C_{60} and figure out how to use it as a starting material. Synthetic chemists should leap on this, rather than spending years and years on trying to make it. It is certainly possible to do so; I have absolutely no question about this in my mind. But I don't see why one should.

R. Stephen Berry, 1995 (photograph by I. Hargittai).

33

R. Stephen Berry

R. Stephen Berry (b. 1931 in Denver, Colorado) is James Franck Distinguished Service Professor of Chemistry at The University of Chicago. He is most famous for the Berry Pseudorotation. He received his A.B., A.M., and Ph.D. in 1952, 1954, and 1956, respectively, from Harvard University. He has been at The University of Chicago since 1964 after having spent some time at the University of Michigan and at Yale University. His interests include the dynamics of atomic and molecular clusters and the thermodynamics of time-constrained processes. He is also interested in a variety of issues concerning science and public policy and has held an appointment in the School of Public Policy Studies at The University of Chicago. He is member of the National Academy of Sciences (1980); Fellow of the American Academy of Arts and Sciences (1978); and Foreign Member of the Royal Danish Academy of Sciences (1980). He has held a great number of distinguished lectureships, including the Hinshelwood lectureship at Oxford University; the Lövdin lectureship at Uppsala University; and the Sackler lectureship at Tel Aviv University. He is the Centennial Speaker of the American Physical Society (1999). Among his other recognitions, he was recipient of the Alexander von Humboldt-Stiftung Senior Scientist Award (Germany) and the Heyrovsky Medal of Merit (the Czech Republic). Our conversation was recorded during a NATO workshop on clusters, in Erice, Italy, on June 22, 1995.

Let's start with the Berry pseudorotation. How would you formulate it for the nonspecialist chemist?

Pseudorotation, in general, refers to a process in which identical atoms permute among nonequivalent sites, so that the net result of the process is what looks like a rotation of the molecule, if we don't have any labels on the atoms. If you put labels on the atoms you'll see that it's a permutation in general, and a rotation as well. The particular process my name became attached to is a large-amplitude motion. It was probably the first real example to be found of a large-amplitude pseudorotation that scrambles bonds, and I applied it in the motions of the fluorines in phosphorus pentafluoride. This came up in the early days of nuclear magnetic resonance, when, early on, people found that identical atoms in chemically inequivalent sites had different magnetic resonance frequencies. These differences were soon called chemical shifts. It was shown, first by Gutowsky, McCall, and Slichter, that if there was rapid exchange of inequivalent atoms, at that time it was basically inequivalent protons, that one would see an average signal and not be able to distinguish the chemically nonequivalent sites from one another. There was one experiment that came from the University of Illinois, by Herb Gutowsky and Andy Liehr, when Liehr was an undergraduate at Urbana, in which they found a single fluorine frequency for PF_5. But it was well established by electron diffraction that PF_5 was a trigonal bipyramid. So I proposed a mechanism in which the axial pair of fluorines bent away from the linear F–P–F line and moved over to form a triangle with one of the equatorial atoms, and two of the three atoms in the equatorial plane moved out and became new axial atoms. The net result of this process is that PF_5 looks like it is rotated by 90 degrees with the polar axis moving from, say, vertical to horizontal. I proposed that this process occurred fast compared with the nuclear magnetic resonance observation time, so that one would observe only a single Fourier frequency. This, in fact, was only one part of a paper, and to me, at that time, not the most important part of the paper, titled *Correlation of rates of intramolecular tunneling processes, with application to some group V compounds.*[1] The paper was about a method for establishing systematic relations for large-amplitude motions, especially tunneling motions in homologous series. So I was comparing PF_5 and PCl_5, and I also made comparisons of other series, and, essentially, was saying, on what time scale you should do an experiment to distinguish the interconversion/pseudorotation rate. If you knew that rate for one molecule in the series, then you could estimate the rates for other molecules in the series. I don't think that that part of the paper has ever been very influential. What did catch on, of course, was the idea of pseudorotation of five-coordinated phosphorus. By the way, I was not the first person to think of such a mechanism. I found out afterward that the first proposal for such a mechanism had come from John Wheeler and Edward Teller in the late thirties in the context of explaining the behavior of a nucleus, the neon-20 nucleus. At that time, they suggested that neon-20 could be described by the then-popular alpha particle model, as five alphas. They proposed that these five alphas form a trigonal bipyramid and could undergo this pseudorotation process. It turned out that it was not the way that neon behaves at all. This proposal was not quoted in my paper because I didn't yet know about it, but I did quote another proposal that had been made in the molecular context but was never published. People were interested in the CH_5^+ molecular ion. Among these people were my research director, Bill Moffitt and Felix Smith, who at that time was a graduate student with George Kistiakowsky but Felix always liked to work with lots of different people. I don't know who really came up with the idea of pseudorotation for CH_5^+, but it was Felix who first told me about it. It turned out, though, that it was not correct for CH_5^+ either because it is not a trigonal bipyramid. It is more like CH_3^+ and H_2 held together weakly. But I did cite in the paper that Felix had suggested the mechanism to me.

Who then did add your name to the PF_5 pseudorotation?

I'm not sure who added it first. I can think of a few possibilities. It may have been Al Cotton because I suggested in my correspondence with him that this process could go on in some of the transition metal carbonyls. In Volume 100 of *Inorganic Chemistry*, Al wrote a long review. In it he referred to the correspondence that he and I had had, and quoted a part of my letter about it. It may also have been Earl Muetterties because I had some correspondence and discussion with Earl about it. Another possibility is Frank Westheimer. Perhaps the most far-reaching implications of the process were what Frank found. He and his students showed that phosphate ester hydrolysis occurs by four-coordinate phosphorus becoming five-coordinate as it goes into its transition state, then goes through the pseudorotation as five-coordinate, and finally goes back to four-coordinate. It works because the axial positions have weaker bonds than the equatorial positions. Something would come in to form an axial bond and rearrange to form a stronger equatorial bond, something else would then become axial and break off. I think that the greatest implications of the process are probably in the biochemical context that Frank pursued.

An important aspect, from my point of view, was time-scale relationships. The question was one of what you observe and more specifically of how to resolve the problem of the apparent paradox that identical nuclei could occupy observably inequivalent sites. We had been brought up in quantum mechanics knowing that identical electrons or any other identical particles have to be indistinguishable and the wave function for identical particles has to reflect this indistinguishability. But everybody knows that chemistry is based on the distinguishability of different sites in molecules. Of course, the answer to this paradox is in time-scale separations. What we consider a time stationary state of a molecule is a stationary state for all practical purposes, but in some philosophical sense it is not. The interesting question to me then was that if you know that the molecule would find some way to establish the equivalence of identical particles occupying inequivalent equilibrium positions in the geometry, what mechanism would any particular system find, what would be the fastest way for the system to do this? This would show itself in experiments, such as the coalescence of NMR lines, which in a fast experiment correspond to different chemical shifts, and in a slow experiment they correspond to an average. In some ways, this concept has reappeared many times in my work.

Did the whole concept of fluxionality evolve from the idea of Berry pseudorotation?

I think so. The notion of small-amplitude motions leading to pseudorotation appeared first in a paper of Pitzer's on cyclopentane, about 1950. In cyclopentane you have an almost-symmetry axis for the nonplanar pentagonal molecule, so you can see the rotation there right away, and Pitzer aptly called the motion of cyclopentane pseudorotation. Then I recognized almost immediately after I thought about the process, that the rearrangement of the trigonal bipyramid was a three-dimensional, large-amplitude extension of that. The motion, applied over and over and over again, eventually covers the whole unit sphere. The system takes on all possible orientations. It poses an interesting symmetry problem, because this particular kind of pseudorotation involves the infinite set of discrete rotations. So it's a discrete subgroup of a continuous rotation group. Then the interesting question is, how does the pseudorotation couple with rotation? I have had in my mind for many years that Bill Moffitt, when he was examining CH_5^+, was already thinking about that. He said a few words in a conversation once, something like Fermat's marginal note that became Fermat's last theorem, and I think Moffitt thought he knew how to solve this problem of the coupling of rotation and pseudorotation.

Let's get back to the very beginnings. At what point did you decide to go into chemistry and physics?

I was about six-years-old when I very much wanted to have a chemistry set, and I got it as a Christmas present. My family is Jewish but we often spent holidays at my grandparents' in Fort Collins, near Denver, where we managed to celebrate both Christmas and Chanukah. I remember that the Christmas presents would be laid out on Christmas Eve after we children were asleep. I woke up about four or five in the morning, crept out quietly, and scanned the presents. I found the one of the right size and shape, smoothed down the tissue paper and could read the words CHEMCRAFT. I was so excited that I couldn't go back to sleep. I was pretty sure from that time that I was really interested in chemistry. Then, it must have been in early junior high school, I remember having long discussions with one of my close friends about science; we both had our laboratories in our basements. I just thought it was the most natural thing in the world to become a chemist. I also considered physics. In ninth grade, I had to do a report on careers. We had to chose two careers we might like, and I chose chemistry and physics to write about. I remember, though, that I was warned by some people that I'd have a very hard time getting a job in the profession because there was a lot of anti-Semitism. For example, I was advised that it was very difficult to get an industrial position as a scientist if you were Jewish. This was probably accurate about the situation before World War II.

May I jump ahead and ask you, did you eventually experience these difficulties?

No, not in terms of anything professional, not that I could ever distinguish. I think the world had changed so much as a result of the influx of European scientists, of whom many were Jewish and as a result of general liberalization and democratization of views. I could see traces of what it could have done to others. For example, one of the famous faculty members of the

University of Chicago, Chemistry Department, was notorious for his anti-Semitism, during the twenties and thirties, but he had died years before I came to Chicago.

What else influenced you in turning you to chemistry?

It was very easy for me to decide to become a chemist or a physicist, based on the books I'd read. One of the first books that influenced me a lot was de Kruif's *Microbe Hunters.* It was about biology, with a lot of chemistry in it. It was about people like Pasteur, Koch, and others. There were then two more books I especially remember, one was called *Magic in a Bottle* and the other, *Crucibles.* These two were also biographical in structure, but very much oriented toward the scientific contributions that the individuals made.

What did your parents do?

My father was in the real estate business and my mother was a teacher.

What made you choose chemistry over physics?

I knew from the high school textbooks what physics was. It was about ladders leaning against walls, and it was really boring, whereas chemistry was about the structure of the atom and all other interesting stuff, quantum theory, for example.

Do you think that for today's child the choice would be similar? Don't we have today a different perception of chemistry and physics?

We do. At this point chemists probably do a far less expressive job, a less effective job of conveying the intellectual excitement of the subject than do the physicists or the biologists, or the astronomers.

Where did you do your studies?

I did all my undergraduate and graduate work at Harvard. Originally I had planned to go to either M.I.T. or Cal Tech. I wanted to get away from Colorado and I was told these were the best schools to study science. However, in my last year of high school, a physics teacher suggested that I might be interested in entering a contest, something put on by Westinghouse. Part of it was to do a project, which was then my first "official" research project done in our basement. Curiously, it never entered my mind to ask anybody for help. I found something interesting in the library, something that looked like an unsolved problem. I bought the chemicals. One of the things I needed was phosphorus pentachloride. It came in a metal can. When I opened the can, there was this crumbled stuff. I didn't know what phosphorus pentachloride should look like, and I couldn't tell the difference between phosphorus pentachloride and vermiculite. So very carefully I saved this crumbled stuff till I got deeper in the can and found there was a bottle, containing phosphorus pentachloride. I then did the experiments and proposed a structure of a thing that I may have actually never had prepared. A few weeks after I had sent in my project, I got a telephone call saying I had won a trip to Washington as one of the 40 finalists. This was the first feedback I had ever had reinforcing, of course, my interest in science. I went to Washington and, for the first time in my life, I met a real scientist.

By that time I had decided that I was really interested in literature, philosophy, and I wanted to learn about such things. I hadn't put these things together yet with my idea of going to M.I.T. or Cal Tech. At this meeting in Washington, one of the people who was helping to shepherd us around was a winner from a previous contest. He had gone to Harvard and was now working for Westinghouse. He made it very clear to me that you could go to Harvard and get a real good education in science. That was very important to me that you could go to a place where you could really learn about literature and history and psychology and not sacrifice the quality of your science education. So I immediately decided that I wanted to go to Harvard, and I went. This was in the Fall of '48; I got my bachelor's degree in '52. Then I got my Master's degree in '54, which kept scientists out of the Korean War, and got my Ph.D. in February of '56.

Please, tell us about your supervisor.

Bill Moffitt came to Harvard in December of 1952. The story I heard was that van Vleck came back from a Shelter Island conference where he'd met Moffitt, and told E. Bright Wilson, Jr., that "We must have this man." So he was hired as an Assistant Professor and as soon as I talked with him I knew that I really wanted to work for Bill Moffitt. What I tried to do was a combined experimental and theoretical thesis. I picked up an experimental project from an instructor who was leaving and asked M. Kent Wilson to supervise my experimental work and Bill was to supervise my theoretical work. The experimental project was excitation of molecular spectra by threshold-energy electron bombardment.

Bill was quite happy to have me (and the other graduate students who started working with him as soon as he arrived). He was very young, extremely articulate, very witty, very urbane. He was educated at Winchester and Oxford in England, and had a proper Oxford accent. He was Scottish though, and at parties, when there was a fair amount of drinking, as he drank, his accent became more and more Scottish. His family came from Sutherland in the far North and he was very proud of his being Scottish.

He was a kind of intellectual glue for the Harvard Department that they had never had. He went around and he asked people what they did, and he talked to the people about what they did. His colleagues thought very highly of him. He had long and very intense discussions with R. B. Woodward and he was a very close friend of Frank Westheimer, and there was tremendous respect between him and Bright Wilson. He was also very close to George Kistiakowsky, both personally and intellectually. Then at the end of 1958, Bill Moffitt died very suddenly. He had had a heart condition. Apparently, he had been told that he should be very careful not to exert himself physically. He was the most competitive athlete I'd ever played against. He would go harder to reach a ball playing squash than anybody else I'd known, and he died on the squash court. He left a young widow, to whom we had introduced him originally, and a daughter. He was about 35.

The first student who joined Bill Moffitt was Andrew Liehr, who had worked a lot on symmetry problems. He later just disappeared, after he'd worked for quite a number of years at Bell Labs following his Ph.D. Then he went to Mellon Institute, before it was Carnegie-Mellon. One day he left a note for his office maid, saying that if he didn't come back in a year, his office maid could have all the books. And he didn't come back. He had been very productive.

How does it work to be in a department where there are Nobel laureates. Is there an aura around them? How does it affect the every day life of the department?

Those that I know and had known and had had close contact with don't carry an aura. They may carry a responsibility, because they're expected to be spokesmen for science in a special way. I have in mind Chandrasekhar, Robert Mulliken or Jim Cronin; I think they are respected for what they'd done in science. There is a sense that you can't avoid, you recognize that the World has acknowledged what they'd done. There are some that I don't know very well who have enormous egos and there is one whom I can think of who has created a gallery in the department where he is, of his honors and awards. It's a small museum of how important he is. It seems to me symptomatic of the level of insecurity of disbelief, and is a little embarrassing to other scientists.

However, Robert Mulliken, for example, was just as intent on understanding what's right, and just as unafraid to try an idea that could be wrong, after he had the Prize, as he was before.

Stephen Berry talking with students at the Erice NATO workshop on clusters, 1995 (photograph by I. Hargittai).

Let's get back to your career.

I got my Ph.D. in '56 and stayed on at Harvard for a year-and-a-half as a temporary instructor. Then I went to Michigan, in Ann Arbor, as an instructor in the Fall of 1957 for three years. Then Yale offered me an assistant professorship and I went there in 1960, and stayed until '64. At that time, Chicago offered me a tenured job and the question of tenure had not come up at that point yet at Yale. By then, Chicago had become a Mecca to me, in my mind, as a scientific center. Chicago's real fame in science has been post-WWII. A lot of this was because of people who came from the Manhattan project. Enrico Fermi, Harold Urey, and Joe and Maria Mayer and others came after the end of the War. But none of these were at Chicago in 1964; Fermi, Franck, and Szilard had died, and the Mayers, Urey, and Libby had left for California. Still, the aura was there for me, especially because of my contemporaries and people younger still.

But was Maria Mayer a professor at Chicago?

She was in the Physics Department. She did not have a regular professorship because they had an anti-nepotism rule, so both husband and wife could not have appointment in the physical science.

Who decided that he should have it and not she?

I really don't know. Maybe he got the first appointment. She went to all the faculty meetings and she could take part in all the discussions; she just didn't have a vote. However, Joe got double salary, as I was told, because of this.

Her Nobel Prize came in 1963. Was she still at Chicago then?

By then she was in California. She and Joe Mayer and Urey all went to San Diego, and she got her own professorship there.

You stayed at Chicago ever since 1964.

Apart from sabbatical years.

What's your main research interest now?

Most of what I'm doing now is concerned with clusters and other systems that are really on the border of being complicated. Much of it is based on using atomic and molecular clusters as vehicles to study particular kinds of phenomena.

A significant part of my work is connected with the thermodynamics of finite-time processes. The thermodynamics of macroscopic systems is just getting connected to the thermodynamics that we've been using to study clusters, the thermodynamics of small, simple systems. This includes studying their phase transitions, melting and freezing behavior, and determining of what aspects of their phase changes become the freezing, melting and evaporation of macroscopic systems, and which ones are characteristic only of those small systems that don't have counterparts in macroscopic systems. My most recent work is addressing questions that have been bothering me for many years, which is related intimately to the question of how a system decides to become a glass or a crystal, how a protein knows how to fold to the right structure, and how this kind of behavior emerges from the forces between the particles. You can think of this question of behavior as something being on a rough surface. Something can roll down on a rough surface and get caught in a little, shallow well, and then get into a deeper bowl. We can think of these potential surfaces the way we think of the potential energy associated with being on top of a hill; in this case, a rough one with lots of structure on the sides of any basin, or at the bottom of the hill.

Thinking in the context of atomic and molecular clusters or molecules means that, instead of valleys and hills on the Earth, as we experience it, these are hills and valleys and passes in the space of many, many dimensions. The first problem is, how do you think about it? How do you conceptualize the topography of a surface in so many dimensions? Then the question is, how do you extract useful information? How do you decide, what information is worth having? Then, how do you put that information to use?

This sounds like an extension of the pseudorotation problem of 35 years ago.

True. Pseudorotation is a good and very simple example of the kind of thing that can happen. There is a logical path between these projects. In the early 1980s, I decided it was time to go back to the pseudorotation issue. I wanted to relate observed spectra to the pathway that the molecule was using to pseudorotate. We figured out a way by which, in principle, we could do that. When we had developed the method but before we applied it to any real case, I realized that our approach was a powerful tool to address a completely different problem. Some computer simulations people had done in the mid-70's implied that very small systems, atomic and molecular clusters, could exhibit solid and liquid-like forms, and have phase transitions. We saw that we could create an analytic theory which would test these simulations on real molecules, using the information that was generated from the very tools we were going to use to relate rigid and nonrigid molecules to each other. So we started studying clusters. That proved to me so productive that we've never gotten back to spectroscopy and to the interpretation of pseudorotation. But it definitely led me to thinking more and more about complex potential surfaces, and to the question of how do you think about a potential surface of a system that is too complex for you to be able to use all information which you, in principle, could derive from the computer about the system. You can get more information than you care to have. What information is worth having? What should you want to know about these complex potential surfaces? During the past year then, while on sabbatical leave, I seem to have found some ways to understand and find answers to some of these questions, such questions as why do some things choose to form well-defined structures, for example, while other things tend to form glasses.

How does a system test all different kinds of arrangement before finding its minimum energy structure?

It depends on the topography of the potential energy surface. In some topographies, for example those of the clusters of argon atoms or copper atoms, the system is testing and exploring the possible arrangements randomly. By contrast, sodium chloride makes rock salt crystals almost instantaneously. It does that even against statistical odds that are even more unfavorable than for a protein to fold to the right physiologically active structure.

Do you, then, know the time period for various systems to find their most advantageous structure?

Better than that, because we have begun to understand why the system doesn't make a random search. Think of the potential surface again in terms of valleys and hilltops and passes in a mountainous region. The system starts off in a very high valley and it goes over a relatively low pass, and it finds itself coming down into another bowl that's a little deeper. So it can run down into the next valley fast because there is a nice steep way down. And it runs down fast and approaches the bottom. If the next pass were very high, then the system could run out of energy (in the mode that carries it over the next saddle) before it gets to the next top. But if the next pass is not so high compared with how far down the last minimum was, then the cluster can come running down, up a little way, and down again over the next pass. In the copper or the rare gas clusters, the next saddle is going to be pretty high, and you may very well be getting into trouble getting over it, compared with how far down you dropped. On the other hand, in the rock salt case, it appears, there are lots of drops that are much bigger than the next saddle that the cluster has to overcome, on its way to becoming a rock salt crystal. So the cluster runs down very fast into what is a very ordered structure.

Is this the kind of thing that you were dreaming about when you were six?

I wish I could have.

References

1. Berry, R. S., *J. Chem. Phys.* **1960**, *32*, 933–38.

34

What Turned You to Chemistry?

Since its publication in 1926, Paul de Kruif's *Microbe Hunters* has been a major influence on gifted children to choose chemistry for their career. The other major source of inspiration has been the chemistry set. In my experience of talking to great chemists, teachers come in a distant third on this list.

Microbe Hunters is about natural scientists and mostly not about chemists, so the quest for uncovering nature's secrets, rather than something specifically chemical, was the determining factor in its impact on future chemists. Paul de Kruif (1890–1971) published many other books but none as successful as *Microbe Hunters.* He published a very personal book in 1962, *The Sweeping Wind* (Harcourt Brace), which contains the story of his becoming a writer. The

Paul de Kruif, from the dust jacket of *The Sweeping Wind.*

Modified from an Editorial in *The Chemical Intelligencer* **1998**, *4*(2), 3 book is also an account of his life, inspired by the loss of his second wife, Rhea Elizabeth Barbarin, in 1957.

Dr. de Kruif met Rhea, a lab technician, in 1919, when he resumed his work in the bacteriology lab of the University of Michigan Medical School, following his two-year service in Europe during World War I. His research was on the blood-dissolving poison of the hemolytic streptococcus and he also taught advanced bacteriology. De Kruif married Rhea in 1922 after he had left his wife and two sons, whom he did not see again for the next 24 years. Back in 1919, he decided to write popular science, first as a sideline, then, from 1922, full time. At that point he resigned, or was fired by, The Rockefeller Institute for Medical Research in New York, where he had moved in the meantime. At The Rockefeller Institute, he had continued his research and published papers on immunology, with John H. Northrop, who received the Nobel Prize in chemistry in 1946 for crystallizing the enzymes pepsin and trypsin. Northrop was one among many famous scientists de Kruif came across during his career. Jules Burdet of Belgium, Nobel laureate of 1919 in physiology or medicine, was another. Burdet praised de Kruif's writing style and gave him this prophetic advice: "What you should think to do is a roman des microbes."

On the way to becoming a writer, a great opportunity opened up for de Kruif in an apprenticeship to Sinclair Lewis, who invited him to help write *Arrowsmith.* At the time, Lewis was the celebrated author of *Main Street*, and *Babbit* was just about to appear. He was awarded the Nobel Prize in Literature in 1930. *Arrowsmith* is about Martin Arrowsmith, a tough young man hell-bent on becoming a microbe hunter. Looking back on the experience, de Kruif remarked: "Red [Sinclair Lewis's nickname] taught me to let my imagination go."

De Kruif started writing *Microbe Hunters* in 1923, and Harcourt Brace gave him a generous contract for it. The title of the book was suggested by Donald Brace. The book appeared in 1926 with a first print run of 2800 copies. It made the American nonfiction best-seller list by the summer of the same year and soon passed the mark of 100 000 copies sold. *Microbe Hunters* has remained in print ever since. Henry Mencken, who played a pivotal role in Paul de Kruif's becoming a writer, characterized the book this way: "One of the noblest chapters in the history of mankind."

Kenneth S. Pitzer, 1996 (photograph by I. Hargittai).

35

Kenneth S. Pitzer

I learned about "Pitzer strain" as a student but later I have not heard much about it. In February, 1996, my wife and I were visiting in California and drove to Berkeley, primarily to see Jeanne Pimentel and Gabor Somorjai. Jeanne had written a beautiful article about her late husband, George Pimentel for *The Chemical Intelligencer*,[1] which was augmented by a brief summary of George Pimentel's professional activities and awards by Kenneth Pitzer and C. Bradley Moore.[2] Jeanne then organized a meeting for us with Professor Kenneth Pitzer. I recorded a conversation with him, sent him the transcripts, and he corrected them in April, 1996.

Kenneth S. (Sanborn) Pitzer (1914–1997) was born in Pomona, California. He got his B.S. degree from the California Institute of Technology in 1935 and his Ph.D. degree from the University of California, Berkeley, in 1937. He had a professorial career at the University of California at Berkeley, eventually rising to the position of Dean of the College of Chemistry. He had also served as President of Rice University and, for a brief period of time, as President of Stanford University. He did war service in 1943–44 as Technical Director of the Maryland Research Laboratory; then he was Director of Research for the Atomic Energy Commission from 1949 to 1951; and was Chairman of the Commission from 1960 to 1962. He and Leo Brewer revised the original (1923) Lewis-Randall *Thermodynamics* in 1961, and he alone published the third revised edition just a few months before our conversation. He seemed very happy about it.

Kenneth Pitzer was a member of the National Academy of Sciences and many other learned societies and he had received many honors and awards, including the National Medal of Science of the United States; the Priestley Medal of the American Chemical Society; the Gold Medal of the American Institute of Chemists; and the Robert A. Welch Award.

Since the interview on February 27, 1996 had not been planned, I did not do my usual homework, and my questions were disorganized and not very comprehensive. Looking back, though, Kenneth Pitzer let me learn about him quite a bit during that single meeting. Otherwise, I don't think he was known as a very communicative person. It may be characteristic what one of his colleagues of many years wrote me recently, "I learned more about the real Ken Pitzer at his memorial service than I ever knew before so I really knew little of the man."

This was our conversation:

As I understand, you follow the activities of the National Academy of Sciences. In addition to having been elected as a member many years ago, what else has happened to you in the Academy?

I was on the Council of the Academy twice. Once in the sixties when I was President of Rice University, and then, again, in the seventies, when I was back at Berkeley. I was also member and Chairman of quite a number of the committees of the Academy, including the Nominating Committee for the President of the Academy.

Does the United States Government request and appreciate the advice of the Academy?

It's a mixed situation. Sometimes it's very valuable and Congress accepts it. There are then other times when they don't.

Any example?

One recent one is the matter of how you would dispose of low-level radioactive waste, such as used in diagnostic work in hospitals and so on. In California, the proposal was to take a very dry desert site, supposedly separate from the Colorado river and use that. This was studied to a point. Many people were satisfied that it was a good site with no water leakage into the Colorado river; otherwise it was a waste desert land. However, others opposed it. The Academy made a study of it and recommended a few more experiments but it seemed very unlikely that it would change the picture. This satisfied almost everybody in the Congress but the hardened environmental types kept opposing it, including one of the California senators. This finally persuaded the Department of the Interior, under whose jurisdiction this comes, since it's Federal land, to refuse to turn the land over to California for this waste disposal until these additional experiments are done. This is a case where it was hoped the Academy study would solve the problem and convince the people as to what the facts were, and let thing go ahead. However, the results were mixed and while the Academy study was useful, the situation remained as mixed as before.

Another recent example was a rating of the best graduate schools for Ph.D. study among the Universities by Departments. We came out number one on that with Cal Tech almost in a tie, and Harvard was third, and so on. The Academy group that did this study within the Research Council, did it in a thorough manner.

In addition to Rice, you were also President of Stanford in another period. Have you had other positions away from Berkeley?

During World War II, I was technical director of a group outside of Washington, D.C. that developed devices to support intelligence and guerrilla activities behind the enemy lines, including time-delayed bombs. It was in the Maryland laboratory which was taken over by the Office of Strategic Services, or OSS, of which CIA grew out. Our most successful project was our train-wrecking devices, which were used by the French Underground during the Normandy landing of the Allies. They essentially tied up the railways in Northern France so the Germans had no use for the railways for some time there.

After the war I got back into basic science. Then in 1948, I was invited to be Director of Research of the Atomic Energy Commission (AEC). It was a newly established organization to take over the nuclear weapons program. I held this position from January 1949 for two-and-a-half years. I set up the Washington headquarters of the basic physical research organization. In addition to coordinating the former research laboratories of Oak Ridge, Argonne (outside Chicago), and Los Alamos, it was agreed that AEC should go much further in research support for individual Universities, not just in the National Laboratories. I was essentially the one who did that.

Early postwar picture of Kenneth Pitzer (courtesy of Glenn T. Seaborg and E. O. Lawrence Berkeley National Laboratory).

Did you come across Edward Teller?

Yes, but even much earlier than that. One of my earliest and major contributions was a study of the internal rotation of the ethane molecule, which was then very controversial. Edward had been one contributor to that. So we had a common background. Although what he did was correct, what finally settled the problem was our low-temperature heat capacity measurements.

I always enjoyed discussions with Edward. The big controversy that he was in, and I was on the same side, concerned the vigorous program to try to make the thermonuclear weapon. My position was that you don't defend the United States in this troublesome world by intentionally remaining in ignorance on an important subject. I'm sure, in retrospect, a lot of people would agree that that was right. Edward was much surer that it was feasible than most and he was going to be in the middle of it while I was just on the sidelines as it were.

The General Advisory Committee, which was dominated by Robert Oppenheimer, was opposed to it, and claimed that vigorous effort in this area was unnecessary, or ought not to be undertaken unless and until there had been an exploration with the Soviet Union as to whether both sides might forgo it. There is a big history of this.

I disagreed with that, and expressed my views that we ought at least to explore the science. Edward was much more outspoken. From hindsight, we were certainly right. What's coming out of Russia now is quite clear. They were going to go right ahead with that. We got there first but not by as much. It would have been quite a situation in this country if they had gotten there first and it would have been known that we had been intentionally slow about it.

Let's return to your career. When I was a student, we learned about "Pitzer strain."

That terminology was more common in Europe than it ever was in this country. The initial molecule was ethane, and the question was, could the two methyl groups rotate freely, one relative to the other? There was some theory that said that the restriction of rotation would be negligible, as compared to thermal energy at room temperature. There was also some indication that it might be substantial. It was a fortuitous situation for me, when I arrived as a new graduate student on the scene in 1935. After my four years at Cal Tech, I was more experienced than many of the typical first year graduate students. The measurements of heat capacity, heat of vaporization, heat of fusion, and so on, for ethane, were originally made by two would-be chemical engineers and they didn't know what do with them, and their advisor, Professor Giauque was not interested in them either. So one of them, J. D. Kemp, asked me to take a look at the situation, and I did, read the literature, including the paper by Teller. I decided to look into the entropy, and integrated the heat capacity divided by the temperature, with respect to the temperature. I got an unambiguous answer, in the vicinity of 3 kcal/mol for the rotational barrier. It was restricted rotation all right, so Kemp and I got out a Letter to the Editor and then a full paper about it. I pursued it for other hydrocarbons, initially open-chain hydrocarbons.

Please, tell us more about the beginnings of research of restricted internal rotation.

In the period 1930–1935, several papers were published about ethane and other light hydrocarbons. All assumed free internal rotation and most saw no inconsistency with experimental data. Only Teller and Topley in 1935 decided that there was a real problem for ethane and that a high barrier to internal rotation might be part of the solution. But they thought the high barrier so improbable that they did not pursue that option for all properties. It was only in my paper with Kemp that I showed the barrier of about 3 kcal per mole to be consistent with all of the other data for ethane. But it was the new entropy value that unambiguously required the high barrier.

I was aware that the same potential effecting torsion about single carbon-carbon bonds in saturated hydrocarbons applied to cyclopentane and cyclohexane. But World War II intervened and I didn't do anything about the ring molecules until after the War. Cyclopentane had been thought to be flat. In terms of the C–C–C bond angles around the five-membered ring, that's just right for tetrahedral carbon atoms, but that lines up all the hydrogens along the ring, eclipsed, which is the top of the potential barrier. So there would be strain, the Pitzer strain, in cyclopentane. In cyclohexane, the ring is puckered in order to fit the tetrahedral bond angles C–C–C. In the chair form all the hydrogens are staggered, and that is the potential minimum, and there is no Pitzer strain. In the boat form, however, there will be about 6 kcal/mol strain. This work was just after the War, in about 1946.

Did you know about the chair and boat forms?

That was known in the literature. Odd Hassel was dealing with substituted cyclohexanes. We worked this out simultaneously and independently, and largely not knowing about the other's work until it was fairly well along. Depending on what you take as a key thing, he was probably the first on some of it but there were certainly other aspects of it that I certainly was the first on. To be fair to all though, I'd have to refresh my memory on all this.

The one paper on the two rings I did just personally. Then I started having collaborators and we did a thorough treatment of both the five-membered and six-membered rings. It was all done by 1948 when I went to Washington. After the war, that was a very fruitful period as many bright young people returned to their studies that had been interrupted by the War. One of the brightest was George Pimentel in my group.

You worked on the book by G. N. Lewis.

The original book *Thermodynamics* was by Lewis and Randall, and it was Published in 1923. The book became a standard American text for many years, and was still going strong in the 1950s. However, it was also getting out of date and the publisher, McGraw-Hill invited me to do a revision.

Both of the original authors were dead by then. I asked Leo Brewer to be a coauthor and we did the revision and published it in 1961. The second edition did very well for 20+ years. During the preparation of the third edition, Brewer dropped out in part because of his wife's death, I believe. I finished the revision alone and the third edition appeared a few months ago. It is just beginning to be used.

You got the National Medal of Science. Was it for a particular achievement?

No. The National Medal of Science is not for a single thing.

It was a rather long time ago, in 1974.

Well, I was no Spring chicken in '74. I've had actually everything but the Nobel prize, which is rather chancy anyway. I'm not one who looks back and says, what if, although you can't help doing a little of it but I do very, very little. The Nobel prize rules are very restrictive. They can't split it just three ways. They can have one group of two who are collaborators and another one. If they had included me with Barton and Hassel, it would have been a third one, really separate, not collaborating with either of the others. I think that's forbidden by the Nobel rules. One could've argued that I could have shared it with Hassel and Barton could have got it for the more complex molecules later or something like that. But as I said, I spend no time on thinking about that.

In addition to the National Medal of Science, I received the Robert Welch Award, and in terms of being remunerative, at the time it wasn't much less than the Nobel prize and I got it all by myself. I also received the top American Chemical Society award, the Priestley Medal.

What turned you to chemistry originally?

I was inclined that way when I was a freshman at Caltech. When I arrived in 1931, the "old man," the founder of modern Caltech chemistry and modern Caltech, Arthur Noyes, wanted to keep his hands on science but didn't want to compete for graduate students with the younger Faculty. So he would get some undergraduates, including me, after the freshman year. He had a summer home down on the waterfront, a beautiful spot. There was a laboratory which was mainly for oceanographic research but there was also a chemistry lab. Noyes had me down there for some research on +2 and +3 silver compounds. It was quite exciting to get that into publishable new research and get yourself some credit for analytical chemistry at the same time. In any case, I was pretty well settled on chemistry through the influence of Arthur Noyes. Chemistry was also more practical than physics, which would have been the competing field for me, and, remember, this was Depression time. Later I did more undergraduate research, and at the end even a crystal structure determination with Linus Pauling. He even declined to put his name on the paper, so it was published under my name alone. He said, you publish it. I don't mean he refused to be on it. Rather, he said, basically it's a routine structure determination, an interesting one but routine. It's not going to contribute to my reputation any but it may call a little attention to you. So that was the last thing I did at Caltech before I came up to Berkeley. Later I emphasized statistical thermodynamics, if you define this term very broadly, but I have always retained my interest in structures. That even includes aqueous solutions. Although their structure is complicated there are overriding simplifications. I like to mention this because it involves my research director Dr. Latimer. There was a paper about 1936 that involves Latimer, another student of his, and myself on the energy and entropy of hydration of ions. Take an ion from the gas phase and put it in water. It's a very simple model for which Max Born gave the original theory for a charge in a dielectric medium, that is, water. You have to have a radius for the charge. By this time, Pauling had given radii of ions in crystals and if you put in the Pauling radius, you got the wrong answer. This wasn't surprising; after all the dielectric constant of water does not apply up to the Pauling radius. Furthermore, there was no reason to believe that the Pauling radius was right for solutions. It was derived from crystal structures. The question here was about the distance to the dipole of the water molecule. We decided to add an increment that was different for positive ions and different for negative ions. We had to get the right radius for the Born expression and the right energy for hydration. As for the positive ion, the electron pairs will come in right next to it, and for the negative ion, it'll be the proton that'll get close to the ions, all with the appropriate corrections. This worked quite well and the paper is still getting cited today. The approach was quiet general. In order to get a little higher precision, you complicate the model one way or another.

References

1. Pimentel, J. *The Chemical Intelligencer* **1996**, *2*(3), 53–58.
2. Pitzer, K. S., Bradley Moore, C. *The Chemical Intelligencer* **1996**, *2*(3), 58.

F. Sherwood Rowland, 1996 (photograph by I. Hargittai).

36

F. Sherwood Rowland

Sherwood Rowland (b. 1927, in Delaware, Ohio) is Professor of Chemistry at the University of California at Irvine. He shared the 1995 Nobel Prize in Chemistry with Paul J. Crutzen of the Max-Planck Institute for Chemistry in Mainz, Germany and with Mario J. Molina of MIT. They received the Nobel Prize "for their work in atmospheric chemistry, particularly concerning the formation and decomposition of ozone."

Sherwood Rowland received his elementary and high school education in Delaware, Ohio and graduated from high school at the age of 16. His science school teacher used to entrust him with operating the local volunteer weather station for short periods of time. He enrolled at Ohio Wesleyan University in Delaware, Ohio, in 1943. At 18, he interrupted his studies and enlisted in a Navy program, finally graduating in 1948. He went to graduate school at the Department of Chemistry of the University of Chicago, where his mentor was Willard F. Libby (Nobel laureate in chemistry, 1960). He took courses in physical chemistry from Harold Urey (Nobel laureate in chemistry, 1934) and Edward Teller; inorganic chemistry from Henry Taube (Nobel laureate in chemistry, 1983); radiochemistry from Libby; and nuclear physics from Maria Goeppert Mayer (Nobel laureate in physics, 1963) and Enrico Fermi (Nobel laureate in physics, 1938); among others. He finished his Ph.D. thesis in 1952 and started his academic career at Princeton University. He then worked at the University of Kansas between 1956–1964, and has been at Irvine since 1964.

Our conversation was recorded on February 22, 1996, in Professor Rowland's office on the Irvine campus, and it first appeared in *The Chemical Intelligencer*.*

Two years ago, I was attending a meeting on fluorine chemistry in Yokohama and there was a special session on chlorofluorocarbons. Officials flew in from Washington and London, and a government representative came from Tokyo. They gave beautiful speeches about all the regulations, but they only considered North America, Western Europe, and Japan. I was rather puzzled by this lack of concern for the rest of the world.

The chlorofluorocarbons are an adjunct of an affluent society. Take the automobile, for example. Automobiles are not very numerous in Third World countries. To have an air conditioner in the automobile is even more a sign of an affluent society. The aerosol propellants, hair sprays, deodorants, or the cleaning of microelectronics — all these products are driven by the First World. So if you control production in Germany, France, Italy, Japan, the United States, and in a few other countries, you have taken care of about 95 to 98% of all the production.

How did the original realization of the hazards of chlorofluorocarbons come about?

The philosophy of scientific research that I picked up from my Ph.D. supervisor, Willard Libby, famous for carbon-14 dating, was that the excitement and the fun in science comes from doing new things. There are two statements in the English language, "in the groove" and "in the rut," which have physically exactly the same meaning but one means that things are going very well and the other means that you have trapped yourself into doing the same thing over and over again. The boundary between getting into the groove of really understanding something, and then pounding on it too long, is not very sharp. The first time you do an experiment, you don't have quite the right experimental apparatus; the next time you improve it, and the third time you really get it done right. But the tenth time, you don't really have to do the experiment because you think you know what the result should be. If that's the case, then it's not much of an experiment anymore. So the problem is how you would

*This interview was originally published in *The Chemical Intelligencer* **1996**, *2*(4), 14–23

stay in the groove long enough and get out before you are in a rut. I made a conscious effort from the beginning of my career, to think about it every six or eight or ten years and strike off in a new direction without necessarily giving up what we were doing before. My original training is as a radiochemist. Radiochemistry for us meant looking at the chemical reactions of atoms that are produced in nuclear reactions — "hot atom" chemistry. Much of it turned out to be done with tritium atoms which are produced in thermal neutron reactions either with helium-3 or lithium-6. The tritium atoms that come out of those reactions have, in one case, 192 000 electron volts and in the other case, 2.7 million electron volts. So they have an enormous amount of energy, and nothing can stop such a tritium atom in a collision with a molecule. However, when the atom has lost most of its extra energy, a molecular collision can leave the tritium bonded to some other atoms in the form of a stable molecule. If you look at the tritium atom after that collision, you are going to see its last collision, the one out of which it formed a chemical bond. The big uncertainty is just how much energy the tritium atom still had when it entered that final collision. Because the atom frequently has a few electron volts of energy then, the name "hot atom" became attached to the process. Over a period of time we put these atoms in contact with various organic molecules and looked at the physical organic chemistry of those reactions. The first experiment was done with crystalline glucose, and we found that the energetic tritium atom was able to substitute for a hydrogen atom bonded to carbon in glucose, and did so with retention of configuration; i.e., it came out as radioactive glucose, and not some other isomeric sugar. This substitution reaction was previously unknown and is still essentially unexplored. You don't expect hydrogen atoms to substitute into a carbon skeleton, unless the hydrogen atom has a very large amount of energy. Later on, we did this with methane and showed that the activation energy threshold was about 1.5 electron volts for a tritium atom substituting into methane. If you react thermal hydrogen atoms with methane, what you'll see is that it pulls off an atom of hydrogen to form H_2. But if the atom has enough energy, it can also trigger a substitution reaction instead. My first tritium experiment was at Princeton. Then I moved to the University of Kansas in Lawrence. In the analysis, we hooked up a proportional counter to a gas chromatograph, and did radio gas chromatography. We measured the radioactive yields of the various species that would come out. That was going well. The first publication on the hot atom chemistry of tritium was in the mid-1950s. It was almost my first publication out of graduate school and it appeared in *Science*. As a result, the Atomic Energy Commission came to me and said, "We would like to support your work". My research support from the AEC started when I went to Kansas in 1956.

I was raised in Ohio and found the Midwest to my liking. I graduated with my Ph.D. in 1952 at the University of Chicago. I went to Princeton as an Instructor that fall, and was an Instructor for four years, and then went to the University of Kansas as an Assistant Professor in 1956. My situation was also complicated by the fact that when the AEC approached me in 1954, the department chairman at Princeton said no to AEC research support — he said I was too young to have independent support. Then the University of Kansas made me an offer and Princeton matched it — in fact, they bettered it. But I didn't like the negotiations — that they would do for you under pressure what they would not do for you on its own merit — and so I went to Kansas. Kansas had built specifically a lot of space for a radiochemist, and my AEC project started there.

Kansas was a very fertile place for me; I had a lot of good graduate students; we had a lot of research freedom. I stayed there for eight years, and left as a Full Professor in 1964 to come out here to Irvine. This is my 32nd year here at U.C.I. The campus didn't take students until 1965; my main job as founding chairman was to have a functioning department in one year.

In the early 1960s, I decided that it was time to do something different, so we went into photochemistry. We were taking advantage of our special capabilities in radio gas chromatography, using radioactive compounds in the experiments. First we took something that we believed was very well understood, and that was the reaction of methylene with hydrocarbons. It seemed to me that there would be some advantages in studying what happened when you labeled the photochemical fragments with isotopes so that you could see what could not be seen before. However, it turned out that methylene wasn't as well understood as we had first thought. When we looked into it, there were two review articles within a couple of years of each other in the early 1960s; one of which said that methylene was one of the most reactive substances anyone has ever dealt with, and the other one said that methylene was extremely inert. Both of those turned out to be correct. The first one was describing singlet methylene, and the second was describing triplet methylene. As it turned out, with nearly all of our methylene experiments, the most interesting result would not have been seen if the methylene were not radioactive, if we couldn't have traced it isotopically.

Then in the late sixties we started doing some radioactive chlorine and radioactive fluorine experiments. Our gaseous inert source of tritium was the neutron bombardment of helium-3. However, the nuclear reactions for creating radioactive chlorine or fluorine started with target atoms of the same element. Thermal neutrons react with chlorine-37 to make radioactive chlorine-38, and fast neutrons make radioactive fluorine-18 from stable fluorine-19. We were looking for inert sources to hold the stable chlorine-37 isotope, and one of them was monochlorotrifluoromethane, $CClF_3$. We knew it wasn't reactive with most thermal species. However, we actually found that an energetic chlorine atom could react and replace a fluorine atom or replace the chlorine atom, in each case leaving a radioactive molecule. Often the substitution process left the molecule sufficiently excited that it would decompose into fragments. But most of the radioactive chlorine atoms became thermalized.

At the same time, I had been chairman of the Irvine Chemistry Department from the time of my coming here in 1964, and in 1970 I decided that six years was enough, and I would return to full-time research. That autumn I went to a meeting in Salzburg in Austria considering the environmental applications of radioactivity. I didn't have any research of that kind at the time, and was looking to see if there was anything that might interest me. There were two outcomes from that meeting. One was that I shared a compartment in the train from Salzburg to Vienna after the meeting with a man from the Atomic Energy Commission, William Marlowe. His duties included organizing meetings for the AEC to cross-fertilize meteorology and chemistry. The AEC was, of course, interested in applications of isotopes, and this is why they were supporting my research. They also had radioactive fallout in the atmosphere, and that's probably why they were holding these meetings. They needed to know more about the atmosphere, and one way of doing that is to get more chemists interested in it. Marlowe found out two things. One was that I was being supported by the AEC and by that time, 1970, I was just going into my 15th year of AEC support. It started in 1956 and stopped in 1994, so it was long term and lasted 38 years. My support has survived two major changes in the Agency during that period, but the contract essentially went on and survived my transfer from Kansas to Irvine. It was an excellent support, and they gave me a lot of freedom.

Why did they stop it in 1994?

Probably because the Department of Energy is under different kinds of pressure than the AEC was. We submitted our next proposal; it got basically good reviews. It was a continuation proposal with some new aspects. However, the question of basic research and of what has to be supported is very unclear at DOE. In the end, they decided that our work was too long-range. By that time, DOE was no longer our principal support agency; our major funding was coming from NASA.

Returning to the consequences of that 1970 meeting in Salzburg, the other was that I came back with an idea of a new experiment, and it was the following. There was a lot of interest in mercury poisoning at that time, and we managed to get hold first of a swordfish that had been caught in 1946, and this was now 1971. So there was a 25-year-old swordfish. Our first reaction was that swordfish would make great historical samples. Everyone puts the fish up on the wall with that long bill and knows exactly where it was caught. We thought that the mercury would go with the calcium, so we'd just analyze the amount of mercury in the bill. However, it turns out that the mercury is in an organic form and doesn't go to the bill. We got a small swordfish, and the mercury wasn't in the bill but in the flesh. It doesn't concentrate in the bone, so the idea that we could just go to everybody's display and get a little bone didn't fly. But there had been then a swordfish caught in the San Diego area which had had a terrible accident when it was tiny. It had stabbed itself and it grew up with its sword bent back and buried in its own forehead. It was not an ordinary swordfish, and was not able to fish on its own. When they caught it, in 1946, they saved the head because it was so unusual. It was put into a preservation solution in a museum in San Francisco. We also obtained some samples of pipe fish, tiny things that were preserved at the same time. We found that the mercury concentration in the flesh of the particular swordfish was essentially the same, as people were recording for freshly-caught swordfish. Then we got some samples of tuna fish that had been saved at the Smithsonian Institution, in Washington, D.C. All those had the same levels of mercury as fresh tuna, about 0.3 parts per million. The oldest one was from 1878, and all of them had mercury. We published this in *Science* and said that the presence of mercury in these oceangoing fish was not a matter of contamination, it was a matter of working its way through the food chain, because the amount that was there a hundred years ago was about the same as it was now. That was the first scientific outcome of having gone to this meeting on radioactivity and the environment.

Another consequence was that Marlowe invited me to the next chemistry-meteorology workshop in Fort Lauderdale, Florida, in 1972. The second thing Marlowe had found out about me on the train was that I was interested in the atmosphere. I always thought that Libby's carbon-14 data was just a marvelous example of how you can use chemistry and physics to study a particular problem that can spread beyond the initial boundaries. As you know, Libby's discovery revolutionized archeology. But it all started in the atmosphere because carbon-14 was produced there by cosmic rays.

We had actually gotten an air sample when I was still at Princeton and spending my summers at the Brookhaven National Laboratory. We measured the tritium content of the atmospheric hydrogen from this sample, and submitted a short note to *Physical Review*. This journal at that time was edited at Brookhaven, so we just took it down the hall. Only one previous measurement of atmospheric tritium had then been made. The answer we got back was that if you withdraw that paper it'll be fine, but if you insist on publishing it, it'll be classified, and it still won't be published. That was 1953. Looking back, the key here was that if our measurement was correct, then you could ask the question, Why is there more tritium in the atmosphere in the beginning of the 1950s than there was in the late forties? The answer was that the U.S. was working on the hydrogen bomb, and had spilled some tritium into the atmosphere. This would be a clue to other people that there had been work going on in the U.S., involving a lot of tritium. Eventually we published our paper in 1961, when it made no difference because everyone knew that we and the Soviets had hydrogen bombs using tritium. The lesson was that if you get too close to something of military significance, you may not be able to get published.

At the 1972 meeting, there was a presentation by Les Machta who had been with the Health and Safety Laboratory of the AEC working on atmospheric effects, and he reported some experimental results that British scientist Jim Lovelock had made the year before. Lovelock was the first to detect chlorofluorocarbons in the atmosphere. Lovelock actually invented the electron capture detector for gas chromatography, an enormously sensitive technique, and put air through his gas chromatograph. He lived in Western Ireland, and detected chlorofluorocarbons in the air there. What I heard from Machta was that Lovelock had then traveled on a ship headed for Antarctica from England, and found trichlorofluoromethane in every air sample taken on board. Machta was talking about what a marvelous tracer it would be, because it's inert. From my point of view, of course, we had been working with tracer experiments all my life. We had been using the CFCs as an inert source, but we knew that they reacted if there was enough energy around and we knew that they could be photolyzed. A CFC molecule is inert, but not totally. This was sort of an idle thought, but it touched a number of things that we had done before.

So in 1973 when I was writing my annual proposal to the AEC as to what we would do during the next year, I had two parts to it. Part A was to go on doing the nuclear reactor kind of studies with chlorine and fluorine and so on. But there was also Part B, that we would branch out and would do something else, and one of those things was to look at the fate of the chlorofluorocarbons in the atmosphere. Mario Molina joined my research group later that year, and, as I would do with all postdoctoral fellows, I offered him a choice of problems. The CFCs in the atmosphere problem interested him because it was so different from what he had been doing before. But, of course, it was also very different from what I had been doing before.

This topic was just one page in the proposal and there was no mention in our proposal of stratospheric ozone. If you start looking at the fate of most compounds in the atmosphere, one of three things happens to them. Some absorb visible radiation leading to photolysis. For example, if you photolyse molecular chlorine, that's the end of it. It's a green gas, absorbs visible radiation, and decomposes. That takes about an hour in sunlight. Hydrogen chloride, on the other hand, is a transparent gas so it's not interacting with visible radiation. It doesn't interact with the near ultraviolet either, so it doesn't photolyse in the lower atmosphere. But it dissolves in rainwater. Other compounds don't photolyse, don't react with rainwater, but may be oxidized by hydroxyl radical, and this may take months or even years in the atmosphere. So the main possibilities for compounds in the atmosphere are photolysis, solubility, and oxidizability. The chlorofluorocarbons, however, don't absorb in the visible — they don't really absorb strongly until you get to the vacuum ultraviolet — they're really not soluble, and they don't oxidize. Nothing happens to them in the lower atmosphere. Molina and I did a calculation that the average life-time of trichlorofluoromethane would be between 40 and 80 years and its sink would be photolysis in the stratosphere. We had models from other people's work of exactly where the ozone was, and the wavelengths that can be absorbed by the CFCs are also absorbed partially by ozone and partially by molecular oxygen. You put into a model the absorption, the amount of O_2 and the amount of O_3, and you have to have some parameter for the vertical transport. When you get to about 30 km altitude, there is enough ultraviolet radiation at 220 nm and below that the molecules are falling apart pretty rapidly. That's what controls the lifetime in the atmosphere. During most of their time in the atmosphere, CFCs are not exposed to anything that would cause them to be destroyed. It's the transit time up to this dangerous level of 30 km that determines how long the average molecule lasts. So now we knew the fate of the CFCs.

Was this just a thought experiment?

It was basically a thought and calculation experiment by Molina and myself, except that we had to know the absorption cross sections of the CFCs in the range from about 180 to 250 nm. The published measurements stopped at 200 nm, right in the middle of the interesting region.

There was an obvious discrepancy in the publication about CFCs in the UV region. In the place where the cross section was changing rapidly, the graph shifted scale by a factor of 10. We looked at it and found out that the overlap of the left and right hand lines didn't correspond to a factor of 10, so there was obviously something wrong with one or both of the lines for cross section. So we measured the ultraviolet cross sections of the compounds down to 180 nm with a simple Cary spectrometer and then could do the complete calculations. That was the only new laboratory measurement that went into it. We reached that point about not more than two months after we started. At this point, we knew more about what's going to happen to these compounds than anyone else did. We could have published these findings then, but we had a loose end — the decomposition of CCl_3F produced a free chlorine atom. First, we wanted to find out what happens to the chlorine atom. That's what put the fat in the fire, because when you ask what's going to happen to the chlorine atom, you find that it overwhelmingly reacts with ozone to give chlorine oxide. Then you ask, what's going to happen to chlorine oxide?

There are two reactions. One of them is chlorine oxide plus an oxygen atom, which gives you back the chlorine atom and O_2. Usually that oxygen atom would have found an oxygen molecule and made ozone. Instead, it is intercepted by chlorine oxide, and out comes an O_2, not O_3, and the chlorine atom reappears. So you have a chain reaction. Chlorine, chorine oxide, back and forth and $O + O_3$ give two oxygen molecules, and you hadn't gotten rid of the chlorine. Then we put in all the other reactions that you think could happen up there. Chlorine can react with methane, but it's about a thousand times less likely than chlorine plus ozone. Having reacted with methane, the chlorine atom is tied up as HCl, and for a moment that looks like the end of the road. But then hydroxyl radical reacts with HCl and gives you the chlorine atom back, and the chain starts all over again. You end up with chlorine existing as Cl and ClO in the chain reaction, or in reservoir molecules such as HCl, HOCl or chlorine nitrate, $ClONO_2$. Plug that into our calculation of life-times, where the concentrations that we were dealing with now were steady state concentrations, and you could see far more CFCs in the future atmosphere and much more Cl and ClO. Multiply those concentrations, and we realized that we were looking at an ozone removal process that was dominant over the processes in the normal stratosphere. That changed the whole exercise from scientifically interesting into a serious environmental problem.

Did it strike you as such at once?

Yes. First, of course, we thought that we must have made some serious mistake, because the calculations indicated that if we go on releasing the CFCs, the amount of ozone is going to be severely depleted. That didn't seem to make any sense. Where could we have made an error of a factor of a thousand? So we went back and looked at it in different ways. Part of this initial response had to do with the instinctive reaction of a chemist dealing with laboratory experiments in which the concentration of free radicals relative to the other molecules is always very low. There were always lots of other molecules and only a few free radicals in our radiation chemistry experiments in the 1960s. In the stratosphere, however, the free radicals could be 15, 20, 30%. It's a photochemical system, putting a lot of energy in, molecules keep dissociating, and you end up with a lot of free radicals. Once you realize that that's the characteristic of the stratosphere, then you know that your instinct was playing false by extrapolating from the usual laboratory conditions. Now laser chemists have the same situation with high power inputs creating many free radicals simultaneously.

When did it finally hit you that this was indeed a global environmental problem?

Mid-December of 1973.

What did you do then with that knowledge?

The first thing we did was to check what was known about it. Remember, neither of us was then an atmospheric chemist and we were working entirely outside that community. So I decided to go to somebody who was in touch with the atmospheric community and find out what's going on out there. I called Hal Johnston, Professor of Chemistry at Berkeley, and Mario and I went up to see him between Christmas and New Year of 1973. We told him about our work and he told us about the work of Ralph Cicerone and Richard Stolarski, who had been coming from the standpoint of hydrogen chloride from volcanoes and chlorine released from the rocket fuel of the space shuttle. They had found the chlorine chain reaction that we were just talking about, but they had decided that neither volcanoes nor the space shuttle were major chlorine sources for the atmosphere. Stolarski had presented a talk about it at an atmospheric science meeting in Kyoto in September of '73, and they had submitted a publication to *Science* that had been rejected. At that time both Cicerone and Stolarski were at the University of Michigan, and Johnston had seen their preprint. Both of them are Molina's age. They were soft-money people working in one of the departments. Cicerone was trained as an electrical engineer and Stolarski as a physicist. Now, both are very well-known senior atmospheric scientists.

At that point we had confirmation that the chlorine chain reaction was already known, but the whole business involving the CFCs had not come up with anyone. We thus decided that it was time to write our paper. The complication was that I was scheduled to go to Vienna for a sabbatical in early January to spend six months to think about doing something new. So I was one week away from going to find something new and this bombshell had just exploded in front of us. My wife and I talked about canceling the trip but decided, no, I would go ahead and go. Molina and I would stay in contact and I would be talking to people in Europe and he would be in the U.S.

The first three things that I did when I got to Vienna was to find an apartment for my wife and myself, to learn about the tennis team of the International Atomic Agency and that I would be eligible to play on it, and to write the paper that was sent to *Nature* in the third week of January. It took about five months for our paper to get published. There were several reasons. One of them was that finding a referee who was a competent photochemist for the CFC part of it and a competent atmospheric scientist for the other part of it proved difficult. Another complication later was that the person who was handling the manuscript for *Nature* went away for Switzerland one weekend and just never came back. He quit without notice, leaving his desk in a mess. I called *Nature* in early May to find out what's going on, and they told me that they simply didn't know where the manuscript was, let alone its status. In the meantime, Molina and I had submitted an abstract for a meeting in San

Sherwood Rowland and Mario Molina at the time of their working together at the Department of Chemistry, University of California at Irvine (courtesy of Sherwood Rowland).

Diego, California, scheduled for the first week of July, 1974. The people of *Nature* made sure to have our paper out before that. It appeared on June 28, 1974, my 47th birthday, with the title "Stratospheric sink for chlorofluoromethanes: chlorine atom-catalyzed destruction of ozone".[1] By that time, my sabbatical in Vienna was over and I was back in Irvine. We knew the paper was coming out and we arranged for a press conference here at U. C. Irvine. This was then a very small university, and we had not very much in the way of contacts. Nothing we had done before had ever been newsworthy at all, except the mercury, and one of my colleagues had talked about that. So the *Los Angeles Times* published a story about our work, but only in its Orange county edition, and nothing from here went outside Southern California. The English news agency Reuters published a story but didn't identify us by name and didn't identify us as being on the Irvine campus of the University of California, so the story had no follow-up. We thought it was an important article, and it had just absolutely zero news coverage for the rest of the summer. Actually not zero, science writers for newspapers in Toronto and Toledo, Ohio, wrote features on it, but no one followed.

By that time, Mario and I had spent all of that summer writing a much longer, much more complete document. We knew far more by the time the *Nature* article came out than we had in the publication, and the question was how we could get that information out. A 150-page review article normally takes a year or more to get published. So I labeled this review as an Atomic Energy Commission Report and sent it off to the AEC. This didn't require any refereeing. It also wasn't an AEC requirement — it was just a procedure that I made up to get our knowledge into print. Then we gave out hundreds of copies to everyone. There was an American Chemical Society meeting in the second week of September, 1974, and we distributed copies of it there too. ACS had a press conference on the work of several scientists of whom I was one, and the Associated Press story from that went to 400 newspapers with a total circulation of 100 million people. Immediately after that, Cicerone and Stolarski, to whom we had sent preprints back in February, and who had then done some calculations, published their work in *Science*. And the Harvard group to whom we had also sent our AEC report came out with their results, and CFCs and ozone became a big media story. Our long review article was also published then, exceptionally quickly, coming out in January 1975, "Chlorofluoromethanes in the Environment".[2] This was basically just the AEC report with some revisions.

Huge poster on the wall of the Department of Chemistry at Irvine, announcing Sherwood Rowland's Nobel Prize (photograph by I. Hargittai).

Do you suppose that in addition to the scientific discovery, politics has also played a role in your Nobel prize?

I think there is an element of scientific politics in it, but I don't think the awarding of the prize involved nonscientific politics. The question of whether or not this kind of work is fundamental chemistry, or whether it belongs in geosciences or meteorology, has always been there. To us it was always chemistry, but from 1974 until 1988 I did not get very many invitations to speak at chemistry departments. I'd get them from other university departments: toxicology, geology, physics. Part of this circumstance is that there is a very strong feeling within most academic chemistry departments that this kind of work is not chemistry; rather, it is some sort of application only, because it's outside the laboratory. Another aspect is that chemistry has had a strong industrial component since the 1920s, and until 1988, we were directly in opposition to some of the major chemical companies, such as DuPont and Allied. In the 1974 time frame, two thirds of the CFC production was in the form of propellants for aerosol sprays in the United States. About half of the global CFC use was then in the United States. The aerosol spray propellants were mostly chlorofluorocarbons. Some used hydrocarbons, but probably 80% of the aerosol industry was in the front line immediately, and they came out swinging. In the American Western movies, the good cowboys always wear white hats and the bad guys always wear black hats so the little kids know whom to cheer for. We had black hats, as black as you could get. Every month Molina and I would read their publication, *Aerosol Age,* to find out what else they were saying about us. One example that illustrates to me how far they would go was when they had an interview with a person who speculated that we were agents of the KGB, intent on disorganizing American industry.

How much did this hurt you?

It probably interfered seriously with my research group in the sense that the postdoctorals I had during this time all came from overseas because I was no longer getting any applications from Americans just finishing their degrees. How people choose where they are going to work as a postdoc has a lot of things feeding into it. Irvine was a small campus, not very well known; atmospheric chemistry was a doubtful area to many chemists; our work was controversial and being ridiculed by industry, so coming to work here was not what the mainstream chemist would tell his graduate student to do. Molina and the others who were in the group at that time were my last mainstream postdocs from major American schools for the next 15 years. But I had a lot of excellent Japanese and European postdoctorals. The quality of research in my group was quite satisfactory to me, but the applications came from abroad. To some extent I was an outcast in my profession, but I hadn't been cast so far out that it affected my life very much, and the funding of my research was not affected either.

Was Paul Crutzen's work known to you?

Not when we started. His findings on nitrogen oxides were important and he had done modeling calculations on the happenings in the stratosphere. He and Hal Johnston also had difficulties with industry because NO and NO_2 are introduced into the atmosphere by supersonic aircraft. It's not the speed of the supersonic aircraft that raised the problem but that they are designed to fly at high altitudes. This concerned the Concorde in Europe and the proposed supersonic Boeing in the U.S. But Congress failed to put U.S. money into supersonic aircraft for a different reason, mainly for economics. Hal Johnston had raised the question about SSTs putting NO_x into the stratosphere back in 1971, and this was one of the reasons I went to Johnston when we wanted more information about current stratospheric studies.

Now you are a hero of environmental protection. Were you concerned about the environment before all this? How did you view environmental problems before your own involvement in 1973–74?

If you have a bipolar world and the environmentalist is at one end and the industrial polluter at the other, I was certainly on the environmentalist side, but not active in any environmental movement at all. Of course, when we came out with our mercury in swordfish findings, for a while we were looked at as antienvironmentalists.

Let me also add that the first Earth Day was in 1970, and the members of my family were very interested in that. Probably I was influenced by them in thinking about the environment as a potential area when deciding what I might want to do next. We are a close family; my children were then teenagers, and they had become very interested in the environmental movement.

When I was a graduate student, our basic attitude was that if you get out of the laboratory alive, then the hazards are over for today. In other words, the only hazards you have are acute hazards. Most of the time, of course, you don't think of experiments as hazardous or dangerous — you just do them. I worked for 3 years as a graduate student with intense lachrymators — alkyl bromides — in a lab without a hood and had to start wearing glasses because I would get headaches behind my right eye from reading. But my eyes recovered in a few years, and I gave up the glasses until age came into play. But the idea that there may be long-time hazards, especially outside the laboratory, was not very prominent. The hazards of contamination of the rivers and the atmosphere were not on our mind. Some of the changing attitude has to do with the sensitivity of our detection schemes. If the sensitivity for measurement were one part in a million, then no one yet would have discovered chlorofluorocarbons in the atmosphere because they aren't even up to a part in a billion level (U.S. billion, 10^9) and they aren't going to get to it, because of the limitations by the Montreal Protocol. If you can measure to a part in a trillion, then you can see CFCs all very clearly. Even if your measurements are good to one part in a billion, you wouldn't find out about the ozone problem from CFC measurements — you would notice first the mysterious disappearance of ozone over the Antarctic.

I remember a conversation back in 1958 with a former Princeton graduate student who went to work for one of the chemical companies. He told me that he was looking for trace species left over in a process. I asked, why don't you use radioactivity because it's so much more sensitive? His answer was that we don't use it because if we did, we would find some. But if we use our present technique and don't find any, then we won't have to do anything to fix it. That attitude was essentially saying nothing really bad actually happens from chemistry. That attitude has changed because we have seen that very negative things do actually happen. On balance, I believe that advances in chemistry have been very good for society, but not all of it has been positive.

Coming back to our work, I can divide the past quarter century into several time periods. The first one is when Molina and I knew about the CFC problem, but no one else did except a few other scientists. That lasted from late 1973 until September of 1974. Then, for about two years, only a few people spoke out about it, chiefly Molina, Cicerone and me. Most scientists tended to say, I know they're talking about it, but I haven't heard anybody unbiased speaking about it. That ended in September of 1976 when the first National Research Council report came out and said that the science behind this was very well founded.

After all the difficulties, initial disbelief, and even accusations, today you are a Nobel laureate. Do you feel elated?

Obviously, any scientist is elated by recognition with a Nobel prize. I wouldn't describe it as a goal, but rather the feeling that if you ever did anything really significant, such recognition was possible. One would expect that the judgment of the Swedes would persuade some people that what we did was a good thing, and in fact that is the reaction of the overwhelming majority. However, in the U.S. it is just as controversial now, but essentially outside the scientific community. The backlash started in 1990. Legislation is being introduced, just as we speak, in several states permitting the use of CFCs. The immediate right wing political reaction to the Nobel prize was that the Swedes must really be in trouble for environmental politics to dominate things so much. The immediate assertion was, what a gross error the Swedes had made; there must be serious political problems in their country. A person from our Environmental Protection Agency asked me about a year ago, have you written anything up recently? I said, no, but the latest document of the World Meteorological Organization and NASA is out with all these details. And he said, the people that I'm trying to provide answers for don't want to hear that the World Meteorological Organization has said anything because the WMO is part of the United Nations, and, of course, they would be saying this because that is part of the takeover of the U.S. Government by the United Nations. So we are living in a very interesting period.

References

1. Molina, M. J., Rowland, F. S. *Nature*, **1974**, *249*, 810–812.
2. Rowland, F. S., Molina, M. J. *Reviews of Geophysics and Space Physics*, **1975**, *13*, 1–35.

N. N. Semenov (photographer unknown, courtesy of Lev V. Vilkov, Professor of Moscow State University).

37

Nikolai N. Semenov

Nikolai Nikolayevich Semenov (1896–1986) was Professor of Moscow State University and long time member of the then Academy of Sciences of the USSR. He had also founded the Institute of Chemical Physics of the Academy of Sciences of the USSR and served as its first director. He was the co-recipient of the 1956 Nobel Prize in Chemistry with C. N. Hinshelwood of Oxford "for their researches into the mechanism of chemical reactions."

I recorded an interview with Professor Semenov in the middle of September, 1965, at the request of Radio Budapest. The occasion of his Budapest visit was to receive an honorary doctorate from the Budapest Technical University. The conversation took place in Professor Semenov's suite in Hotel Royal. The original recording was in Russian. I then translated the conversation into Hungarian and it was broadcast by Radio Budapest shortly afterwards. The program was judged as very successful and the conversation made its way into a volume, *Rádió és Televízió Évkönyv*.[1] The present English version of the conversation is based on the Hungarian text. Thus the choice of words may have been influenced by the double transformation. I hope, however, that the meaning and flavor of what Professor Semenov had to say have been preserved. It was my first interview ever, and I still remember his kindness, his awareness that the program was meant for a broad, non-specialist audience, and his dedication. It is especially interesting to read today about Semenov's prognosticating, more than three decades ago, the directions of scientific research.

How did you start in science?

I have to refresh very old memories to answer this question. I was still in school, contracted typhus, and could not attend school for quite a while. Since I didn't want to be left behind, I read a lot of books. My favorites were all chemistry books. When I got well, I got to the nearest drug store and bought all the available chemicals there and started experimenting with them. For me it was the greatest puzzle that sodium, this flammable and malleable metal, and chlorine, this extremely reactive gas, formed the innocent table salt. To check this out, I bought a piece of sodium, burned it in chlorine gas, and re-crystallized the precipitate. It was a white powder which I poured over a big slice of bread and it was table salt indeed, the best kind. Although I was not aware of it, I hit right on one of the fundamental and at that time yet unsolved problems of science. It became possible to be solved only many years later, using the electron theory and the atomic arrangements in molecules.

My interest in chemistry kept deepening until I read in a book that the future of chemistry was in physics and for a good chemist it is mandatory to know physics very well. Since my goal was to become a good chemist indeed, I signed up for the Faculty of mathematics and physics of the university. I was enormously impressed there by Academician Ioffe and started working for him from my sophomore year. Our main interest was the electron-molecule collisions. We were very excited by Niels Bohr's new theory which explained so many things.

When I graduated from the University, I immediately became Ioffe's assistant. At the same time there were tremendous changes in our country. There was a revolution, civil war was on its way, and so was foreign intervention. Many thought at that time that under the circumstances scientific research would be suspended. Due to Lenin's foresight, however, things took a different turn for science. He convened the scientists scattered all over the country, created conditions for independent work, and secured the necessary funds for international monographs and journals.

This was a heroic era indeed. It was moving and uplifting to see the thirst for science of our impoverished, tortured, and liberated people. We received letters from the most remote corners of the country. If somebody read or created something of interest, he let us know at once. The scientific institutes themselves turned often to the public for recruiting new coworkers. Although we could hardly pay them, people were joining us *en mass*. Let me just mention one example, Cherkov, now a Professor, who had abandoned animal husbandry for science, and who was not alone. Everybody was willing and ready, even if living merely on bread and water, to participate in creating the new science. It is also true, though, that the young researchers were given greater independence than is customary today. But we were all very young. Even Ioffe, the oldest among us, was not yet forty.

I was put in charge of an independent laboratory in 1920, and was already directing 60 coworkers by the early thirties. Progress was breathtaking and new institutes were mushrooming. Among them Pavlov's physiological institute, Ioffe's physical institute, institutes for radio-physics, radioactivity, and others. The technical physics institute was very strong and I was in charge of its Department of Physical Chemistry. The reactions abroad were interesting. For quite a while they could not appreciate what was going on in our country. Only when the spectacular results appeared, the atomic bomb, the atomic power station, the first Sputnik, only then did we finally receive their recognition.

Three generations of a school on stamps: A. F. Ioffe (1880–1960), Semenov's research adviser (Soviet stamp, 1980); N. N. Semenov (Russian stamp, 1996); and N. M. Emanuel' (1915–1984), the most famous of Semenov's pupils (Soviet stamp, 1985).

However, it was a long, very long way before we reached that destination. In the 1920s, England and Germany were the great scientific powers. Our Soviet science was, to some extent, provincial, in spite of the many excellent scientists. American science was in a similar situation at that time. Today Soviet and American science are at the top. Without belittling the American achievements, their development was, no doubt, greatly accelerated by the tremendous import of outstanding European scientists. We, on the other hand, had to do everything ourselves but, I think, we have solved our task with great success.

Soviet picture postcard from 1953 displaying "Stalin Prize winner academician, N. N. Semenov (physicist)", by the artist, A. M. Gritsai. Courtesy of Lev V. Vilkov, Professor of Moscow State University.

Please, tell us about your research, especially the work that eventually led to the Nobel Prize.

During one of our experiments we made an interesting observation. The phenomenon of luminescence of phosphorus, that it emits light in oxygen, is well known. We were interested in the changes of this light emitting ability of phosphorus with decreasing pressure. It was an unexpected experience that at a certain low pressure phosphorus ceased to emit light, and its oxidation reaction stopped. There was thus a lower pressure limit below which the reaction would not go. Curiously, there was a similar upper limit of pressure as well, above which the oxidation reaction would not go either. This was strange since all the conditions, including the composition of the reaction mixture, were present and yet, the reaction would not go.

We duly published our observations, and soon enough received a letter from one of the best known physical chemists of that time, Bodenstein of Berlin. He thought that our observation was a result of the imperfection of our experimental apparatus. It was obvious that we had to repeat our experiments, but our new results only confirmed our previous observations. In the continuation of this research, we were varying other conditions to monitor the changes in the phosphorus luminescence under the impact of these other conditions. Thus, for example, we added some inert gas to the reaction mixture and observed that the luminescence reappeared above a certain amount of inert gas. The same thing happened when we enlarged the volume of the container in which the reaction was taking place. As a result of many similar experiments, we finally came to an understanding of this phenomenon in terms of branching chain reactions.

Before, however, I tell you about them, I need to tell you something about the chain reactions without branching.

The chain reactions without branching were discovered by Bodenstein in photochemistry. In order to start a chain reaction, there must be an active center in the reaction volume, created by light or heat. This active center then reacts with another particle. If there is a mixture of chlorine and hydrogen gas, one of the molecules may dissociate under the impact of some external influence. Let's thus suppose that one of the chlorine molecules dissociates, resulting in two chlorine atoms. When one of these chlorine atoms collides with a hydrogen molecule, a hydrogen chloride molecule forms plus a free hydrogen atom. At this point, the hydrogen atom carries on the reaction and it continues as a chain until two hydrogen atoms, or two chlorine atoms, meet and form a H_2 or a Cl_2 molecule which means the end of that particular chain. Thus there is a distinct beginning and a distinct end of this chain reaction. The oxidation of phosphorus is a much more complex process.

In the process of the oxidation of phosphorus, active centers can form not only under the impact of external light or heat but also by the large energy liberated by the collision of two particles. The excess energy can break an oxygen molecule, producing two oxygen atoms, and every oxygen atom is the beginning of a new chain. Thus there are branches of the chain giving rise to newer and newer branches and all these are multiplied, resulting in an avalanche of chains. This process is accompanied by light and often even by explosion. These reactions are the branched chain reactions. If the pressure is very low, there will be very few branches before the active particle reaches the wall and the reaction stops. On the other hand, with increasing pressure, larger container, addition of an inert gas, the probability of having enough branches to occur increases and the reaction continues rather than terminating the chain. This mechanism provided an explanation for the observation that had been a puzzle for a long time.

Eventually, we also determined that this mechanism characterizes many other chemical reactions, such as polymerization, cracking hydrocarbons, and combustion. Our work has

Cyril N. Hinshelwood (1897–1967). Photograph courtesy of Keith J. Laidler, Professor Emeritus, University of Ottawa.

generated a lot of interest all over the world, and in particular in England, where Professor Hinshelwood later found an explanation of the reaction stoppage at the high pressure limit as well. The branched chain reactions are still being studied by many scientists in different places.

Which research directions do you find the most promising in your field, as we are looking at the development of science in the mid-1960s.

Physical chemistry has been around for a long time in that physics has been applied in chemistry in the most fruitful ways. Earlier it meant primarily the application of classical physics, especially thermodynamics. Thus we have seen such interaction for quite some time. It is also well known that there was a revolution in physics from the beginning of this century. Following the discovery of the electron and the quanta, great strides have been made into the structure of the atom, and quantum mechanics has been created. All this has made a strong impact in chemistry. Scientists have uncovered the secrets of the nature of the chemical bond. I could finally understand the formation of the innocent table salt from two aggressive materials, sodium and chlorine. Considerable work has gone into another area of physical chemistry, dealing with the problems of the speed of chemical reactions, that is, chemical kinetics. There has been an increasing demand to find out more about the interactions between molecules and electrons and between molecules and quanta. New and powerful physical techniques have been introduced to investigate the structure of molecules. An increasing number of scientists have addressed themselves to these problems. We named this area first "electron chemistry" and we even published a book under this title. However, in Germany the name "chemical physics" had been adopted and this is what has become accepted all over the world. There are many institutes of chemical physics, including the Institute of Chemical Physics of the Academy of Sciences of the USSR, and I have been its director from its foundation. From the very beginning, we have built very good relations with foreign institutions and scientists, including Bodenstein who, after our second publication, accepted our findings.

It is increasingly difficult to delineate the various sciences, to draw the border lines between physics and chemistry or between chemistry and biology, and who needs these borderlines anyway? I remember we used to say about the difference between physics and chemistry that physics deals with dirty materials but operates with clean techniques, whereas chemistry deals with clean materials but operates with dirty techniques. This would not even stand as a joke today because physicists very often use super-pure substances,

Soviet stamp (1981) commemorating the 50th Anniversary of the Institute of Chemical Physics of the Academy of Sciences of the USSR, whose director, Semenov, served to the end of his life. Today it is the N. N. Semenov Institute of Chemical Physics of the Russian Academy of Sciences.

for example, in atomic energy research and in radio-techniques, whereas there is an ever broadening application of physical techniques in chemistry. Thus it is increasingly difficult to find a sharp dividing line between physics and chemistry, except, perhaps, in high-energy physics, but even there a periodic system of the elementary particles seems emerging not unlike the periodic system of the elements.

Let's consider another field, biology. It is mandatory to get to know in great detail all those substances that determine the various phenomena related to life. It is no longer possible to carry out serious research in biology without chemistry and without the participation of chemists. The living matter is highly developed not only from the point of view of its structure but also from the point of view of chemistry. The biological structure is a higher order chemical and physical structure. This is the area where we should be expecting the next chemical revolution. In the 1820s, Wöhler showed the possibility of making organic substances from inorganic materials when he synthesized carbamide. Before Wöhler, the ability to create organic substances was attributed to some vital force. This discovery gave a tremendous impulse to the development of organic chemistry, leading to the synthesis of plastics and many other entirely new classes of compounds.

However, the investigation of the processes in the organism is a highly complex task. The chemical reactions in the living matter happen at room temperature, whereas in industry often extreme conditions are needed to maintain them. These processes consist of large series of small steps each of which need large activation energies. The synthesis of proteins is a good example of how different the reactions in the living organism may be. Insulin has been produced in an extremely lengthy procedure in the laboratory, but it takes only a few minutes in the organism. The high speed is made possible by the catalysts operating in the organism. They are called enzymes. We should learn about them and when we do, it will mean the beginning of another industrial revolution of unprecedented magnitude.

Let me conclude with an example. There is much effort going on in utilizing solar energy. However, the efficiency of the various solutions pales in comparison with the efficiency of the photosynthesis in the living organism in which solar energy also plays a decisive role. If we could only learn how nature does photosynthesis in the living organism, it would lead us to high-efficiency utilization of solar energy. Uncovering the secrets of biological structure would at least as much facilitate progress in other fields as in biology.

Taking very distant perspectives, it seems to me that there are two most important directions in scientific research. One is the elementary particles. In addition to working out their classification, another important task is to uncover the internal relationships of the enormous number of elementary particles. This would lead us into the details of the fundamental organization of matter. The other direction is the investigation of matter of higher organization — crystals, semiconductors, and so on — and in the final account, the investigation of the most organized of all, the living matter.

***

A few biographical data on N. N. Semenov on the basis of the introductory article by A. E. Shilov in *Nikolai Nikolayevich Semenov 1896–1986* (in Russian), compiled by L. V. Shit'ko and A. F. Abashkina. Nauka: Moscow, 1990.

1896	Nikolai Semenov was born in Saratov. His father was a civil servant. While in high school he started doing chemical experiments at home. He realized soon that the future of chemical theory lay in the advancement of physics.
1913–1917	He studied in the Faculty of Physics and Mathematics of St. Petersburg University, graduating in 1917. A. F. Ioffe was a great influence on Semenov and his starting of his research career.
1916	His first papers were published in the *Journal of the Russian Physical Chemical Society*, on the collision of slow electrons with molecules.
1918–1920	He taught in Tomsk; then, from 1920 at the Leningrad Polytech, where he became Professor in 1928.

1920 He got his first research appointment, eventually becoming director of the Institute of Chemical Physics of the Soviet Academy of Sciences, a position he held until his death. The Institute, a huge research organization, moved to Moscow in 1943.

The branched chain reactions were discovered by his co-workers in his laboratory in 1926–27, and he developed the theory of chain reactions in 1928–29. Chain reactions and reaction kinetics have remained his main interest throughout his life.

1929 Semenov was elected Corresponding Member of the Soviet Academy of Sciences and in 1932, incredibly young, he became Full Member of the Academy.

He was Member of the Commission of Defense of Leningrad during its siege in World War II.

1944 He organized a new chair of chemical kinetics at Moscow State University and remained Professor and Chair to the end of his life. [A chair, like Semenov's, in the Chemistry Department of Moscow State University could, by itself, be as large or considerably larger than a chemistry department of any American university.]

1947 Semenov joined the Communist Party of the Soviet Union. Throughout his life, he held a variety of positions, primarily in the Science Academy.

1956 Nobel Prize in Chemistry, shared with Sir C. N. Hinshelwood of Oxford, U.K., "for their researches into the mechanism of chemical reactions."

1958 Foreign Member of the Royal Society (London).

1963 Foreign Member of the National Academy of Sciences (USA).

1978 Foreign Member of the French Academy of Sciences. Semenov was elected to many other academies and received many honorary doctorates.

1986 Nikolai Semenov died in Moscow.

References

1. B. Lévai, Ed., *Radio and Television Yearbook*, Magvetö: Budapest, 1967.

George Porter, 1997 (photograph by I. Hargittai).

38

George Porter

Baron George Porter of Luddenham (b. 1920 in England) is Professor and Chairman of the Centre for Photomolecular Sciences, Imperial College of Science, Technology and Medicine in London, U.K. He received the Nobel Prize for Chemistry in 1967 with R. G. W. Norrish, Cambridge, the other half going to Manfred Eigen, Göttingen, Germany, "for their studies of extremely fast chemical reactions, effected by disturbing the equilibrium by means of very short pulses of energy".

Professor Porter is Fellow of The Royal Society (London) and served as its President from 1985–1990. He is Foreign Associate of the National Academy of Sciences; Foreign Member of the USSR Academy of Sciences; Member of Pontifical Academy; the Japan Academy; the Academia Europaea; and many other learned societies. His many distinctions include the following medals of The Royal Society: Davy (1971), Rumford (1971), Michael Faraday (1991), Copley (1992). Since 1989, he has been in the Order of Merit of the U.K., which is restricted to 24 members. He was knighted in 1972 and was created Baron (Life Peer) in 1990. We recorded our conversation in Professor Porter's office at Imperial College in London on September 11, 1997. The interview has appeared in *The Chemical Intelligencer.**

*This interview was originally published in *The Chemical Intelligencer* **1998**, *4*(4), 30–36 © 1998, Springer-Verlag, New York, Inc.

For the first time in my life, I'm having a conversation with an English Lord. However, you were not born a Lord.

Good Lord, no! In that case I would be a different kind of Lord, an hereditary Lord and they are probably going to dispense with the hereditary Lords within a year or two. But I am just an ordinary Lord, sort of elected. It's fun and its committees, like the Select Committee on Science and Technology, are interesting.

Were you present in the House of Lords on December 10, 1991, when buckminsterfullerene was discussed?

Yes, the question was asked about its possible use, and nobody was sure and then Lord Russel said, "My Lords, can one say that it does nothing in particular and does it very well?" quoting the words of Gilbert and Sullivan about the House of Lords itself.

There are lots of knowledgeable people in the House of Lords. Some may be unknowledgeable about science but very knowledgeable about other things. In any case, they are streets ahead of the House of Commons. Famous musicians, authors, scientists, diplomats along with the politicians — you'll find them represented in the House of Lords.

What was your family background?

I was born and brought up in Yorkshire. It was a poor world at that time, a time of depression. Ours was a mining area, mainly unemployed when I was a child. I went to an ordinary tin hut school. Some of my little friends there were going around without shoes and socks. My father was a builder and we were a bit better off, but not well enough for me to go to public school [i.e., private school]. But I went to a grammar school and our grammar schools in England were very good, until the Labour Government, just after WWII, destroyed them on egalitarian grounds.

My father was also a Methodist lay preacher. This background had an absolutely negative influence on me. I went to chapel every Sunday as well as several weekdays. But I became a confirmed atheist by the time I was about 13. I started doing science when I was 10 and developed a scientifically inquiring mind. I asked awkward questions, and I read Thomas Paine and many other people who were questioning religion. Fortunately, it didn't cause much conflict with my family because sensibly I didn't really discuss it with them. I just don't like upsetting people. If their belief makes them happy, the best thing is to keep quiet about it.

But I don't agree with people who say that science and religion are totally different things and that they can exist side by side. Of course they can't. Religion is a dogma. You have to believe, you're not supposed to ask questions, which is exactly what science is about.

You started becoming very successful very early. Did your parents witness this?

Unfortunately not. My father died when I was 19. He died as a result of a chest complaint, arising from digging air raid shelters in our house in 1939. I was doing quite well, although my school was not high level academically. Nobody from my school ever went to Oxford or Cambridge. Most of my friends left school to work and support their families. My father saw me go to Leeds University at the age of 17, and he was thrilled about it. I graduated with a Chemistry degree at 20. Had he lived a few more years, he would have seen a few other things that he would have been even more thrilled about.

How did you decide to become a chemist?

It was at the age of 9 or 10 that I was given a chemistry set. I'm a great believer in chemistry sets; they used to be terrifically exciting. It was something you could do by yourself in the kitchen, which I did. There was the magic of colour changes and bangs from our hand-made fireworks, on November 5, Guy Fawkes day (the fellow who tried to blow up Parliament). We chemists, aged 10, used to make our own fireworks. But you can't do that now.

Safety?

So-called safety, but it may be very dangerous because it means you get no practical chemical experience. We had a very good master of chemistry, Mr. Moore, who first made a chemist of me. But the teacher I remember best was our mathematics teacher, Mr. Tomkins. I mentioned this once on a TV programme and the BBC found him. He was just about retiring as Headmaster of a very good school and we were able to meet after some 30 years. Nowadays I don't distinguish between chemistry, physics and biology, I do all of them. At present, I am working mainly in biology, biophysics, biospectroscopy, biochemistry and photosynthesis. I have 10 students and collaborators here at the Imperial College of Science, Technology and Medicine. Most of them were physics graduates and they are now

BBC Television Lecture on the "Laws of Disorder," 1965 (courtesy of George Porter).

employed as lecturers in biochemistry or chemistry. Ours is a very interdisciplinary subject.

I've heard about your activities of popularizing science. During a Christmas lecture at The Royal Institution, you had a beautifully laid table and after repeating three times the sentence, "I believe in Isaac Newton", you pulled out the tablecloth with everything intact on the table.

Yes — illustrating the law of inertia. I would like to add something to that story. Those Christmas lectures were watched by the Royal Family, encouraged by the Duke of Edinburgh, at Buckingham Palace at Christmas. In the New Year, The Queen kindly invited me to lunch there, with her family and a dozen other people. Princess Ann was there; I've always liked Princess Ann, she is a bit of a tomboy. She was about 17 at that time. She said, "I liked your demonstrations, especially the one where you cleared the table." We were walking towards lunch at the time and as we entered the splendid drawing room, with its large table set for 12 people and loaded with silver and gold plate, Princess Ann said, Go on, try it now!

Like every other professor, I had to give an inaugural lecture when I became Professor at the University of Sheffield. I took a lot of trouble and I got together about 20 or 30 spectacular and interesting demonstrations. It went down extremely well, and someone told Sir Lawrence Bragg about it. He was then the Director of The Royal Institution and he invited me to lecture there to 350 schoolkids. Then I gave a discourse to The Royal Institution, quite a formal event, and a number of press and television people were there. It was about entropy and it was called, "The Laws of Disorder." I gave six one-hour lectures at the time when there was one television channel only, and they were in prime time. That was how I got involved in presenting science to the public, which I greatly enjoyed.

With Sir Lawrence Bragg at The Royal Institution, 1968 (courtesy of George Porter).

Did you get through to the non-scientists?

I think I did. I remember coming into the London Airport, very late one night, into a large, empty, hanger-like space. There were some other porters (those who carry luggage!) and one of these toughies shouted, "What's new in thermodynamics?" I did about 200 programmes and became quite well known to the public around the early 1960s. However, instant fame on television doesn't last long.

This was before I went to The Royal Institution in 1966 where I stayed as Director for 20 years. I had a large research group and was very involved in the lectures for the public and for young people in the theories made famous by Davy and Faraday. Then, in 1985, I became President of The Royal Society. At a rather advanced age, I was invited to come here to Imperial College to be Professor. I have a nice new laboratory here. I am now thinking of taking "early retirement" at 78. It doesn't mean I'm giving it up but I really can't expect Imperial College to support me forever.

I am all for going on as long as I can, but there comes a point where you slow down a little bit. People of my age don't often do anything very bright or original themselves, although they can sometimes be useful when working with younger colleagues.

You started very early.

Yes, especially counting the five years I'd served in the Navy. I was very lucky. Upon graduating at 20, I became a Naval Officer. That year, I went to the University of Aberdeen where they had a very good school of natural philosophy as they call it in Scotland. I was doing radiophysics. That was what the war effort wanted. During the war, I spent much of my time at sea in the Atlantic trying to keep radar sets going and learning about electronics. There were about a hundred ships in that convoy and I was the only person properly qualified in radar. I also ran a naval radar training school in Belfast.

I had a lot of time at sea to think over what to do after the War. Maybe I should make money or I should do something useful or try to write books? Soon I decided unquestionably that what I wanted to do was science. Not just to have fun, which I did, but to do scientific research, make a really big discovery. Not that I thought I was very brilliant; I wasn't. But I felt there was nothing so worthwhile in life as trying to advance our understanding of the world.

By then I was 24, and I decided that I must move quickly: I wrote to three professors in Cambridge. It had to be Cambridge, the centre of science. The three professors were a physicist, a physiologist and a physical chemist. Finally, I chose the physical chemist, R. G. W. Norrish, FRS. I

Lieutenant Porter in 1943 (courtesy of George Porter).

had a wonderful time in Cambridge and stayed for 9 years. I decided to do something rather ambitious and quite expensive, but I was able to beg most of the apparatus from the Navy.

I was lucky in many ways but especially because the idea for my research came by combining experience of radar and photochemistry. People in Cambridge Chemistry were interested at the time in free radicals, which only lived for a thousandth of a second or less. Many people even doubted their existence, and I wanted to see them. I'm a simple soul and I don't really believe in things like free radicals unless I can see them.

The way I did it was, instead of using a continuous source of light, an army search light, for example, to use the very powerful flash lamps that had been developed during the war; they were put into bombers, the bomber bay being filled with condensers to store the energy, and flown over to photograph Germany. I brought one of these lamps back from the makers in Lancashire, I got the condensers from the Navy, and obtained an enormous flash, far bigger than I needed. I overdid it but the condensers were war surplus and free.

The second idea, about a year later, was to use two flashes, one as a pulse and the other as a probe. This was the optical equivalent of radar. Radar uses pulses of electromagnetic radiation, radiowaves. Those pulses were a few millionths of a second long, corresponding to the microseconds in the latest radar sets that we were using before I left the Navy. The idea was to send out one pulse and then to probe it, using the velocity of light to change distance into time.

George Porter with Professor R. G. W. Norrish (courtesy of George Porter).

So it wasn't only equipment you got from the Navy but also ideas?

Absolutely. There is nothing more rewarding than linking two quite different subjects.

How long did your association with Professor Norrish last?

I stayed in the department for 9 years. The work for which I got the Nobel Prize was part of my Ph.D. Thesis, but I had to work another 15 years before anybody took much notice.

Did you do anything as important as your Ph.D. work later?

Not really. I applied the technique in biology, in chemistry and in physics. Some of these led to quite exciting results, but everything I have done since was made possible through the flash photolysis technique.

One of the first applications of flash photolysis was to make short-lived substances, called free radicals, and record their structure and study their chemistry. I detected quite a number in the first few months of my research. A most satisfying example was the free radical ClO, chloric oxide, unknown

Receiving the Nobel Prize from King Gustav of Sweden in 1967 (courtesy of George Porter).

before that time. It forms very easily by shining light on a mixture of chlorine and oxygen, but it only lasts for a few thousandths of a second. I was able to do some fascinating chemistry with it. Norrish used to bring round lots of visitors who might give money to the Department. Many of them inevitably asked, "But what use is it?" I was a young, pure scientist and very proud of it, so I replied that it was no use at all, and for 20 years it wasn't. But it turned out, in the end, that ClO was the key molecule in the atmospheric ozone problem. We had most of these reactions worked out before their importance in the reactions of the chlorofluorocarbons was realised.

The first applications of flash photolysis made possible millisecond time measurement of chemical reactions. Within a year we could go to microsecond measurement and then nanosecond flash photolysis became possible with the laser, which replaced our flash tube. It is much more powerful, has shorter pulse times and is coherent. The laser has a very narrow beam and is almost monochromatic.

Back in the microsecond days, did you envision or dream about the day when nanosecond, picosecond, let alone femtosecond scales would become a possibility?

I spent more time thinking about that than anything. It was the obvious thing to do. We knew we were on the edge of much faster things. Before lasers I tried hard to make a nanosecond flash lamp, and we did make some; and published work using them. Sparks, for example, can be used for nanosecond reactions. Unfortunately, they don't have much energy and are not coherent.

The laser transformed my work. First the nanosecond laser, then quite quickly afterwards, in another five years, came the picosecond laser, and then the femtosecond.

Is this the end?

Yes. As far as molecular chemistry and biology are concerned you can't follow usefully any event in times less than a femtosecond. Because of the uncertainty principle, it becomes impossible to measure energies accurately on such a short time scale. Even the lifetime of the transition state is longer. The rate of going through the transition state is kT/h, where k is the Boltzmann constant, T is the absolute temperature and h is the Planck constant. This corresponds to a time about 50 femtoseconds which is the end of the time scale for chemistry. In nuclear physics you can push it back further, because much bigger energies are involved than those in chemistry.

Lord Porter with wife Stella (Lady Porter) at home at the Royal Society, Carlton House Terrace, London (courtesy of George Porter).

However, there are many interesting things to study within the available time scale. I've mentioned the transition state which is, for many physical chemists, the ultimate thing in studying reactions. One of the early workers who advanced this concept originally, M. G. Evans, was one of my teachers at Leeds who greatly inspired me. He himself studied under Michael Polanyi at the University of Manchester. I met Michael Polanyi in my first year as an undergraduate, at the age of 17. I was given the daunting task, as the secretary of the student chemical society, of proposing a vote of thanks to Michael Polanyi for his lecture. I didn't really understand the lecture very well but I managed somehow to say what a marvellous lecture it was, and that even I could understand some of it. I met him many years later when his son, John took me along to dine with him at the Athenaeum club after a Faraday Society meeting. By this time he had become a social scientist.

Would you care to mention an example of your current work?

For the last ten years, I have studied the primary events of photosynthesis. Surprisingly, some of the fastest reactions of all occur in the leaf of a plant. When the leaf absorbs light, the whole photochemical photophysical process is over in less than a nanosecond. The chemical reaction is produced by electron transfer from chlorophyll through membranes in the leaf. When the electron is lost from chlorophyll, oxidation occurs and oxygen is produced from water. Then the electron goes to the other side of the membrane, there is reduction, and hydrogen is produced or, more usually, carbon dioxide is reduced and sugars are produced. This is a wonderful area right now because in just the last two or three years not only have we been able to do femtosecond spectroscopy rather precisely, but the biochemists have been able to isolate the reaction centres, and through the techniques of electron microscopy, electron diffraction, and X-ray diffraction, to obtain the structures to nearly atomic resolution. The green engines of the leaf at this resolution are even more beautiful than the leaf itself.

The first of these structures was the reaction centre of a photosynthetic bacterium for which a German group was awarded the Nobel Prize in 1988. But full photosynthesis with splitting of water occurs only in the green plants with their two photosystems and the oxygen evolution is engineered in one of these, called Photosystem 2. My colleagues here at Imperial College, led by Jim Barber, are at the threshold of obtaining the structure of this reaction centre at medium resolution and should have it at atomic resolution very soon.

The primary events occurring in the reaction centre are the transfer of electrons across the membrane, resulting eventually in oxidation of water and reduction of carbon dioxide to carbohydrates. But, before this happens, the light has been harvested by antennae, consisting of many chlorophyll and other pigment molecules and the structures of these antennae have also been established recently by groups at Imperial College and Stuttgart led by Werner Kühlbrandt.

All this structural work is of the greatest importance and interest especially to us, because it is complementary to the kinetic flash photolysis studies of the fast reactions of energy and electron transfer, which are themselves being elucidated by the photomolecular sciences group of David Klug, James Durrant and other colleagues at Imperial College.

These processes of photosynthesis are perhaps the most important of all fast reactions in the femtosecond and picosecond time scale and it is very gratifying that the rapid structural advances of the biochemists have appeared almost simultaneously with the ultimate triumphs of picosecond and femtosecond flash spectroscopy, which are essential to resolve the detailed kinetics of photosynthesis.

I can think of no more gratifying way to conclude this race against time than to be able to elucidate what are, after all, not only the fastest but also the most important of all chemical reactions, the reactions which provide all the food, fuel and energy for life.

Ahmed H. Zewail, 1997 (photograph by I. Hargittai).

39

AHMED H. ZEWAIL

Ahmed H. Zewail (b. 1946 in Egypt) is Linus Pauling Professor of Chemistry and Professor of Physics and Director of the NSF Laboratory for Molecular Sciences at the California Institute of Technology. He is best known for his work in the field of femtochemistry, leading to the paradigm change in our thinking about the dynamics of chemical bonds and reactions. Dr. Zewail has received many awards, including the King Faisal International Prize, the Wolf Prize, the Robert A. Welch Award, the Peter Debye Award, and the E. Bright Wilson Award. He is a member of the U.S. National Academy of Sciences and many other academies and societies. Our conversations took place on February 14 and 15, 1997, in Pasadena, California. Below is a somewhat shortened version of the complete conversation that appeared in *The Chemical Intellingencer*.* Ahmed Zewail was awarded the Nobel Prize in chemistry in 1999 "for his studies of the transition states of chemical reactions using femtosecond spectroscopy."

Please, place femtochemistry in perspective.

Time resolution in chemistry has witnessed major strides in over a century. Svante Arrhenius' introduction of the famous theoretical expression (1889) for the speed of chemical reactions gave us a clue of the time scale of rates, and Henry Eyring and Michael Polanyi's (1931) microscopic, theoretical

*The complete version of the interview was originally published in *The Chemical Intelligencer* 1997, *3*(3), 20–31 © 1997, Springer-Verlag, New York, Inc.

description of a reaction made chemists think of the atomic motions through the *transition-state* on the vibrational time scale. But the focus naturally had to be on what could be measured in those days, namely the slow rates of reactions.

One of the first successful experiments involving subsecond times was made with flow tubes. Two reactants were mixed in a flow tube and the reaction products were observed at different distances. Knowing the speed of the flow, one could translate this into time, tens of milliseconds, representing an advance from the time scale of seconds achieved with radiative glows using phosphorscopes. Then came the stop-flow methods that reached the millisecond scale. The stop-flow method is still used today in biological kinetics.

The next stride in time resolution in chemistry came when R. G. W. Norrish and George Porter in England and Manfred Eigen in Germany developed techniques reaching the microsecond time scale. Porter and Norrish developed the method of flash photolysis, few years after World War II, using electronics developed at the time. They produced a burst of light of a millisecond duration or so and created radicals in the sample, and, using other light, they recorded the spectrum of these radicals. They achieved kinetics on this time scale and observed some relatively stable intermediates. Eigen developed the relaxation method which reached the microsecond and close to the nanosecond scale. By disturbing the equilibrium of a solution by either a heat jump, a pressure jump, or an electric field, the system shifts from equilibrium. This is the point of time zero. Then the system equilibrates, and its kinetics can be followed. Manfred Eigen called these reactions "immeasurably fast" in his Stockholm lecture. There was a feeling that this time resolution was the fastest that could be measured, or that needed to be measured for relevance to chemistry. At about the same time, shock tubes were used to provide kinetics on similar time scales.

What was not anticipated was the laser. It could generate nanosecond pulses of light, and, with mode-locking, it could provide picosecond and femtosecond pulses. In the late sixties and seventies, a number of research groups in the U.S. and Europe were using the picosecond technology for the study of fast chemical kinetics of nonradiative processes in solutions. I recall the exciting results reporting the photophysical rate of internal conversion (Peter Rentzepis), the picosecond chemical reactions of caging and electron transfer (Ken Eisensthal), and the rates of intersystem crossing (Robin Hochstrasser) and vibrational relaxation (Wolfgang Kaiser) in solutions. At about the same time, molecular-beam studies of reactions were being developed, and, although I was not a member of this community earlier in my career, beams later became part of our effort in femtochemistry.

When I arrived at Caltech 20 years ago, there were two things that helped the development of femtochemistry. One was that I had done picosecond studies at Berkeley. The other was that I have had an interest in the concept of coherence, about which I would like to say a little more later. For looking into the *dynamics* of how a bond is broken and how a bond is made, you must introduce a time scale that is shorter than the vibrational period of the bond, and here coherence is important. We had the fortunate idea of coupling ultrafast lasers to molecular beams to study *isolated* molecular dynamics. This was in analogy with studies made earlier by us on coherence of small molecules in the gas phase, but now extended to large molecules and reactions.

Extending the 1976 work, we observed an unusual phenomenon in 1980. We started with anthracene, a large molecule, and we energized it. We might have expected the energy to flow all over the place due to the large number of vibrational modes of this molecule, as if the molecule is in a "bath" of many modes. How fast will the energy redistribute in this molecule? This was the initial question. We expected a decay of the energy from the initial mode excited and its increase in other modes. This would be the time constant, measured in real time, which we could assign to energy redistribution in the molecule, hitherto unmeasured directly. My students, Peter Felker, now a Professor at UCLA, and Bill Lambert, now at AT&T, observed something striking during the experiment. There were oscillations; that is, the energy departed from the initial mode, but it came back too. I knew we had to be very careful with this striking observation, and many control experiments were made, even with my own hands. There was no doubt it was there. We immediately published our first paper describing the phenomenon, which would later become the coherent wave packet of reaction dynamics. Essentially, we prepared molecules at time zero in phase in a given vibrational state and then observed the system going into another vibration and coming back, without all the multitude of states being involved. That observation was the real trigger that opened my eyes. From this point I knew if we could study isolated chemical systems, we should be able to see the vibrational and rotational motions of molecules and reactions.

In a couple of years, we achieved a time resolution which enabled us to see not only the vibrational redistribution but also the rotational motion and reaction rates as well. From the period of rotation, we could get the moments of inertia and the molecular structure. From microcanonical rates, we tested theories at the most fundamental level. In the meantime, the lasers kept improving and their pulses became shorter: first sub-picosecond, then about 90 femtosecond, and then 6 femtosecond. Chuck Shank and colleagues provided these short pulses from unique designs of dye lasers, and I immediately saw to it, in the early eighties, that we upgrade our lasers, which had already been interfaced to molecular beams for the studies I mentioned above. Funding was provided by the Air Force, which had volunteered to support this transition in our instrumentation. Thus, "Femtoland I" was born and housed in the same room where Linus Pauling's X-ray apparatus once was. Currently, we have up to Femtoland VI.

In 1985, we were able to develop the methodology of probing reaction dynamics with femtosecond resolution, and in 1987 the first paper reporting the observation of transition states, "Real-Time Femtosecond Probing of Transition States in Chemical Reactions" was published. It is remarkable that this was the first time a bond breakage process was seen in real time, even though chemistry is defined as the process of bond breaking and bond making. Also in 1987, we published our work on real time observation of bond making in the paper addressing "the birth of OH from the reaction of H + CO_2." This was followed in 1988 with the paradigm case of sodium iodide. This case illustrated the power of observing the dynamics of covalent/ionic bonds and the transition states.

Linus Pauling and Ahmed Zewail (photo courtesy of California Institute of Technology).

In the last decade, the scope of applications went beyond expectations, from elementary to complex chemical reactions, from the gas phase to clusters to dense fluids, and to solutions and biological systems. In retrospect, the femtosecond time scale was just right for seeing the fundamental dynamics of the chemical bond, and that is why the applications have been and will continue to be impressive in many areas.

What is the shortest time that may still be useful to resolve for studying chemical reactions?

Chemical reactions are characterized by the motion of the nuclei. The nuclei move with a speed of the order of one kilometer/second and for a bond length of one angstrom, the time scale is about 100 femtoseconds. If we can monitor the molecular vibrations and molecular rotations in real time, we can learn essentially all that we need about the dynamics of chemical reactions. These fundamental limits define the chemical as well as biological time scales.

When you move to the sub-femtosecond time scale, electron motions become interesting. This scale may give us the possibility of watching the electron moving around in the benzene ring. On this scale, physics also becomes important rather than chemistry.

What would you single out as your most important contribution to chemistry so far.

Perhaps, for the "Dynamics of the Chemical Bond," the development of femtochemistry, building on the key concept of coherence in chemical dynamics. The concepts have changed the way we think about nuclear motion and dynamics in chemistry.

Would you elaborate on the concept of coherence?

When we speak about rates and kinetics in chemistry, we usually think of the number of molecules that had reacted in a unit of time, or of the number of molecules that exist in certain states of a system. Thus, we speak of average behavior of the ensemble, with no coherence among the molecules. However, with laser light we can actually prepare molecules in a coherent state, and in this state the molecules will know of each other's existence, phase, and orientation. In a sense they are "glued" together, so when they move, they all move together. We call this a single-molecule trajectory. This is the world of dynamics, because the system of all molecules becomes in-phase at time equal to zero.

Coherent preparation and probing of molecules and reactions is central to the field of dynamics and, in fact, explains why chemistry was ahead of physics in this field. The state is called wave packet in quantum mechanics. In 1926, Lorentz wrote to Schrödinger suggesting that it would not be possible to prepare such a state. As mentioned before, we can now create such coherent state in molecules and reactions quite easily.

These concepts make us think of reactions as they proceed in real time and as a single-molecule trajectory, just like a classical visualization of motion. Actually on the femtosecond time scale, the classical description of motion is a remarkably valid picture and dynamics is at the atomic resolution.

Your book *Symmetry through the Eyes of a Chemist* is full of examples of coherence in chemistry. In the structural sense coherence is a search for pattern. This is spatial coherence. There is then the temporal coherence which we just spoke about and which was foreign to chemistry 20 years ago.

Would you care to give examples showing how your research has impacted chemistry?

Chemistry is bond breaking and bond making. The development of techniques to observe these processes is very exciting by itself because we can see the phenomena — "Seeing is believing." Processes often appear complex because we look at them on an extended time scale during which many steps in the process are integrated over. On the femtosecond scale, these steps are resolved and the complexity of the process breaks down into a series of simpler events. Now, for the first time, we have the methodology and resolution for seeing the nuclei moving in real time, and we can speak and think in the same terms in which the reaction takes place. It is a new concept to actually see it while it is happening.

I would like to mention some examples of areas where such observations were critical for the way we think about the phenomena. They span elementary reactions, complex reactions, solvation dynamics, and femtobiology, and may provide the scope of different phenomena, concepts, and applications of femtochemistry.

Consider the concept of bond breaking in elementary reactions. We may ask the fundamental question: Will we ever be able to localize the positions of the nuclei in space to observe their motion? Having in mind the uncertainty principle, can we speak of dynamics with a resolution of a tenth of an angstrom? The work on sodium iodide showed that indeed we can localize the nuclei and have atomic resolution in dynamics. The principle of coherence was established and the dynamics of the chemical bond was seen as the bond changes its structure from being covalent to being ionic. Such a process was not observed before, and we now know the dynamical picture and the extent to which quantum effects can help or impede such observations. This is fundamental and, in view of previous theories predicting the opposite behavior of localization, is important. Incidentally, this curve crossing (covalent/ionic) is ubiquitous in chemistry — electron transfer, S_N2 reactions, acid-base reactions, etc., — and to see it in real time was a thrill.

Take another example of elementary, but bimolecular reactions. One such reaction is relevant in the upper atmosphere and in combustion. A hydroxy group reacts with carbon monoxide and produces carbon dioxide and a hydrogen atom, or the reverse. In the reverse reaction, the hydrogen atom approaches carbon dioxide to take away an oxygen. How does it happen that a H–O bond forms and a C=O bond breaks? How long does it take? The duration is extremely important. If it happens in 10 femtoseconds, it means that the hydrogen strips the oxygen instantly, even before the whole system has a chance to vibrate or rotate. In this case the key players are the electrons. On the other hand, if it is a picosecond event, it may be that the hydrogen complexes to CO_2, taking some time so the nuclei can move. The molecule can then rotate and vibrate, and energy redistribution becomes a determining factor for product formation; that is, the product is the result of nuclear motions. By isolating the transition-state/intermediate HOCO in real time, we can establish the mechanism, and, equally important, test theoretical predictions at the most fundamental level.

What about the complex reactions in organic and inorganic chemistry?

Femtochemistry has been extended to the world of organic and inorganic chemistry to unravel the dynamics and mechanisms of complex reactions that chemists have been studying for many decades. These include Diels-Alder reactions, Norrish reactions, elimination, pericyclic, proton- and electron-transfer, isomerization, and tautomerization reactions. The femtosecond resolution makes it possible to study the elementary steps in these complex systems. The same approach was extended to reactions in clusters, that is, finite numbers of solvent molecules, to study acid-base and other reactions and to solutions and biological systems.

Take, for example, the Diels-Alder reaction or the pericyclic reaction of two ethylenes, the classic Woodward-Hoffmann description. Two mechanisms have been suggested: one in which the reaction proceeds in a concerted manner — that is, the two reactants come in a well-timed and synchronized manner — and the other, through the formation of a biradical species in the transition-state region. This biradical vibrates and rotates and eventually the second bond forms. The mechanism is crucial to our understanding of how C–C bonds are formed and broken and to stereochemistry, reaction channels, and reaction yield. We examined these reactions, observed their femtosecond dynamics, and defined concertedness in relation to the actual nuclear motions. We also elucidated the origin of stereochemical retention and provided a united picture of the energy landscape.

Could you give an example of condensed phase chemistry?

The cage effect, a phenomenon of considerable importance in solution-phase chemistry, is another case. This occurs when you break a chemical bond in a solvent and there is a finite probability that the two atoms of a molecule, or fragment, will recombine. This process has been studied for 60 years, and we teach this concept in text books. The dynamics of this process is fascinating and involves many steps. If you observe with picosecond or nanosecond resolution, you will be integrating over the times for the separation of the two atoms, their interactions with the solvent, and their return to each other and the re-establishment of the bond. Following recombination, they are vibrationally hot and the effect includes the cooling-down period as well.

With femtosecond resolution we were able to observe the different steps, and we did such studies in solvent clusters and in different phases from gases to solutions. We could monitor the nuclear motion, observe the departing nuclei and their recombination, and could even measure their rate of cooling. Furthermore, we could truly check the supposed mechanism — whether the atoms really move away from each other and find the solvent cage, and whether they move back as was supposed, and whether there was a connection between the phases of their motion. We also raised the question about the necessity of the solvent cage. We found that even one solvent molecule is capable of caging the atoms in "bimolecular" solvent–solute encounters. The collision with the solvent molecule takes most of the energy of the departing atom, and it is trapped. So it is not as if it had been bounced back from a wall. The process is not random; there is coherence in the whole event and the emerging mechanism is now based on direct observations.

What have you learned about electron and proton transfers on the femtoscale?

We have been interested in these processes because of their fundamental importance. Take the problem of charge transfer studied by Robert Mulliken, Joel Hildebrand, and many others. The change in color of solutions of iodine defined in those days what was called "regular or irregular solvents." We studied the famous benzene-iodine solution but now at a microscopic level. The electron jumps from benzene to iodine. Now there was an I_2^- on top of the benzene$^+$. What we resolved on the femtosecond scale was the nuclear motion and found the unexpected mechanism: there are two channels for the reaction once the I_2^- is made. One is that the electron hops back to the benzene, attracted by the positive charge there. This leaves behind an I_2 but of very high energy, so it breaks again and gives iodine atoms. This process takes 300 femtoseconds. In the other channel, the I_2^- vibrates on top of the benzene$^+$, and it kicks out an iodine atom and becomes I^-–benzene$^+$. We can separate the two channels, determine the two different kinetic energies of the iodine atoms originating from the two different channels, and give the geometry of the transition state. Thus we could understand the dynamics of the two processes and can shine some light on the nuclear motions involved in this old problem. Note that the final product in both channels is iodine atoms, and without the time resolution we simply cannot isolate the different pathways and their dynamics. We have similarly studied intramolecular electron transfer and both intra- and intermolecular proton transfer.

Femtobiology?

The progress is impressive, and many of my colleagues have made it possible. For example, rhodopsin, the "molecule of vision," has a bond, called the 11-*cis* bond, about which it twists. The dynamics of this twisting motion was studied on the femtosecond scale at Berkeley and other institutions. We now know that in the primary process of vision it takes 200 femtoseconds for the molecule to twist. The rhodopsin molecule is huge, yet the quantum yield around the 11-*cis* bond is about 80%. This means that nature did not make this molecule to first get the energy, redistribute it around, find this particular bond, and make a twist around it using the remaining energy with a yield of about 5%. The process is coherent and remarkably mimics the description we gave for sodium iodide. So we have the dynamics of a two-atom system being central to the primary process of vision. This unification defines the importance of dynamics in chemistry and biology.

The photosynthetic reaction provides a similar example. Its primary process is an electron transfer in a protein. As the electron is transferred, the nuclei around it are moving in a coherent way, at least on the time scale of the transfer. It is even called the coherent electron transfer. Nature made it in a way that it is not using all the atoms of the protein and distributing the energy and wasting it. The electron transfer goes in a certain direction and involves a small coherent vibrational phase space. These results from work in France, Sweden, and other countries have changed our way of thinking of how to describe these processes in a dynamical fashion on the scale of the atom.

How about the dynamics of less drastic changes, weaker interactions, those beyond the covalent bond?

This is another area that interests me since it relates to supramolecular chemistry and molecular recognition. We have already mentioned our study of acid/base reactions, where hydrogen bonding with bases, such as ammonia or water, plays the key role. The question is how does this recognition of the lone pair of nitrogen, or oxygen, by the hydrogen influences the acidity and stability of the species. More recently, we extended the study to model base pairs with two hydrogen bonds — two nitrogen lone pairs and two N–H bonds. This makes a double hydrogen-bond structure, just as you see in some DNA base pairs. These are weak interactions with large-amplitude motion and have low barriers, yet they are of enormous importance in establishing the structure and in mutations. Because of the barrier, the system may tunnel through the barrier, which takes a much longer time than going over the top; hence the probability of mutation is relatively low. Löwdin called this quantum genetics. We wanted to understand this phenomenon, so we energized the molecule and found that the two hydrogen atoms moved nonconcertedly in the model base pair. We determined it to be a quantum tunneling effect, and we measured the rate. We replaced the hydrogen by deuterium, and it drastically changed the tunneling probability. These changes appear on the femtosecond scale. The phenomenon was related to the molecular structure, and we proposed the nature of the intermediate and transition-state structures, confirmed later by ab initio calculations.

How can we assess the validity of the Born-Oppenheimer approximation at the femtosecond time scale.

Your question gets to the heart of the importance of the time scale. The Born-Oppenheimer approximation fundamentally comes from the following idea. The electrons move much faster than the nuclei. In this sense we can speak about electron excitation and promotion while the nuclei are relatively fixed in space. The approximation is based on the separation of time scales of the electron and nuclear motions. At the level of femtochemistry, the Born-Oppenheimer approximation is still valid in that the 10^{-14} s order of magnitude of nuclear motion is well separated from that of the electron motion. However, at the level of 10^{-16} s time scale, this separation becomes fuzzy and even the electronic states of the atoms and molecules are ill-defined.

Let's now make a big jump and talk a little about you, rather than about your science. You spent your first 22 years in Egypt.

Egypt gave me wonderful opportunities. I grew up near Alexandria in a family of good values and hard work. My father worked for the government. I was the only son in our family and I have three sisters. I was given special attention and was expected to become an engineer or a doctor. I especially value that I was taught the importance of good interactions with people, to be likable and to be liked. At home I was always taught moderation and doing what is sensible; we are raising our children in the same spirit. I got my university education in Alexandria, a place of ancient traditions in science and, of course, the place of the Great Library. It is surprising that I became a scientist because this was not in the family tradition.

How about religion?

My family is Moslem, as is probably 90% of the population in Egypt. We have always been moderate, and during my time in Egypt we spoke of being Egyptians, independent of religion, color, or region; many of my teachers at the University were Christian Copts and the students were of different backgrounds and from different regions. Tolerance has always been characteristic of most Egyptians, and, in my view, this is also why the civilization has survived for such a long time, more than 5000 years.

Your wife and children?

The present Zewail family (photographed by I. Hargittai).

I met my present wife, Dema, in 1989 in Saudi Arabia, after ten years of being a bachelor following a divorce. I went to Saudi Arabia to receive an international prize, the King Faisal Prize. I got it in physics. Dema's father was being awarded the prize in literature at the same time. This is how we met. I have altogether four children, two from my first marriage. One of them, Maha, is a graduate of Caltech and is now studying genetic engineering for her Ph.D. in Texas. The other, Amani, is about to go to Berkeley. My first wife was a student of Alexandria University and she is a loving mother. Dema had studied in Syria, got a medical doctorate there, and then a Master's degree at UCLA in public health. We have two young boys, Nabeel and Hani, and Dema devotes lots of efforts to them. She keeps us a happy, cultured, and caring family.

How was your education in Egypt?

I received a very traditional education. It was important to get an education but it was irrelevant what direction I would choose later. There was no pressure from my family, for example, to get into business and to make money. We have a proverb that I heard all the time: "A degree in your hand is like the power of wealth in your pocket." It protects you, makes you a better human being, makes you culturally rich. This was the tradition. Achieving good grades and getting into a good university was the driving force among my classmates. The hope was to become somebody some day. There was nothing fancy in our school, but there was serious emphasis on lots of homework and discipline.

I don't know what triggered my interest in becoming a scientist. I went to school and I was getting A's in science. During the precollege years, I set up some science experiments at home in my room. I had some test tubes and would heat some substances, such as wood, in them. I could have created an explosion, but I didn't; I was lucky. I was intrigued by these experiments, especially when observing the substance changing from solid to a burning gas. My mother didn't mind my experiments; I think she thought that it was part of my studies.

Learning was everything. When I went to visit the University of Alexandria for the first time and I saw its beautiful, ornate buildings, the shrine of knowledge and learning, I had tears in my eyes. I was emotional about the "house of learning" and by then I knew that one day I was going to become a scientist.

When I entered the University of Alexandria, I started with the sciences. The first year I could take four subjects, chemistry, physics, mathematics, and one elective, and I chose geology. The second year you could drop one of these four. After the second year, the top few students were allowed to specialize in one subject. There were seven such "special students" in our class of about 500, and I was one of the seven. I chose "special chemistry." I graduated with distinction and first-class honor. As a special student, I had privileges, such as taking classes in the professors' offices, borrowing books from them, and they knew us by name. I got a fellowship, which was like a salary. When I graduated there was a position waiting for me since I was the so-called first student of the university. The idea was that I would go away somewhere, get a doctorate, and return to my position at the University of Alexandria. For my doctoral studies, I went to the University of Pennsylvania. My research supervisor was Professor Robin Hochstrasser, who introduced me to state-of-the-art scientific research. I had a productive four-and-a-half years there with a dozen or so papers, but just as important, I was exposed to a large number of interesting research problems.

During your doctoral studies, did you ever feel that having graduated from Alexandria put you at a disadvantage?

Certainly. I knew the basics of chemistry, the periodic table, the inorganic salts, etc. However, I was lacking knowledge about quantum mechanics, lasers, and other modern things. The first course I took was called 501,

Introduction to Quantum Mechanics and Spectroscopy. I was worrying so much that I would not be able to compete, but in the end I was one of the two who got an A in the whole class.

What happened then?

I had every intention of returning to Egypt. However, I wanted to postdoc for two years to learn the art of being a professor before returning. I also wanted to bring home a big American car. I went to Berkeley to work with Professor Charles Harris. This gave me a very different experience. Berkeley was a new world. Almost everybody was working on frontier problems. The laboratories were richly equipped, there were lots of famous people, and the majority of graduate students were of high quality.

When the time came to think about returning to Egypt, Charles had a talk with me. He saw in me more than I saw in myself. He suggested to me to apply to the top universities in America. At the end I had about six or eight offers, from places such as Harvard, Chicago, Caltech, Northwestern, Rice. It was a magnetic force for staying.

Your final choice was Caltech.

When I came for interview here, I was so impressed by Caltech's determination to make me succeed. It was evident that they were going to

Ahmed Zewail in the laboratory (photograph by I. Hargittai).

do it right. They made the commitments and they kept every one of them to the last detail. My name was painted on my parking spot before I arrived.

There was one more thing. At Chicago, there were many excellent faculty in chemical physics. At Caltech, in my general field, I was alone, and would start to rebuild modern experimental chemical physics. I was intrigued by the idea of coming to Caltech with nobody around in my general area. I have found this interesting throughout all my career. I don't like crowded areas, so in this sense Caltech was also attractive for me.

I came in 1976; I remember it very vividly, it was in May. Our first paper was submitted in August and published in December of the same year. I had ordered some of my equipment already while still at Berkeley, so the equipment was arriving as I was coming in. I was able to start my experiments the moment I set foot in my new lab, and they worked. After a year and a half, the chairman came to see me to discuss the tenure process. I was tenured in 1978. It was very exciting; lasers were being introduced into molecular science to exploit phenomena that cannot be achieved with normal light!

However, in 1978 I realized that I could make a better contribution to real-time molecular dynamics if I could isolate these molecular systems in a beam. At that point, I didn't know anything about molecular-beam machines, but I was, once again, ready to learn. This transition was fruitful and has played an important role in our studies of the "dynamics of the chemical bond." In 1990, I was given the first Linus Pauling Chair. Interestingly, Pauling's group's effort at Caltech was centered on the "structure of the chemical bond."

Whom would you single out as an important influence on you?

Outside Caltech, Richard Bernstein. Dick was a first-rate experimentalist, and one of the founders of molecular-beam scattering. He had high integrity, worked all the time, had a commitment to science and family, and was especially helpful to younger colleagues.

My relationship with him was my good fortune. I was a newcomer in molecular reaction dynamics when he happened to be visiting Caltech. First he was a colleague, then a supporter, and, ultimately, my friend. He took two sabbaticals with me. He was in the "tenth excited state" about femtochemistry. He always believed in the "unlimited opportunities" of this field, and these are his words. At that time, hearing this from a well-established and well-respected senior was a tremendous boost for me. You

Richard B. Bernstein (1924–1990) had a distinguished career, including appointments at the Illinois Institute of Technology, University of Michigan, University of Wisconsin at Madison, University of Texas at Austin, Columbia University, the Occidental Research Corporation, and the University of California at Los Angeles. He was a pioneer in molecular beam scattering and reaction dynamics studies. His last book, titled *Molecular Reaction Dynamics and Chemical Reactivity*, was co-authored with R. D. Levine. In addition to the Zewail interview, there are references to Richard Bernstein in the conversation with Robert L. Whetten in this volume. (Photograph courtesy of Ahmed Zewail).

hear about how difficult it is sometimes to make a breakthrough in other fields, especially if you are an outsider. You may be seriously hindered by hostile attitude or simply by being ignored. Dick Bernstein saw to it from day one that this was not the case, and he made prophetic predictions about the field. During his first sabbatical, we did collaborative work here. Every day when he was in town and I was in town, there would be a pot of coffee and we would sit in my office and talk. It was a fantastic experience talking with somebody of some 30 years of experience and seeing that I was stimulating him. During his second sabbatical, just before he died, we built a new apparatus and wanted to study the femtochemistry of surfaces, and we had all kinds of plans to work together. What can I say? He is a hero.

When you go somewhere where people don't know about science, do you experience reservation because you are also an Arab from Egypt?

I feel one thing, my own feelings. When you approach other humans, it's either that you have this label in the back of your mind and live with

Ahmed Zewail on an Egyptian stamp (1998). According to the Caltech News Release of August 5, 1998, Dr. Zewail said: "I am particularly pleased as this honor comes from my country of birth and that I could be in the company of stamps honoring the pyramids, Tutenkhamen, and Queen Nefertiti."

it — you are an Arab, you are a Jew, you are a Christian — or you don't. If you do, you will always have some insecurity about it, and people will see that. Even if they did not have the intention of offending you, you'll be sensitive. Throughout my life I have not had this feeling of pressure at the back of my mind, but this does not mean that I was not upset with unfairness. I'm proud of being an Egyptian, proud of being an American, proud of being an Arab, proud of being a scientist.

When I came to the United States, I was seeking to receive a better Ph.D. education. The people at the University of Pennsylvania had not seen too many Egyptians before in this situation. It was different from someone coming from Britain, Germany, or France. They didn't know my background and my culture. I was 22 and for me it was also a culture shock. I grew up in the Middle Eastern culture. Friends are very important. You can borrow money from friends without writing it on paper. You can visit without calling. If you are having a crisis, your friends will spend hours with you and talk to you. When a good friend left Alexandria for Cairo for a week, we would go to the train station, kiss the guy, and might even have tears in our eyes.

The culture in the United States looked different to me, and work was the number one priority. There was an episode early after my arrival in 1969. It was snowing and I was wearing light shoes and I fell down. Every car passed by me. It was not that they were bad people, it was just that people minded their own business. In Cairo, traffic would have stopped, they would bring a chair in the middle of the street, and somebody would rush to me with tea with mint.

Speaking of the culture, I now laugh at some incidents in the beginning. When good friends are together in Egypt, we sometimes use an Arabic saying, "I kill you" in verbatim translation. It's just teasing, of course. My English was very poor at the beginning, and often I would just translate words and expressions. This is how I told one of my fellow graduate students, "I kill you". I immediately noticed that something terrible had happened. In his mind I just came from the Middle East, and he was sure I meant literally what I had said.

Another difference was the student/professor relationship. In the culture I came from, the professor is respected like a prophet of knowledge. There is a saying that "If somebody teaches you as little as just a letter, you owe it to him to become his slave." I arrived at the University of Pennsylvania, and I considered my professor to be a prophet of science. I found out that he liked coffee so I bought a coffee machine, got the coffee, made the coffee, and brought it for him in his office. All the other graduate students thought I was really weird. They were sure I wanted something special from our professor. When I arrived in the U.S. I gave the professor a gift that my parents had selected and wrapped for him, and I saw a similar reaction. By the way, I now like it when my students do these things if they wish!

I ignored the cultural difficulties and spent all my energy on learning. I only wanted to get the best education I could get. For four years in Philadelphia I did not learn much about the city, but just knew my way around. I was, and still am, very excited about science. It was a paradise of science. In Alexandria we had difficulties in finding journals, chemicals, equipments, etc.; here I could have anything. In Philadelphia, I made friends; some remained friends for over 25 years.

There is a strong Jewish presence in science. Have you had any special experience that you would like to mention?

I noticed this when I started applying for a job. But my origin may have even been beneficial. There was a natural curiosity to meet an Egyptian scientist. In academia, I have also been aware of the Jewish tradition of scholarship. I know that Jewish families educate their children to these values and to reason. I think the majority of people could appreciate me for what I was. Many of my friends are Jewish scientists, and they can value me as a scientist and as a human being. Once we reach this level, we don't let politics ruin or interfere with our relationships and we can agree or disagree about political issues. There were exceptions, of course. I remember one job interview in a distinguished university. I entered the office and this person, obviously very emotional, asked me, "For heaven's sake, why don't you go back to your country?" On the other hand, there was Stuart Rice of the University of Chicago, who was the first chairman to offer me a job, and it was a very good offer. There are others, who might not want me to mention their names, who have been truly helpful and scholarly in their attitude. So it all depends on how individuals feel and behave and on how I feel and behave.

Is there anything you would like to add?

I would like to emphasize a point. Whenever I go to get an honorary degree, a prize, or give a special lecture, people are always generous and praise me for what I have done, and I don't want to be modest about this. However, I think we should not forget two things. One is that, however good you are, you have to be *in the right place at the right time* too. The other is the institutions what make us what we are. For me they are my parents, Egypt, and the United States and Caltech. Credit for the success of research in "femtolands" must be given to the many talented members of our group at Caltech, the scholarly and helpful support we received from colleagues, and the research impact of femtochemistry by colleagues all over the world.

Name Index

Page numbers in bold refer to interviews.